ELEMENTARY
STATISTICAL
PROCEDURES

Clinton I. Chase

Indiana University

ELEMENTARY STATISTICAL PROCEDURES

THIRD EDITION

McGraw-Hill Book Company

New York St. Louis San Francisco Auckland
Bogotá Hamburg Johannesburg London Madrid
Mexico Montreal New Delhi Panama Paris
São Paulo Singapore Sydney Tokyo Toronto

ELEMENTARY STATISTICAL PROCEDURES

Copyright © 1984, 1976, 1967 by McGraw-Hill, Inc. All
rights reserved. Printed in the United States of America.
Except as permitted under the United States Copyright
Act of 1976, no part of this publication may be
reproduced or distributed in any form or by any means,
or stored in a data base or retrieval system, without the
prior written permission of the publisher.

234567890HALHAL8987654

ISBN 0-07-010677-0

This book was set in Baskerville by Santype-Byrd. The
editors were Christina Mediate and James R. Belser; the
designer was Elliot Epstein; the production supervisor was
Diane Renda. New drawings were done by ECL Art.
The cover photograph was taken by Joseph Schuyler,
courtesy Stock Boston.
Halliday Lithograph Corporation was printer and binder.

Library of Congress Cataloging in Publication Data

Chase, Clinton I.
    Elementary statistical procedures.

    Includes index.
    1. Statistics.   I. Title.
QA276.12.C45   1984   519.5   83-18780
ISBN 0-07-010677-0

Contents

8   Regression: Predicting One Variable from Another   161

9   Inference and Probability: Basic Concepts   181

10   Probability and Inference: The One-Sample Test   207

APPENDIXES

Preface

The objectives of this revision remain the same as those of previous editions—to present basic statistical methods used in the social sciences in a direct, nontechnical manner. The aim in the present edition has been to improve the vehicle through which the ideas are presented by a more careful sequencing of ideas, by introducing new material in smaller increments, and by expanding some topics for a more complete coverage.

The major changes have been to divide the discussion of correlation into two chapters (one dealing with correlation and one with regression), to expand and rearrange the material on probability, and to add a chapter on post hoc tests following analysis of variance. Almost every chapter has been expanded to include topics that experience has shown will elaborate and provide application for the central idea of the chapter. In all these changes the focus has been the same—to present only the fundamental content but with greater detail, clarity, and application.

The pattern of the book has always been to present a topic, follow with exercises to allow students to apply their skill to the topic, and then move on to the next topic. At the end of each chapter a more extensive set of problems across all topics is provided to serve as review and stimulate enough practice to keep the ideas alive by constant application. Since the exercises within chapters are there primarily to enable students to get an immediate estimate of their grasp of the current topic, practice was considered to be of secondary relevance; but since the problems at the ends of chapters were intended for review, practice, and integration, many more of them have been added.

Two features are new. Each chapter concludes with a set of key terms for student review and self-test. Also, a flowchart of the operations, decision points, and consequences is provided at the end of many chapters. Although the charts are designed primarily to review the procedures introduced in the chapter, they also serve as a bridge to electronic computing by familiarizing students with the logic and use of computer flowcharts.

Application to research has always been an incidental concern of this book; however, the third edition moves this topic more centrally into focus. Basic experimental design is presented in the early part of the book, and the ideas are used from there on. The general plan of an experimental study and how statistics relate to the plan should be fairly well acquired by students before they have finished the book.

As in previous editions, certain ideas are presented incidentally before they are presented formally. For example, students have a set of sample means and calculate the standard deviation on this set some time before the idea of the standard error of the mean is introduced. These "advance organizers" are intended to smooth the transition from one idea to another.

The appendixes provide ancillary aids. The review of mathematics in Appendix A, presented for students who have been away from formal study of algebra for some years, is limited to processes used in the book. It is concise and provides self-tests. A brief introduction to computers (Appendix D) is given, but with many designs of personal computers and a variety of larger facilities bursting onto the scene, no introduction can adequately address all beginning students. Therefore, only general ideas applicable to the widest audience have been included.

During the years of the first and second editions the comments of students have been valuable as a base for revising the book, and I am greatful for their assistance. My appreciation also extends to a number of other people; only a few will be mentioned here. Susan Hillman and Hing-Kwan Luk helped with the review and revision of the problems in the text, and the copy editor, Elizabeth P. Richardson, contributed enormously to the form, style, and readability of the text. Harriet Miller also provided valuable aid in putting the manuscript together. I would also like to express my thanks for the many useful comments and suggestions provided by colleagues who reviewed this text during the course of its development, especially to Robert Karabinus, Northern Illinois University; Dennis W. Leitner, Southern Illinois University; and Bernard J. Schneck, Bloomsburg State College. Help from these and other people has gone far in making this a more usable book. Lastly, the willingness of Pat and Steve Chase to alter their personal plans to accommodate my writing schedule I acknowledge with gratitude.

Clinton I. Chase

1

Definitions and Concepts

Modern societies survive on data. Every business must know the characteristics of its clients; intelligent consumers need facts about competing products; informed voters must have statistics on economic trends, voting records, and population characteristics. Sports fans everywhere are masters of data consumption: batting averages, percentage of passes completed, probability of making a foul shot, and hundreds of other performance records are dutifully recorded by sportswriters and poured into the ears of the public.

The scientific community relies on data even more. While the physical scientist is recording the readings of dials and gauges, the social scientist is counting people in various categories, altering conditions under which people perform various events, and assessing the impact of the alteration. All these activities produce data which must be put into some kind of order before they yield the information hidden in this mass of numbers.

Some compilations of data are merely expository since they are intended to support no argument, but others are clearly devised (and often cleverly contrived) to support a particular point of view. How much of this information can we trust? What do we need to know about how data are legitimately manipulated?

These questions lead us to the topic of this book. The popular idea of statistics as large compilations of data is *not* the topic we shall study. Although all sections of society need data, data are useful only to the extent that we can put order into the masses of numbers that social and scientific agencies collect. We therefore need methods of managing these masses of data to answer the questions we have in mind. It is these methods which are the topics of the chapters ahead. This book is not on statistics but on *statistical methods*.

Basic Concepts

The techniques described in successive chapters will do one of only two things. They will describe a set of data, or they will provide a basis for making a generalization about a large group of individuals when only a selected portion of such a group has been observed. Thus certain procedures are called *descriptive methods* because they point up a characteristic of the group being observed. Other techniques are called *inferential procedures* because they allow us to make inferences about large numbers of individuals when only a small sample from the larger group has been observed.

Look at some examples of these two methods. I have in my files IQs of all children in my sixth-grade classes in Lewiston, Idaho. What do these data tell me? Just thumbing through the files I cannot really say. But statistical methods can tell me what the typical IQ is, or between what two IQs the bulk of the students fall. These are descriptive methods in that they summarize a mass of data into a few simple ideas.

On the other hand, suppose that we want to say something about a large group of people—all ten-year-old middle-class boys, for example—but can reasonably study only a very small sample of this population. Certain methods allow us to speculate about the nature of a population once we know something about a sample from that population. These are inferential techniques because they allow us to make reasonable projections about a large group of individuals when we have studied only a small portion of the group.

We need to identify the terms *statistic* and *parameter*, but to do so we must distinguish between samples and populations. A *population* is any group of individuals all of whom have at least one characteristic in common. We typically think of populations in terms of census data, such as the population of a nation, a state, or a city. Certainly these concepts fit our definition, since they each describe a group of people who have a common characteristic—locality of residence. But populations can be defined on bases other than residence. For example, all women with naturally red hair or all men with beards or all people with annual incomes above $100,000 are populations, and the defining characteristic has nothing to do with residence. Although populations may be large, e.g., all persons within the city limits of New York on a given day, or small, e.g., all persons who serve in forest lookout stations in Clearwater County, Idaho, or all persons over one hundred years of age in California, statistical procedures typically assume very large populations.

In research in the social sciences, as well as in other disciplines, it is often impossible to study entire populations. Instead we select smaller portions of the population and from them make inferences about what the population is like. These smaller portions of a population we call *samples*.

In choosing samples our hope is to get a segment of the population which looks just like the entire population, i.e, is representative of it. There are several techniques for selecting representative samples, but the one on which the inferential techniques in successive chapters rely is *randomization*. For a sample to be truly random, all individuals in the population must have an equal chance of being selected each time a selection is made. In this way the selection of one individual has nothing to do with determining who else will be selected in the sample.

In actual practice the individuals we observe are often not truly random samples from the parent population. Therefore, we go to some length to compare the characteristics of the sample with known characteristics of the population so that we can decide whether our sample looks sufficiently like the population to make further assessments. Suppose that we know something about the expected varieties of IQs for all six-year-olds, we know something about the numbers of people in various socioeconomic levels, and we know something about the numbers in various ethnic groups in the city. With these types of information we can randomly select children in socioeconomic levels, within ethnic groups, etc., equal to the proportions in the population. This procedure produces a *stratified random sample*.

When we have selected a sample from a population, we usually compute certain characteristics of that sample, e.g., an average for some trait. Characteristics of samples are called *statistics*, but when we are dealing with populations, such characteristics are called *parameters*. Thus, if we define our population as all people taking elementary statistics at Eks University, the average height of these people will be a parameter. But if we put all the names of these students into a barrel, stir them well, and, blindfolded, draw out 10 names (replacing each name once it was drawn out—why?), the average height of this group of 10 people will be a statistic. Statistics tell us something about samples taken from given populations; parameters tell us something about populations.

Mathematical Models

Statistical procedures are derived from mathematical models of what is presumed to be "reality." For example, most of us have experienced grading on the curve. This method is based on the belief that human talents are distributed in the population in such a way that many people cluster around average, a few are above average and a few below, and a very few are very much above average and a similar-sized group very much below. If we believe that human traits are indeed distributed in this manner, mathematicians can fit to this model of the world a curve showing how proportions of the population will vary with talent level. Statisticians then exploit this curve in describing groups of people and in making inferences about populations from samples from those populations.

The well-known bell-shaped curve is not the only model we shall exploit, however. Statisticians use a variety of mathematical models of the world's phenomena as a foundation for developing their methods. It will often help students to understand a procedure if they attend first to the model then to the procedure.

Mathematical models not only tell us how given characteristics are expected to be distributed in the world but also provide a concept of the scale on which varying amounts of the condition might be measured. In the real world we typically measure less precisely than the mathematical model prescribes.

Stevens* has dealt with this problem in describing four classes of measurement scales. The *nominal scale* merely assigns numbers to categories to identify them, e.g., serial numbers on bicycles. The number does not tell us about varying degrees of quality; it merely distinguishes one style or model from another. Numbers worn by athletes similarly distinguish one player from another but tell us nothing about the size or quality of the player.

The *ordinal scale* is more sophisticated. Larger scores reflect more of the quantity or quality, but units along the scale are unequal in size. The order of winning a race is an example. The difference in speed between the first- and second-place runners is not necessarily equal to the difference between the second- and third-place runners. A teacher-made arithmetic test is also an ordinal scale. The difficulty level of one problem is not presumed to be equal to the difficulty level of other problems.

Interval scales have equal units and are therefore more precise than ordinal scales, but we do not know where true zero is on the scale. The Fahrenheit thermometer is an interval scale. We have equal units along the scale, but zero on the thermometer does not indicate an absence of all heat. Standard scores on an intelligence test are also on an interval scale.

The *ratio scale* also has equal units, but zero on the scale means an absence of the quality being assessed. Measures of length and of weight are ratio scales.

Stevens† has argued that because of the nature of various scales, not

---

* S. S. Stevens, On the Theory of Scales and Measurement, *Science,* **103:**667–680 (1946).
† S. S. Stevens, Measurement, pp. 18–36 in C. West Churchman (ed.), "Measurement: Definitions and Theory," Wiley, New York, 1959.

all statistical procedures are appropriate for all kinds of measures. If we have collected data using an ordinal scale, Stevens' view has been that any technique requiring arithmetic manipulations of these scores, e.g., averaging, is inappropriate. How can we add units that are unequal in length? Lord* has argued, however, that mathematical procedures are appropriate for any number sets, and Burke† has pointed out that if our statistical methods correspond with the mathematical model of the population with which we are dealing, we need not worry much about the nature of the measurement scale. The fact that we measure clumsily does not change the nature of the population and should not deter us from using the statistical procedures indicated by our concept of the population.

The question now arises: If we use crude measures (like ordinal scales), will our data indeed reflect the character of the mathematical model, i.e., will the data allow us to make conclusions that are reasonable in terms of the model? Labovitz‡ has shown that fairly crude (ordinal) data may well lead us to the same conclusions about populations as more sophisticated (interval) data. If this is the case, scientists should strive to refine their measuring techniques (for a variety of reasons) but should not refrain from employing complex mathematical procedures on data simply because their measuring devices do not fit the requirements of Stevens' interval or ratio scales.

Discrete and Continuous Variables

The data we apply to our mathematical models come from giving numbers to variables. A *variable* is a specified condition that can take any of a set of values. For example, height is a variable, and so are mental age and the strength of right-hand grip for a sample of ten-year-olds.

Statistical workers deal with many kinds of variables; all these kinds can be sorted into two categories. In the first category we have data obtained by counting indivisible units, such as children in a family or errors on a true-false test. We call a succession of these units a *discrete series*. The data are always collections of whole numbers, since at no time do we have a part of a unit. If we are going from house to house tabulating the number of children in the family, each report is of a

* Frederic Lord, On the Statistical Treatment of Football Numbers, *American Psychologist*, **8:**750–751 (1953).

† Cletus J. Burke, Measurement Scales and Statistical Models, pp. 147–159 in M. H. Marx (ed.), "Theories of Contemporary Psychology," Macmillan, New York, 1963.

‡ Sanford Labovitz, Some Observations on Measurement and Statistics, *Social Forces*, **46:**151–160 (1967); The Assignment of Numbers to Rank Order Categories, *American Social Review*, **35:**515–524 (1970).

whole number of children. There is no such thing as a fractional part of a child. Families are made up of two, three, four, etc., children, never three and a fraction or two and a fraction. If there are two children in a family and another child is born, the number of children changes from two to three without passing through a series of fractional parts between those two numbers. The series is a discrete one.

On the other hand, many kinds of data come in units which are divisible into an infinite number of fractional parts. This is the second category. To expand an observation from one unit to the next we must pass through a large number of fractional parts of that next unit. For example, a child growing from 43 to 44 inches in height passes through all fractions of that forty-fourth inch before actually reaching a full 44 inches. Numbers obtained from a succession of units each divisible by an infinitely large number of times constitute a *continuous series*. If we have driven 2 miles, to go 3 miles we must pass through all fractional parts of that next mile; if we walk a block, we must walk through all fractional parts of that segment of the street. All such measurement kinds of data represent a continuous series.

At times we may wish to visualize discrete and continuous data progressing together. For example, a certain mile race may mean that the runners must go around the race track four times. Although the actual distance of a mile represents a continuous variable, if we count the runner's progress by laps around the track, we are in the realm of discrete data. Either a runner has completed one, two, three laps or not. What if a runner has completed 2.67 laps? Shall we count him as having completed only two laps? But he is really nearer to completing three laps than two. When we are counting events in a discrete series paralleled by a continuous one, we typically mark off units at the halfway point between two successive numbers. For example, if we are counting inches, the unit labeled 4 inches would run from 3.5 to 4.5. The unit labeled 2 inches would run from 1.5 to 2.5, and so on, as shown in Fig. 1-1. In other words we shall think of any given number in a continuous series as the midpoint of a unit. The unit may be small ($\frac{1}{8}$ inch, 1 ounce) or it may be large (1 mile, 1 decade), size being a relative matter. In any case, the numbers we use to label the units will represent the midpoint of the unit. Thus, if our unit is $\frac{1}{2}$ inch, the number 3 would represent the third half inch in a series of $\frac{1}{2}$-inch units, and it

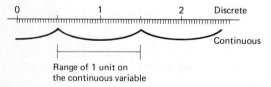

Range of 1 unit on
the continuous variable

Figure 1-1

would be the midpoint of that segment of length that runs from $\frac{1}{4}$ inch below the third half inch to $\frac{1}{4}$ inch above it. If our unit is the mile, 3 would represent the distance from 2.5 to 3.5 miles, etc.

Thus, in statistical methods we often have occasion to deal with a discrete series of units paralleled by a continuous series. When this occurs, the units in the discrete series label the midpoints of units in the continuous series and the units in the continuous series are thought of as beginning a half unit below the labeled point and extending a half unit beyond it.

Experimental Design

The scientific community has wide interest in statistical methods because they help scientists not only describe their data but also make projections from sample data to populations through controlled research on relatively small samples. Scientists in all disciplines seek answers to questions that will generalize to the broad population from which their sample came. Suppose, for example, that a particular method of learning mathematics appears to have promise. The method is tried out on a group of elementary school children while a similar group of children continues to use the standard method. A year later we compare the mathematics skills of the two groups. If the special-method group achieves notably higher scores than the standard-method group, we must first make sure that the difference between groups is not just as likely to have occurred by chance between samples from a population. If we indeed have an important difference in favor of the special method, we want to generalize the finding to all children like those on whom we ran our study.

It is through procedures like this that scientists test new methods of learning, new procedures for psychotherapy, and new drugs, and work to discover "laws" of learning, personality, physiology, and social behavior. Procedures of this sort are called *experimental designs.*

The special vocabulary tied to these designs will be used in later chapters of this book. Here are some basic terms.

Experimental group This is the group in an experiment which gets a special treatment of some kind. In the example above, the experimental group is the one that received mathematics instruction by a special method.

Control group This is the group that gets no special treatment. It is used to estimate what happens to a random group of people over the course of the experiment as a result of "normal" conditions. We do not want to report positive findings for our special treatment if its results are no better than normally expected during the time of the study. The performance of the control group provides information on the normal expectation.

Independent variable This is the special condition injected into the study which the researcher feels will have a special effect on the performance of the people in the experimental group. The independent variable in the example was the special method of mathematics instruction.

Dependent variable This is the outcome condition that we are assessing. In the example above it was mathematics achievement. If I am testing a new drug for preventing headaches so that one set of people gets the new drug and the other gets an inert powder, the occurrence of headaches would be my dependent variable. Because there may be a difference between the dependent variable as named and the dependent variable as measured, we must add two more terms.

Conceptual variable This term refers to the broad construct which carries a trait name, like anxiety, arithmetic skill, or intelligence; but before we can apply statistical procedures to analyze the impact of our independent variable, we must measure the variable, i.e., assign increasing numbers to increasing amounts of the variable so that the more of it there is the higher the number. In the example above the experimenter was observing arithmetic skills (the conceptual variable), but to do so had to give a test on arithmetic. The content of the test defined what arithmetic skills were in fact observed. Hence, the conceptual variable must have a definition in terms of how it is measured.

Operational definition This is the definition our measuring device confers on the conceptual variable. It is the observable (countable) event that represents the conceptual variable. For example, in the mathematics study, mathematics skill (the conceptual variable) was operationally defined as scores on the final arithmetic test. Operational definitions tell us how the conceptual variable was quantified. This is especially important in interpreting the results of research. For example, if in the above arithmetic study the test used to operationalize the dependent variable had been made up of 50 word problems in which the verbal part was long and complex, students might have improved their arithmetic skills in the experiment without showing improved test results because they had problems in reading. Here the operation reflects the conceptual variable poorly.

Operational definitions can also be used to the advantage of people who want to present a specific picture of social conditions. Suppose Joe Poore is arguing before the city council that more aid must be provided for the unfortunate members of the community. His charts show that 20 percent of the city's people are in poverty (his conceptual variable). This is an impressive argument; council members shift uncomfortably in their seats. But council member Mary Smart, who took a course in statistics in her college days, discovers that Poore has counted in "poverty" all those who are making $1 above the minimum wage or less (his operational definition). Such a definition would allow him to inflate his poverty figures and consequently strengthen his argument. Whenever we see data that reflect social conditions or experimental findings, we must ask: What is the operational definition of the dependent variable?

All these ideas will be focused in a simple experiment asking: Does exposure to *Sesame Street* have an effect on achievement of innercity kindergarten children? To pursue the answer we randomly select two groups of kindergarten children from 10 schools in downtown Chicago. One group (the experimental group) will watch *Sesame Street* programs daily for 1 hour; the other group (the control group) will have 1 hour of television cartoons. The independent variable is the television programming. After 6 months of this routine, reading readiness (the conceptual dependent variable) is assessed for both groups. Since the Metropolitan Reading Readiness Test (MRRT) is used, reading readiness is operationally defined as performance on this test. Table 1-1 shows the layout of our study.

There are other experimental designs, but they are essentially variations on Table 1-1. For example, there can be more than one experimental group, because the experimenter wants to observe more than one aspect of the independent variable. In the above study, one experimental group might be assigned to *Sesame Street* TV and a second to 1 hour daily of number and letter workbooks while the controls watched cartoons.

Other variations involve compensations for the fact that random assignment is not always possible. This usually entails pretesting to note (and later account for) differences between groups on important variables before the independent variable is imposed. For example, suppose we studied children who already watch *Sesame Street* compared with a group who do not watch it. Are there socioeconomic differences between these groups that might affect our results? Since our samples are not randomly selected from a defined population, generalizing results will be difficult, even with sophisticated statistical procedures.

Table 1–1   Layout of Study

Group	Assigned	Treatment	Measurement
Experimental	Random	Yes (*Sesame Street*)	Yes (MRRT)
Control	Random	No (cartoons)	Yes (MRRT)

The Role of Statistics in Experimental Design

The interpretation of the results of experimental procedures is tied intimately to statistical procedures. When one randomly selects two groups from a population and looks at their average weight, height, IQ, or any other variable, the two samples will not provide identical averages. Some differences between samples are expected because samples do not match populations exactly. This variation of samples from the population is known as *sampling error*.

In the arithmetic study we expect sampling error to cause the average score (descriptive statistic) of one group to be a little different from that of the other group at the outset. How different? Statistical procedures will tell us what to expect, so that if the average score of one group exceeds the average for the other by more than the limits revealed by statistical methods, the experimenter may wish to conclude that the differences between the groups were greater than chance, i.e., beyond sampling error. First we need *descriptive statistical procedures* to describe our samples; then we use *inferential procedures* to provide the criterion against which we can decide whether our groups differed because of sampling error or real differences in achievement. Without such procedures it would be risky to make generalizations to populations once the data from samples have been observed. Statistical methods are essential for the correct application of experimental methods.

Summary

Not everyone today is a statistician, but everyone, almost without choice, must be a consumer of statistics. Knowledge of basic methods of manipulating data is essential to intelligent behavior in a quantitative world.

This book discusses methods of dealing with data—statistical methods. Methods that tell us how to describe a body of data are called descriptive statistics. Techniques that allow us to make judgments about a large group of individuals when we have observed only a carefully chosen segment of the total group are called inferential statistics.

All individuals with one or more characteristics in common can be defined as a population. Characteristics of a population are called parameters. A collection of individuals who represent a portion of a given population is called a sample. Characteristics of samples are called statistics.

Conditions observed by social scientists provide numbers. These numbers represent a discrete series, in which each unit is an indivisible whole, or a continuous series, in which a given unit may be divided an infinite number of times. A discrete series is reported only in whole numbers, such as the number of students in a class, whereas a continuous series may involve combinations of whole numbers and fractions.

Although statistical methods have many uses, they are especially important in the design of experimental studies. Experimental studies typically begin with a sample of individuals who are given a special treatment (the experimental

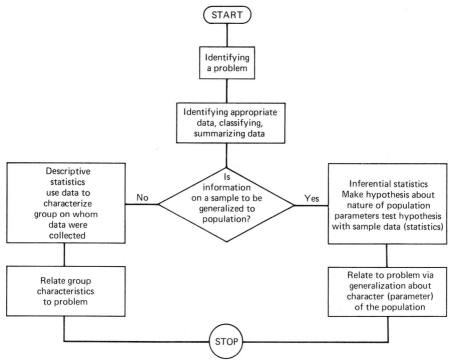

Figure 1-2

group) and a second sample that gets no special treatment (the control group). After the treatment we assess the condition we were treating (the conceptual definition of the dependent variable) with some kind of test or counting procedure (the operational definition of the dependent variable). We then apply statistical procedures to see whether the two groups look like random samples from one population or look quite different from each other. In the latter case we hope to make some conclusions about the treatment relevant to our population, an inferential use of statistics.

The flowchart in Fig. 1-2 lays out different types of data and relates them to research questions and to types of answers the data will provide.

Key Terms

descriptive methods	ratio scale
inferential procedures	variable
statistic	discrete
parameter	continuous
population	experimental group
sample	control group
random sample	independent variable
stratified sample	dependent variable
nominal scale	conceptual variable
ordinal scale	operational definition
interval scale	sampling error

Problems*

1-1 List four populations other than those tied to geographical boundaries such as city limits or state boundary lines. For each of these populations describe how you might draw a random sample.

1-2 Label each of the following conditions as a parameter or a statistic and state why you have so labeled it.
 (a) The average age of all red-haired children in the sixth grade in Centerville Public School
 (b) The proportion of children who are left-handed among a sample of 40 children stopped at random on the playground of Rogers Elementary School
 (c) The average weight of 10 girls whose names have been drawn from a hat containing names of all girls in the senior class in Centerville High School
 (d) The range in height (the difference between the shortest and the tallest) of all the males who were convicted of a felony in the State of Oregon in 1982

1-3 Decide whether the following variables represent discrete or continuous series.
 (a) The number of dogs caught by the dogcatcher in a given city on December 12, 1983
 (b) The time it takes a rat to run through a T maze
 (c) The number of children whose fathers are college professors at Eks University
 (d) The amount of change in height for a given child in a 6-month period
 (e) The number of birthdays a sample of 10 elementary school children have had

1-4 Professor Wyzee has just done a study to test the hypothesis that meditation reduces anxiety. He randomly selected 20 students from a group-therapy class and assigned them to the meditation group; 20 others remained in group therapy. Meditators were given training in meditation and asked to practice it twice a day for 2 weeks. At the end of 2 weeks both groups were given a test of anxiety and test scores produced by meditators were compared with those produced by group-therapy participants. In terms of experimental design:
 (a) What name would be given to the meditating group?
 (b) What name would be given to the group that stayed with the group-therapy class?
 (c) What is the therapy method called?
 (d) What term is applied to anxiety in this study?
 (e) What is the operational definition of anxiety?

1-5 The superintendent of schools in Eckswye School District believes that small kindergarten classes (12 children) will result in better reading readiness at the first-grade level than normal-sized classes (20 and more children). He randomly assigns children to large and small kindergarten classes until he has 10 small classes and 10 large classes. At the end of the school year he gives these children the Metro Reading Readiness Test and compares the 10 average test scores made by small classes with the 10 average scores made by large classes.

(*a*) Which children make up the experimental group? The control group?

(*b*) What is the independent variable?

(*c*) What is the dependent variable?

(*d*) How is the dependent variable operationally defined?

(*e*) The superintendent in this study concluded that small classes are superior to large ones in developing reading readiness. Is this a descriptive use of statistics or an inferential use? Why?

2

Frequency Distributions and Graphic Representation of Data

The collection of data is central to most research. These data are derived from the operational definition of the conceptual variable in which we are interested. If we want to know whether training in relaxation (the independent variable) alters anxiety (the conceptual dependent variable), we can administer a personality inventory (the operational dependent variable) to a group of people some of whom have had relaxation training (the experimental group) and some of whom have not (the control group). The test data will then be the focus of our analysis.

If we want to know the relationship between marital happiness (our conceptual variable) and the size of the town of residence, we tabulate the number of divorces per 100 population (the operational dependent variable) in cities of various sizes over a year's time. The divorce rate, as we have defined it, provides the data we shall analyze.

All researchers are collectors of numbers who seek ways to put order into the masses of numbers they collect.

Frequency Distributions

One of the most fundamental techniques for putting order into a disarray of data is the *frequency distribution*, a systematic procedure for arranging individuals from least to most in relation to some quantifiable characteristic. Suppose we are observing the weights of a group of eleven-year-old children. We can see better what our data tell us if we record the weights (in pound units) in a column with the lowest weight

15

in the group at the bottom of the column, the next possible lowest weight second from the bottom, and so on, until we have the largest weight at the top of the column. Next, for each weight we record the number of children in the group who weigh that amount.

We end up with two columns, one showing the units on the scale of measurement (in this case pounds) and the other showing how many people were found within the limits of each unit along our scale. An example is given in Table 2-1. Here we have the weights of 20 children, first in an unordered group just the way the data might have been collected, and then in a frequency distribution. You can practice this frequency distribution by first making a column of weights beginning at the bottom with 97 pounds and then tabulating beside the various weight values the number of children who were at that weight.

Frequency distributions are constructed primarily for two reasons: (1) they put the data in order to enable a visual analysis of the results of the measurements to be made, and (2) they provide a convenient structure for simple computations, as we shall see in later chapters. By scanning the frequency distribution in Table 2-1 we see that most of the weights, three-fourths of them in fact, lie between 102 and 106 pounds and that half of them fall between 102 and 104. We could not have observed these simple facts (and others like them) quickly from the unordered data, but when the unordered data are arranged into a

Table 2-1   Fictitious Weights of 20 Children

Unordered weights			Ordered weights	Frequency of occurrence
			Frequency distribution	
97	108	102	110	1
103	106	104	109	0
105	110	101	108	1
103	105	104	107	0
106	102	102	106	3
99	103	104	105	2
106	103		104	3
			103	4
			102	3
			101	1
			100	0
			99	1
			98	0
			97	1

frequency distribution, many interesting results can quickly be seen simply by scanning the distribution.

In building frequency distributions, we go through three basic stages:

1 We choose a given step to represent points on our scale. The step represents a specific quantity of the condition being observed. In Table 2-1 the step was 1 pound. But we could have made our measurements to the nearest $\frac{1}{2}$ pound and arranged our scale in $\frac{1}{2}$-pound steps, we could have gone to an ounce scale, or we could have made our scale in 2-pound steps. Steps should not be confused with units of measurement. For example, we could use a pound as our step but measure to the nearest ounce. The pound would be the step, the ounce the unit of measurement.

2 We then arrange our score values from the lowest to the highest in terms of the selected step.

3 We tabulate the number of individuals whose scores come within each step on the scale.

As an example, let us take height as the characteristic being observed and make a frequency distribution following the three stages described above. We might (1) prescribe 1-inch steps, although we could measure to the nearest $\frac{1}{4}$ inch. We would then (2) arrange our scale units, and if we are observing height among 30 high school senior boys, a scanning of the data could show the scale might begin at 60 inches and run— inch by inch—through 79 inches. We would then (3) tabulate the number of boys whose height fell within the range of each of the inch units. The result would be a frequency distribution for height for this group of individuals. It might look like Table 2-2.

In Table 2-2 we have the steps on our scale in the left-hand column and the number of boys in each step in the right-hand column. A step in the scale, in this case 1 inch, is called a *step interval* or simply an *interval*; the number of individuals in a given step interval is called the *frequency* (*f*) for that interval.

Several decisions must be made before we can construct a frequency distribution. One of these decisions deals with how to classify people whose measurements involve fractions of steps. Where do I tabulate the boy who is 68.75 inches tall? Where do I include the one who is 71.25? Before I begin to construct a frequency distribution, I need a few basic rules that will tell me how to classify these cases so that I shall be consistent in my classifications of all persons whose measurements involve fractions of units.

When the frequency distribution is based on a scale with a defined unit of measurement and individuals are measured in fractional parts of the defined units, we usually handle them in the following manner.

Table 2-2 A Frequency Distribution of Height for a Sample of 30 High School Senior Boys

Scale in inch units	Number of boys (f)
79	1
78	0
77	1
76	0
75	2
74	1
73	0
72	2
71	2
70	4
69	5
68	3
67	0
66	3
65	2
64	0
63	1
62	2
61	0
60	1

If for any individual the measurement contains a fraction which is less than half a unit, it is included in the table under the last whole unit in the scale *before* the fraction. Thus, a boy who is 68.25 inches tall would be tabulated at 68 inches in the frequency distribution of Table 2-2; a boy 72.38 inches would be tabulated at 72. Fractional parts that are equal to or more than half a unit represent amounts which are nearer to the next larger unit on the scale than to the previous unit and are therefore tabulated under that next larger unit. Thus, a boy who is 68.75 inches tall would appear in Table 2-2 under the unit labeled 69 inches.

This leads us to determining the *real limits* of a given step interval. For a given point on a scale the real limits begin a half unit before that point and extend up to a half unit beyond that point. Therefore, in Table 2-2, the real limits of the unit listed as 70 are actually from 69.5 to 70.5, as shown in Fig. 2-1. Any measurement which falls from 69.5 to 70.5 would be tabulated within the class interval of 70. Thus, in Table 2-2 the four cases that are in the interval labeled 70 may range anywhere within the real limits of the interval 69.5 up to 70.5.

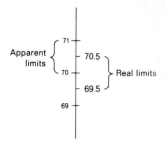

Figure 2-1.

We must identify the real limits because many of our data come from continuous variables which parallel discrete units. The *apparent limits* identify the discrete units and are the midpoints of the ranges of the continuous variable that parallel the units. For example, when I measure height, I have all possible amounts of height, but if I decide to measure to the nearest inch, a discrete unit, I must have a rule to tell me in which discrete unit a given measurement belongs. Suppose I put my data on height into a frequency distribution with 1-inch intervals. I pick out one interval to inspect. The interval is labeled 70 for 70 inches. Joe is $69\frac{3}{4}$ inches tall. In which interval does he belong? According to our rule, the unit of which 70 is the midpoint begins half a unit, i.e., $\frac{1}{2}$ inch, below 70 and extends half a unit above 70, or 69.5 to 70.5. Therefore Joe belongs in the interval labeled 70, although he is slightly shorter than 70 inches.

A second decision to be made in regard to frequency distributions is the number of measurement units that will be included within a step interval. In Table 2-2 our unit was 1 inch, and we included only 1 unit per step interval. However, suppose we had decided to use $\frac{1}{4}$ inch as our unit. Then by using 1 inch as our step interval, we would have been including four measurement units within each interval ($\frac{4}{4}$ inch equals 1 inch, the step-interval width). Many measurement units are made up of smaller parts, so that when we use these units, we are in effect using groups of the smaller parts that make up the unit. For example, the unit used in measuring a football field is the yard. But 1 yard is a group of 3 feet, or it is a group of 36 inches. Agreeing that 3 feet make up 1 yard is pretty much an arbitrary arrangement. We might do as well if we all agreed to use, say, 2 feet or 4 feet instead. The point is that measurement units are often made up of smaller parts, and how we group those parts into larger units is largely an arbitrary matter, workable only as long as we all agree on the grouping method.

With this thought in mind, let us turn again to Table 2-2. There are quite a few class intervals which have no frequencies, making our distribution uneven. Suppose instead of using 1 inch for our step interval we all decided to use 2 inches per interval, just as we could decide (if everybody agreed) to use 2 feet for a yard. In the table this would not spoil the integrity of the inch as the basic unit of measurement. It influences only our method of grouping the inches, just as grouping feet into a yard does not disturb the adequacy of the foot as a unit of measurement. Now using our 2-inch interval, let us make up a second frequency distribution from the same data as before. Our new frequency distribution will look like Table 2-3.

The data in Table 2-3 are listed in 10 step intervals, instead of the 20 in Table 2-2, but the distribution has no breaks in it, as it did before when six intervals contained no frequencies. The result of compressing distributions by including several measurement units per interval is known as *grouped data* and is especially helpful when the range of scores is wide, wider even than in Table 2-2. The method reduces the number of intervals to simplify handling them in subsequent computations and makes them meaningful in terms of the number of cases that come within each interval.

The question now arises: How many intervals are a "good" number to work with: As a rule of thumb something near 15 intervals is usually quite satisfactory, but we would certainly not want fewer than 10 and usually not more than 20. If we use too few intervals, the shape of the distribution may become distorted; too many intervals may not compress the data enough to be of advantage in subsequent manipulations.

The number of intervals to be used also provides us with the basic information we need for estimating how many measurement units will

Table 2-3 A Second Frequency Distribution of Height for a Sample of 30 High School Senior Boys

Intervals based on inch units	Number of boys f
78–79	1
76–77	1
74–75	3
72–73	2
70–71	6
68–69	8
66–67	3
64–65	2
62–63	3
60–61	1

go into each interval. The procedure, represented by formula (2-1), goes like this:

$$\frac{(\text{Largest value}) - (\text{smallest value}) + 1}{15} = \text{number of units per interval}$$

(2-1)

First, we subtract the smallest quantity in our distribution from the largest and add 1 to that difference. This gives us the total range of scores. Then we divide the result by 15 and round the answer, whenever possible, in the direction that will make the number of units in an interval an odd number. (The reason for this will become evident later; for now it suffices to say that we shall want to find the midpoint of an interval, and if the intervals have an odd number of units in them, the midpoint will always be a whole number. If the interval width is an even number of units, the midpoint of an interval will always end in a fraction.)

Suppose in a group of 30 six-year-olds the highest IQ is 125 and the lowest is 90. How many IQ points should go into each interval? First, we subtract 90 from 125 and add 1. This comes out to 36. Dividing this by 15, we find an answer of 2.4 units. An interval 2 units long would result in 18 intervals; 3 units per interval would result in 12 intervals. Since we agreed to go with an odd number of units in an interval, we choose the latter width, 3 units of IQ per interval.

We are now ready to portray a set of data by putting them into a frequency distribution. The steps in the procedure are summarized with an example in Table 2-4.

Another question arises from grouping more than one measurement unit into a step interval: What are the real limits of intervals that contain more than 1 unit?

Nothing is changed when we decide to group several units into one interval. We still are measuring in terms of the basic unit, even though we have grouped several of these units into a single interval. Therefore, the *real limits* are based on fractions of the *unit of measurement*, not upon fractions of the step interval. For example, the first interval in Table 2-3 is 60–61. Its real limits begin $\frac{1}{2}$ unit before the designated beginning point of the interval and end $\frac{1}{2}$ unit beyond the designated ending of the interval. Thus, the real limits of the interval designated 60–61 are from 59.5 to 61.5. The real limits of the next interval begin at 61.5 and go on to 63.5, etc. A graphic representation is shown in Fig. 2-2.

It is well to remember in working with frequency distributions like Table 2-3, where measurement units have been grouped within intervals, that by so grouping we do not abandon our unit of measurement. The last unit in the interval 60–61 is still 60.5 to 61.5. Grouping

Table 2-4   Construction of a frequency distribution

1. Locate the largest and smallest values

2. Determine the number of units per interval, i.e.,

$$\frac{(\text{Largest value}) - (\text{smallest value}) + 1}{15}$$

3. Arrange intervals in order with smallest score values at the bottom and largest at the top

4. Tabulate scores within appropriate intervals

5. Change tabulations to frequency of occurrences

6. Sum frequencies; the sum should equal number of scores originally recorded

---

Example   Given the following scores on a 50-item spelling test taken by 25 children in grade 8:

30	40	26	43	30
47	22	16	34	33
30	27	36	24	18
27	12	37	28	40
6	49	15	32	34

Step 1:   49 and 6

Step 2: $\dfrac{(49 - 6) + 1}{15} = 2.93$

3 units per interval will be used

Step 3:

Interval	*Tabulation*	*f*
48–50	1	1
45–47	1	1
42–44	1	1
39–41	11	2
36–38	11	2
33–35	111	3
30–32	1111	4
27–29	111	3
24–26	11	2
21–23	1	1
18–20	1	1
15–17	11	2
12–14	1	1
9–11		0
6–8	1	1
	Sum =	25

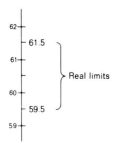

Figure 2-2

it with another unit (59.5 to 60.5) does not change this unit in any way so far as its limits are concerned, just as grouping inches within a foot does not change the nature of any of those 12 inches. Thus in determining real limits of an interval, the basic unit of measurement still is used in setting real limits regardless of how many units are grouped into a single step interval.

Exercises*

2-1 On a group of 50 ten-year-olds we have IQ scores which we want to put into a frequency distribution. The highest IQ is 144 and the lowest 78.
(a) What will our scale *unit* be?
(b) How many units will be used in an interval?
(c) How many intervals will this make?
(d) What are the apparent limits of the first interval?
(e) What are its real limits?

2-2 We have the weights of 75 high school freshmen. The heaviest weighs 192 pounds 8 ounces and the lightest 89 pounds 14 ounces.
(a) If our scale unit is 1 pound, how many units should we include in the interval?
(b) How many intervals will this make?

2-3 Given the following IQ data, construct a frequency distribution:

114	116	97	95	112		85	94	128	123	106
80	93	118	113	104		121	101	138	122	112
118	112	109	144	105		125	129	133	103	

2-4 Using the data given in Exercise 2-3, construct a frequency distribution using five intervals. Now compare the shape of this distribution with that of the one constructed in Exercise 2-3. Does this give you some idea why using too few intervals is not desirable? (Be on the lookout for more on this topic later.)

*Answers to Exercises are given in Appendix E.

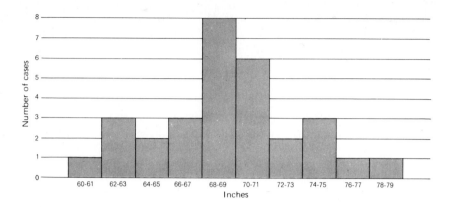

Figure 2-3

Graphic Representations of Data

There are many opportunities to present frequency distributions in graphic form and several standard procedures for graphing data. The choice of procedure depends upon what characteristic we wish to emphasize.

The Histogram

One way frequency distributions can be presented is in the form of a *histogram*.* The procedure is first to construct a frequency distribution and then within the *real limits* of the first interval to construct a bar whose length represents the frequencies in that interval. The same procedure is repeated for each succeeding interval until all intervals are depicted as adjacent bar graphs. A histogram of the data in Table 2-3 is shown in Fig. 2-3.

In Fig. 2-3 the frequencies in a given interval can be interpreted either by the height of the various bars or by their areas. For now we probably should think of the frequencies in terms of the height of the bars, but later we shall have occasion to interpret a distribution in terms of relative areas of the histogram associated with given segments on the base line of the distribution. If we think of a given interval in terms of the area of its bar and the relationship of the frequencies in, say, two intervals as the relative magnitude of the two areas, we shall be prepared for a later chapter, where the area in a graph which is cut off between a given segment of scores assumes primary importance.

The histogram represents the data in terms of distinct and discrete

---

*The Greek word *histos* refers to something vertical, e.g., the mast of a ship or the lengthwise threads in the loom. Karl Pearson, the English statistician, originated the term *histogram*.

segments (intervals) and emphasizes the relative magnitude of each of these segments. If this is the characteristic of the data we wish to illustrate, the histogram is the technique to use.

The Frequency Polygon

We often wish to reflect the continuous nature of our data. In this case we would like to illustrate differences in frequencies from interval to interval as they occur in a steady incremental fashion. We can do so by placing a point at the center of the top of each bar in Fig. 2-3 and connecting the points with lines. The resulting figure is called a *frequency polygon*, which for our data would look like Fig. 2-4.

In building a frequency polygon the points we plot reflect interval frequencies and are plotted at the midpoints of the respective intervals. In Fig. 2-4 the interval 66–67 had three cases in it. The point is opposite 3 on the vertical scale (the *ordinate*) and is at the midpoint of the interval 66–67.

Smoothing

We often graph data from a sample of people who are supposedly typical of the population from which the sample came. For example, we may wish to show the distribution of IQ scores for ten-year-old children, but we have scores on only 100 cases. If we put these data into a frequency polygon like Fig. 2-4, we would probably see irregular peaks in the graph since our sample does not amply represent the entire population. Such irregularity might well diminish if we could graph data from the entire population; i.e., the graph would show a minimum of irregularity, or, become *smooth*.

Figure 2-4

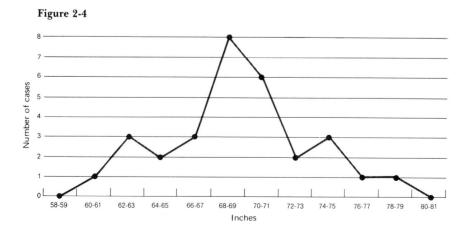

Luckily, there is a simple technique for smoothing frequency polygons. We do a kind of averaging between adjacent intervals in the distribution. The smoothed frequency for a given interval is the average of the frequencies of that interval and the intervals on each side of the one being considered.

From Fig. 2-4 let us determine the smoothed frequency for the interval 62–63. We work with the frequency of that interval, 3, and the frequencies of the interval below and above it, which are 1 and 2. The sum of the three frequencies is 6, divided by 3 intervals is 2. The smoothed frequency for the 62–63 interval is therefore 2. This procedure is repeated for all intervals, and the smoothed frequencies are then graphed.

Sometimes students are uncertain about handling the first and last intervals in a distribution, since the first interval contains no frequency below it, and the last no frequency above. The process is the same as that used in constructing the histogram. The interval below the first one in which we had frequencies is certainly there but simply has no cases falling within it. Its frequency is 0, and so is the frequency of the interval beyond the highest one in which we have cases. Therefore, in smoothing our first interval 60–61 as it appeared in Table 2-3 we begin with the frequency of that interval, 1, add the frequencies of the interval below, 0, and the one above, 3, and divide by 3. The smoothed frequency for the interval 60–61 is 1.3.

How do we handle interval 58–59, the one just below our first frequencies? Again, the intervals 56–57 and 58–59 are certainly real, even though they have no frequencies. We handle the interval 58–59 like any other. Its frequency is 0, the frequency of the interval below is also 0, but the frequency in the interval above is 1. We add 0, 0, and 1, and then divide by 3, and our smoothed frequency for the interval 58–59 is .33. The smoothed frequency for the interval 56–57 is 0 (why?).

We deal with interval 78–79 and 80–81 in a similar fashion. The frequency in the interval 78–79 is 1. The interval below it also has a frequency of 1; the interval above has 0. The average is .67, the smoothed frequency for the interval 78–79. The frequency in the interval 80–81 is 0, in the interval below it is 1, and in the interval above it is 0. Adding 0, 1, and 0 and dividing by 3, we get a smoothed frequency of .33 for the interval 80–81. What is the smoothed frequency for the interval 82–83?

Figure 2-5 shows the same data as Fig. 2-4, except that smoothing has been applied, reducing the irregularities seen in Fig. 2-4. Since marked irregularities in graphs are often a result of properties of the sample upon which the frequency polygon is based and not of characteristics of the population from which the sample comes, smoothing usually gives us the best picture of what the population

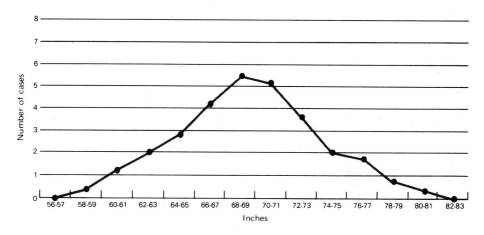

Figure 2-5

distribution would look like. Therefore, if we are using a sample of cases to represent the entire population, it is advisable to use the smoothing procedure. For example, where IQ scores were available for 100 ten-year-olds, if we are attempting to illustrate the shape of the distribution of IQs for all ten-year-olds, smoothing would be advised for these data.

Skewness and Kurtosis

Sometimes only a few frequencies fall below the central pile up of frequencies, but many continue to string out far above it; or vice versa. Outcomes such as these are shown in Fig. 2-6. These unevenly spread

Figure 2-6

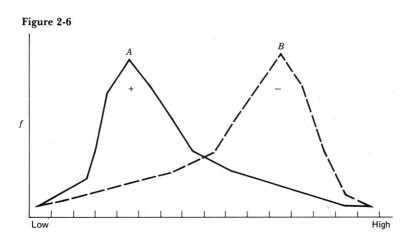

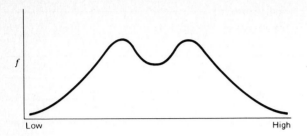

f

Low   High

Figure 2-7

score distributions are said to be *skewed* or characterized as having *skewness*. If the tail of the curve extends far above the central peak, the distribution is said to be *positively skewed* (curve *A*, Fig. 2.6); if the tail extends far below, the distribution is said to be *negatively skewed* (curve *B*, Fig. 2.6).

A number of circumstances can cause skewness. For example, giving a very easy test to a group of students may result in a negatively skewed distribution of scores. Many students will get high scores near the maximum, but there will be a trickling off of scores from the less capable students on the lower end. Sometimes skewness is an artifact of the units on our scale. If I do not discriminate well between persons on the lower end of a continuum but do discriminate between persons on the upper end, scores will bunch up in the lower intervals and will spread out in the upper ones. This will produce a positively skewed distribution.

Some distributions of scores produce a regular, bell-shaped curve similar to the one in Fig. 2-5, but sometimes the curve is more peaked near the center or flatter in the center than Fig. 2-5. These variations from flatness to peakedness, referred to as *kurtosis*, will be discussed in Chap. 6.

Occasionally a distribution has two nonadjacent intervals both containing a large number of frequencies. This produces two high points in our frequency polygon (Fig. 2-7). We call such a distribution *bimodal*; the meaning of the term will become clear in Chap. 3 when we discuss the mode.

As previously noted, the data in Fig. 2-5 piled up near the center of the distribution, with few cases appearing at either extreme of the scoring range. This is characteristic of many kinds of data collected in the social sciences; in fact, if enough cases are included, this type of graph, even without smoothing, begins to look bell-shaped. This bell-shaped curve is called the *normal curve*; it can be plotted by using a specific formula which will be given later when we have seen how to compute and use the elements that go into it.

Models

The normal curve is one mathematical model of how quantities of a given characteristic are distributed in the world. Although it is useful in helping us decide how often a given value will occur, other mathematical models also tell us about what quantities to expect in reference to given characteristics we wish to study. One such curve deals with ratios of two observations; if observations of a given event are from two samples of persons in a common population, the ratio of the two observations should be near 1.0. For example, if you measure the height of two samples of six-year-old girls, the ratio of the average height of sample A to the average of sample B should be near 1.0. Since samples do not always match populations exactly, some variation in ratio size will occur due to sampling error. Nevertheless we expect only a few samples to deviate noticeably from 1 and very few to deviate conspicuously from 1. Our curve of expected numbers of sample pairs that correspond with different ratio values might look like Fig. 2-8, a

Figure 2-8

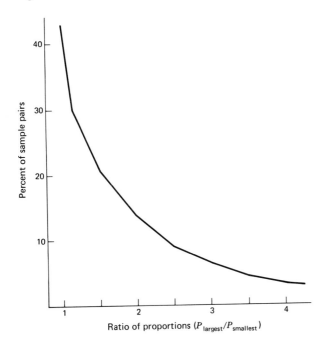

graph of ratios of statistics we shall learn about later. If our ratios consistently fall far into the tail of the curve, we may conclude that something other than sampling error is at work: possibly our samples are coming from different populations.

Social scientists use mathematical models like the normal curve to help them study the world and make decisions about it. We shall indeed have occasion to use some curves like the one in Fig. 2-8 in the chapters ahead.

Exercises

2-5   A group of 40 children's records were screened to see at what age in months each child learned to ride a tricycle. The results were as follows:

32	37	21	23	41		32	29	32	28	29
28	30	31	37	31		26	35	34	36	25
33	30	32	29	30		35	22	16	25	38
27	35	37	31	33		45	39	42	31	33

(*a*) Arrange these data into a frequency distribution.
(*b*) Construct a histogram of the data.
(*c*) Construct a frequency polygon from the data.
(*d*) Construct a smoothed curve of the data.

2-6   Which procedure of Exercise 2-5, (*b*) or (*c*), best illustrates the data as continuous? Why?

2-7   Which graph of Exercise 2-5, (*c*) or (*d*), is probably most like the curve for a sample of 1000 instead of only 40 cases? Why?

Grouping Errors

Arranging data into intervals which are larger than the unit of measurement has some conveniences but is not without problems. We assume that within a given interval the frequencies are equally distributed across the interval. For example, suppose we have an interval of 60–64, a 5-point width, and this interval contains 10 cases. We hypothesize that the data look like Fig. 2-9, with two frequencies in each of the 5 units in the interval. Often this is not the case. It is very likely that one unit in the interval will have more frequencies than another, e.g., six persons having a score of 61 and only one person having each of the rest of the scores in the interval, thus causing an

Figure 2-9

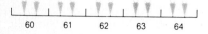

60   61   62   63   64

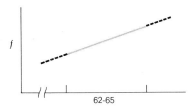

Figure 2-10

imbalance of frequencies at one end of the interval. We are not especially concerned with this problem, however, because across many intervals the effects of the imbalances tend to average out.

However, there is a problem in grouping scores into intervals which is tied to the assumption of frequencies being equally distributed throughout the interval. Looking at most distributions of scores, we see that beginning at one end of the distribution and moving toward the middle of the score range, the frequencies tend to increase (Fig. 2-5). Now let us chop out the rather wide interval 62–65 from Fig. 2.5 and see what that portion of the curve would look like. Clearly in Fig. 2-10 the frequencies in the lower half of the interval are not equal in number to those in the upper half, nor would we expect this condition to be averaged out in the next interval because Fig. 2-5 shows the frequencies in successive intervals steadily increasing as we approach the center of the distribution. Our assumption of equal spread of frequencies within an interval probably is not entirely correct and does in fact lead to some small errors in computations, especially if the intervals are wide. This is one reason we hesitate to have too few intervals, since the fewer the intervals the wider they must be and the greater our grouping error will be.

Cumulative Frequencies

Occasionally we want to describe a given interval in a distribution in terms of the total number of cases that have appeared up to and including that interval. What we are saying about our distribution is that from interval X down, so many people have appeared. For this purpose, we illustrate our data by using a cumulative frequency graph; i.e., each interval includes not only its own frequency but also the sum of all frequencies below it.

The data in Table 2-3, which have been converted into the graphs in Figs. 2-3 to 2-5, are here incorporated into the cumulative frequencies in Table 2-5. Here the first interval is composed of all the cases below and included in that interval. Since no cases came in the interval below, we include only the frequencies of the 60–61 interval, or 1 case. The

Table 2-5 Construction of Cumulative Frequencies Based on the Heights of 30 High School Senior Boys

Intervals based on inch units	Number of boys in each interval	Cumulative frequency	Cumulative percent
78–79	1	30	100
76–77	1	29	97
74–75	3	28	93
72–73	2	25	83
70–71	6	23	77
68–69	8	17	57
66–67	3	9	30
64–65	2	6	20
62–63	3	4	13
60–61	1	1	3

next interval is tabulated as containing its own frequencies plus all those below it, or 3 plus the 1 in the first interval, or 4 cases. We proceed in this manner through the last interval, 78–79, where there is only 1 case, which is added to the frequencies of all intervals below, becoming 30 cases.

Since frequencies themselves are often less informative than percentages of the total cases, we often see cumulative frequencies converted into cumulative percentages, as for the data in Table 2-5. It is probably more meaningful to say that 57 percent of the boys were 69 inches tall or less, than it is to say that 17 boys were in this range. The value of 17 is relative to the total number, and this relationship is reflected in the percent.

Cumulative frequencies and percentages can also be graphed to provide a pictorial representation of the distribution. The data in Table 2-5 are graphed in Fig. 2-11 as cumulative percentages. Here the data appear to have a "lazy" S shape. This characteristic shape of cumulative percentage data is often referred to as an *ogive*.

Figure 2-11

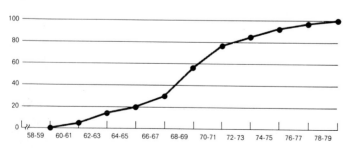

Exercise

2-8 From the data given in Table 2-4 a frequency distribution was constructed. From this frequency distribution:
(*a*) Compute cumulative frequencies and cumulative percentages.
(*b*) Make a graph of the cumulative percentages.

Bivariate Graph

Sometimes we want a graph to show the relationship between two conditions which we have observed. For example, how do height and weight change together? How does student achievement change with changes in amounts of training for teachers? When two variables are plotted together, the result is called a *bivariate graph*. In this type of graph two measurements must be made for each individual observed. One variable is scaled along the *ordinate* (the vertical axis), and the other along the *abscissa* (the horizontal axis). A point is placed on the graph where a given subject's ordinate and abscissa value intersect. This point is called the *coordinate*. An example will illustrate this procedure.

The following fictitious data represent hours of study and number of items correct on an arithmetic test for four sixth-grade boys.

Hours of study	1	1.5	2	2.5
Test score	10	14	18	22

A bivariate graph of these data, with hours studied recorded on the abscissa and test scores on the ordinate, will look like Fig. 2-12. Here the graph shows how one variable is expected to change when there is a change in the other one. The data in Fig. 2-12 show that for each hour studied after the first hour, the test score is expected to increase 8 points. From the graph we could also determine expected test scores for various fractional parts of hours. For example, $1\frac{3}{4}$ hours of study should produce a test score of 16.

Figure 2-12

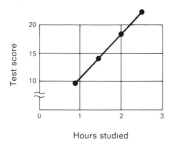

Hours studied

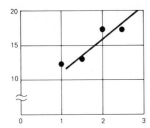

Figure 2-13

As presented in Fig. 2-12, test scores are said to be a mathematical function of hours of study; as one condition changes, a predictable change occurs in the other.* For most human traits, however, relationships are not as precise as those in Fig. 2-12, partly because our measuring tools are imprecise and partly because variables do not change together with complete regularity. Nevertheless, we use a functional relationship between two variables to predict one from the other, realizing that a margin of error exists. Suppose the points in the above problem looked like those in Fig. 2-13. We can still plot our line through the dots, coming as close to each as a straight line can, and from this line predict test scores from hours studied. Our predictions would not be as exact as they were in Fig. 2-12, but from these data we could determine the range of our error. If this device will prove predictions more reliable than mere chance, even with our range of error, it will serve a purpose. In a later chapter this idea will again appear. At that point the predictions of one variable from another will be referred to as *regression*, and our plotted line will be a regression line, but for now Figs. 2-12 and 2-13 are bivariate graphs.

In Figs. 2-12 and 2-13 the data fit a straight line. We say the variables show a *rectilinear* relationship. But sometimes the straight line does not fit the data as well as a curve does, in which case we say the relationship is *curvilinear*.

Summary

Several simple techniques for bringing order to data are tied to the frequency distribution. Data can be grouped into segments (class intervals) that are longer than the unit of measurement without distorting the basic character of the score distribution. By arranging class intervals from the lowest score value to the highest and tabulating the number of cases in each interval we construct a frequency distribution.

Frequency distributions can be graphed in several ways. The bar graph, with adjacent bars representing intervals and their frequencies, is called a histogram. If the midpoints of the upper end of each bar on the histogram are

*Students unfamiliar with the idea of functions should consult Appendix Sec. A-4.

located and a continuous line is drawn to connect these points, the resulting figure is called a frequency polygon. Frequency polygons are often rather irregular (sawtoothed) in shape. Smoothing techniques can be applied to the data to reduce this irregularity.

Sometimes instead of how many individuals fall at a given point on our frequency distribution we want to know how many people come below a given point. For this purpose the cumulative frequency table is used; it is constructed by adding the number of cases in an interval to the total number of cases that have preceded this interval. A typical graph of cumulative percentages takes an S shape and is called an ogive.

The relationship between two variables can be shown graphically by putting the scale of one variable on the ordinate and the other on the abscissa and locating the coordinate points. With this method we show the change in one variable that is expected with a given change in the other variable, a procedure which provides a basis for predicting one condition from another.

Key Terms

frequency	smoothing
interval	skewness (negative, positive)
frequency distribution	kurtosis
apparent limits	bimodal curve
real limits	grouping errors
grouped data	cumulative frequency
histogram	ogive
frequency polygon	bivariate graph

Problems

2-1 What are the real limits of the following intervals (assume that we measured to the nearest whole number; i.e, our unit of measurement is 1)?
(*a*) 1–5   (*b*) 56–58
(*c*) 28–32   (*d*) 101–103

2-2 Give the width of the interval and the number of intervals for each of the situations in the following table (assume each problem has a unit of 1):

	Highest score	*Lowest score*	*Interval width*	*No. of intervals*
Example	40	8	3 points	11
(*a*)	56	26		
(*b*)	101	32		
(*c*)	17	−2		
(*d*)	116.25	11.75		

2-3 The following data, collected by a school psychologist, represent raw scores on a clerical aptitude test.

20	25	26	48	44		43	40	42	29	39
23	26	24	47	45		28	29	41	38	36
27	44	42	43	29		37	34	31	33	30
42	43	28	41	29		36	35	30	32	31

(a) Make a frequency distribution beginning with the interval 20–21.
(b) Make a frequency polygon from these data.
(c) How would you describe this distribution?
(d) Construct a smoothed curve from the frequency polygon.
(e) From the frequency distribution construct cumulative percentages. What percent of the group falls below a score of 36? Below 34?

2-4 Construct a second frequency distribution from the following scores made by children on a 50-word spelling test. Make the first interval begin with a score of 13.

48	37	25	28	27		41	14	17	20	18
35	51	36	35	16		28	20	21	30	25
26	33	30	32	20		24	20	33	23	32
24	21	29	30	38		20	29	44	21	32
15	20	23	29	30		13	17	41	18	20

(a) Is this distribution positively or negatively skewed?
(b) Is the distribution bimodal or unimodal?
(c) If we smoothed this distribution, what do you think would be the effect on skewness? On modality?

2-5 Five children made the following scores on two spelling tests:

Child	A	B	C	D	E
Test 1	8	7	6	4	2
Test 2	6	6	3	2	1

(a) Construct a bivariate graph from these data.
(b) If a pupil had a score of 5 on test 1, what score would you predict for test 2?

2-6 I have taken a random sample from the voter rolls in Lake County and found the proportion of females in the sample to be .75; that is, 75 percent of the sample is female. I take a similar sample in Jefferson County and find the proportion of females there to be .50. The ratio of my sample proportions is .75/.50 = 1.5. From Fig. 2-8:

(a) What percent of pairs of random samples should give me ratios of 1.5?
(b) If I repeat the study in Monroe and Posey counties and find a ratio of 3.5, what percent of pairs of random samples is expected to give me a ratio of 3.5?
(c) Would you say that differences that produce ratios as large as 3.5 are so unusual that they are not expected among "typical" samples of voters?

3

Measures of Central Tendency

Probably the most striking thing about people is that they are all different. In a typical social gathering no two people are likely to be exactly the same height and weight, nor are they apt to be identical on any other common measurement. Yet differences between them in a given trait are not without limits. For example, we rarely find an adult who is as tall as 7 feet or as short as 4 feet, the height of most people ranging around a point located centrally between these extremes. Because so many measurements cluster near the middle of the distribution, we say scores have a *central tendency.*

Since so large a portion of the group clusters near this central level, we think of that point as representing the typical characteristic for the group. This most typical point is what we compute when we find an average. Averages and similar values are called *indicators of central tendency* since they identify that point in a distribution around which other scores seem to group.

We are often faced with problems which demand knowledge of typical conditions. If I am going to buy a set of reading books for my fifth grade, I need to know the reading skills of the typical (or average) fifth-grade child; if I am studying heart disease and diet, I must determine the average number of attacks per thousand population for people on diet A compared with the average number for people on diet B; if I am explaining divorce rates by socioeconomic conditions, I must determine typical education and income levels for people having high divorce rates compared with people having low divorce rates; if I have calculated the average number of sexual experiences for a sample of fifteen-year-olds, I can hypothesize what the average is for the population from which the sample came. We use indicators of central tendency extensively in building and altering our belief system. This is also true for scientists as they attempt to build a data base to explain, predict, and control natural phenomena.

The above examples point to three principal uses of indicators of central tendency:

1   They indicate the amount of a given condition which is typical for a defined group of individuals.

2   They provide a basis for comparing a condition in one group with the same condition in a second group.

3   They allow us to estimate a typical condition for many individuals when we have measurements on only a portion of the total number of those individuals.

The chapters ahead will provide specific statistical procedures for each of these applications of indicators of central tendency. In this chapter we shall deal only with the first, the descriptive use of central tendency to indicate the typical quantity of a given characteristic.

You may think that "average" is the only indicator of central tendency, but in fact three indicators are widely used by statisticians, the *mean*, the *median*, and the *mode*. Since each defines the point of central tendency in a slightly different way, we shall deal with them separately. Keep in mind, however, that each is an attempt to identify a typical condition for a defined group of individuals.

Exercises

3-1   List five specific situations in which you have been involved where knowledge of the typical or average condition was important in describing a group of people.

3-2   List three specific situations in which you have been involved where averages were useful as a basis for comparing groups of people.

3-3   List three specific situations where we may wish to estimate an average condition for a large group of people when we know the average of a sample from that large group.

THE MODE

Suppose we want to find the central point in a distribution of scores around which the bulk of the data seems to congregate. A quick scan of the distribution shows that several scores near the center of the range

Table 3-1 Procedure for Locating the Mode

	X	f
1. Put the data into a frequency distribution, where X = score and f = number of people	21	1
	20	0
	19	4
	18	12
	17	19
	16	14
	15	9
	14	10
	13	4
	12	2
	11	1

2. Locate the score value that has the largest frequency

Mode value = 17

3. If data are grouped in intervals:
   (a) Locate the interval with the largest f, here 25–27
   (b) Locate the midpoint of this interval by dividing the score point width of the interval by 2, and adding the result to the lower limits of the interval

X	f
37–39	1
34–36	4
31–33	3
28–30	12
{25–27	18
22–24	13
19–21	8
16–18	10
13–15	3
10–12	1

Score point width = 3
$3/2 = 1.5$
Lower limits of interval = 24.5
Mode = 24.5 + 1.5 = 26.0

have relatively high frequencies. From these we might select as the most typical score that one scale value which appears more frequently than any other. We call this point the *mode*; it is defined as that scale value which occurs most frequently* in a distribution. Since scores usually pile up around a central point, it can be assumed that the point of central tendency is that value which accumulates the greatest frequency of cases. This is the point we call the mode. The procedure for locating the mode in a distribution is given in Table 3-1.

Because the score point which identifies the mode depends for its

* Mathematicians often call this the *crude mode*.

location on only a fraction of the total number of observations in a distribution, the mode in one sample of data may not correspond with the mode from a second sample from the same population. For this reason the mode is an unstable indicator. Nevertheless it still has value because it is easily calculated and consequently is a quickly obtained indicator of central tendency.

Now that we have dealt with modes, we should think back to Chap. 2, where we looked at characteristics of frequency distributions. One distribution was identified as *bimodal*. Now we understand this as a distribution in which two nonadjacent intervals have many frequencies, i.e., there are two intervals either of which could be identified as the mode.

The Median

If research data tend to cluster around a value centrally located between the extreme values, why not locate the exact central point and label it as most typical for the characteristic being observed? We can begin at the lower end of the distribution and count up until we find the point below which exactly 50 percent of the cases fall. This will be our central point. It will also be the *median*. The median is simply that point in the distribution which divides the total observations into two equal parts. It is not influenced by how far extreme scores may range in a given direction, because the range of these scores does not change the point that divides the distribution into two equal-sized groups. For example, the median in the following two sets of scores is the same. In each case 6 divides the set of scores into two equal-sized groups.

Sample 1	Sample 2
1 3 5 6 7 9 10	1 3 5 6 7 15 21

The fact that the second distribution has two scores (15 and 21) that deviate far beyond any score in the first distribution does not alter the value of the median, which is 6 for both samples. The performance of sample 1 is not identical to that of sample 2, but the median is not sensitive to the difference since widely deviating scores do not influence the midpoint in a distribution.

The data that researchers collect are not always additive because the distance between points is not the same length on all segments of the scale. For example, the same number of seconds does not elapse between the runners who are first, second, third, fourth, and fifth in a race. Such data only provide us with a basis for ranking individuals; therefore, we may not wish to add quantities involved since not every step on the scale is equal to all other steps. The median, because it is determined simply by a counting procedure, is appropriately used with these data. Other techniques involving even the simple arithmetical process of addition may be less appropriate.

The median is a more stable figure than the mode; two random samples from a population are more likely to agree on medians than on modes. It is, however, a less stable characteristic than the mean, as we shall soon see.

Locating the median in a distribution of scores is a relatively simple procedure, although not as simple as finding the mode. If ungrouped data are arranged in order of magnitude, as in samples 1 and 2, the median can be found by counting up to the point below which there are $(N + 1)/2$ cases, where N stands for the total number. For example, if the ordered scores are 1, 3, 4, 5, 7, then $N + 1$ is 6 and 6 divided by 2 is 3. The third score, 4, is therefore the median. But data only rarely come in such neat order, with only one case at each score value. Usually a frequency distribution will appear with a number of cases at each interval, especially in the central portion of the distribution. Therefore, the following more detailed procedure for identifying the median must be used.

The age data in Table 3-2 represent the patrons of a neighborhood hardware store during one day's business. What is the median age for the patrons? Our first step is to find the number of cases that will make up half of the group, since the median is the point dividing the distribution so that 50 percent of the cases fall above it and 50 percent fall below it. The median will then be the age which is exactly at that dividing point.

To locate this point we divide $72/2 = 36$ cases and begin at the

Table 3-2   Ages for Store Patrons

Ages	f
80–84	4
75–79	6
70–74	9
65–69	12
60–64	11
55–59	9
50–54	8
45–49	8
40–44	4
35–39	1
	$N = 72$

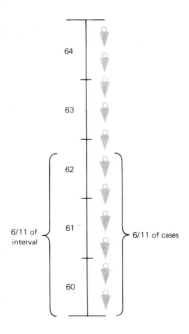

Figure 3-1

bottom of the distribution to count cases up to the interval containing the thirty-sixth case. This is the interval 60–64 (real limits 59.5–64.5). Up to this interval we have accounted for 30 cases; we need 6 more from the interval to make up our total of 36. These 6 cases must come from the 11 which are in the interval. In other words, we need $\frac{6}{11}$ of the cases in the interval to add to the 30 cases below the interval.

We now make an assumption to help our arithmetic. We assume that the cases in any interval are evenly distributed across the range of the interval. Therefore, if we move into the interval to get $\frac{6}{11}$ of the cases, we also move up into the interval $\frac{6}{11}$ of the width of the interval. In our data how many units is that? Since the interval is 5 units wide (60–64), we move

$$\tfrac{6}{11}(5) = 2.73$$

units into the interval.* This is illustrated in Fig. 3-1. Study it carefully before reading on.

Now we locate the median. We had 30 cases up to the age level of 59.5, the lower real limits of the interval. Since we moved 2.73 more units into the interval, we add 2.73 to the lower limits of the interval

$$59.5 + 2.73 = 62.23$$

* You did this in algebra class and called it *linear interpolation*.

The median age in our sample is 62.23 years, which means that 50 percent of the group will have ages below this point and 50 percent will have ages above it.

Table 3-3 shows what we have done to the distribution shown in Table 3-2. We are looking for the point that divides the total number of cases into two equal parts. We found that this point comes within the interval 60–64, but we need a portion more of the interval in order to have 50 percent of the cases below the chosen point. In Table 3-3 segment A of the distribution is short of the 50 percent point by 6 cases, or $\frac{6}{11}$ of the next interval. This needed additional amount is segment C. Since we need segments A and C to make 50 percent of the cases, we also need scale-value segments B and D in order to identify that scale point which gives us the 50 percent split in cases. If we take a given fraction of the cases in interval C, we also take a similar fraction of the width of the interval D. This fraction of the interval added to the top of the preceding interval (59.5 in Table 3-3) will identify the median point for us (59.5 + 2.73 = 62.23).

A second example with a different kind of data may be helpful as an illustration. In Table 3-4 the following steps are necessary to find the median:

1   We find $N/2 = 25/2 = 12.5$ cases. The score point below which 12.5 cases comes is the median. The next steps will identify that score point.

2   We find the interval containing case 12.5; it is the interval 100–109. It will contain the median score. We have 7 cases up to that interval.

3   We find the portion of the interval needed to complete 12.5 cases. This is 5.5 cases. We have 7 cases up to that interval; adding 5.5 to this gives 12.5.

Table 3-3   Reproduction of the Data in Table 3-2

Ages	f
80–84	4
75–79	6
70–74	9
65–69	12
D 60–64	11 } C
55–59	9
50–54	8
B 45–49	8 } A
40–44	4
35–39	1
	$N = 72$

Table 3-4   *Distribution of IQs for a Fourth-Grade Class*

	IQ	f	
	140–149	1	
	130–139	2	
	120–129	4	
	110–119	3	
(5.5/8)(10 points)	100–109	8	←5.5/8 needed
	90– 99	3 ⎫	
	80– 89	2 ⎪	
	70– 79	1 ⎬	7
	60– 69	0 ⎪	
	50– 59	1 ⎭	
		$N = 25$ $N/2 = 12.5$	

4 We need 5.5/8 of the cases in the interval to complete 50 percent of the cases; therefore, we also take 5.5/8 of the score width of the interval to identify the score point which cuts off that 50 percent.

5 The value 5.5/8 of 10 (this interval width) is 6.88, and we add this to the range already accounted for below the interval, or 99.5 plus 6.88. The median, then, is 106.38.

Exercise

3-4 A teacher-rating form is given to students in Introduction to Psychology classes. The score is the number of favorable points checked by a student in reference to his instructor. Instructor X received the following rating scores:

7	12	11	9	18		20	9	15	16	13
7	6	8	9	12		16	13	12	13	18
19	17	15	12	21		14	13	14	15	10

Instructor Y received the following ratings:

13	5	6	16	7		15	15	10	11	8
8	15	8	15	6		15	8	9	12	7
14	13	9	13	14		15	13	14	9	14

(*a*) Compute a median for each instructor

(*b*) In scanning the distributions of ratings for each instructor, do you believe the medians would be a good basis for comparison of X and Y? Why?

Now let us investigate what we have done. $N/2$ tells us how many cases will be below the median in a given distribution. To get this number of cases we usually have to divide up an interval to add a few cases to make the total of $N/2$. we do this by subtracting

$$\frac{N}{2} - f_b$$

where f_b is the sum of all frequencies below the interval we are dividing. This gives the number of cases we need out of the interval in question. To find what portion of the cases in the interval to which this is equivalent, we divide

$$\frac{N/2 - f_b}{f_w}$$

where f_w is the frequency within the interval. Now we have the proportion of the cases in the interval we need to make up 50 percent. We also take the same proportion of the interval score width

$$\left(\frac{N/2 - f_b}{f_w}\right) i$$

where i is the score width of the interval, and add it onto the lower real limit of the interval

$$\text{Mdn} = l + \left(\frac{N/2 - f_b}{f_w}\right) i \tag{3-1}$$

where l is the score value that identifies the lower real limit of the interval. The result is the median (Mdn). Formula (3-1) follows the steps we went through in our logical procedure for calculating the median and in fact takes us through those logical steps with each application. Table 3-5 lays out the procedure for using formula (3-1) to compute the median.

Exercises

3-5   Apply formula (3-1) to the problem you solved in Table 3-3 to see whether you get the same median with the formula as you did without it.

3-6   Repeat Exercise 3-5 for the data in Table 3-4.

Table 3-5　Computing the Median by Formula

Formula (3-1)	What it says to do
$$\text{Mdn} = l + \left(\frac{N/2 - f_b}{f_w}\right)i$$ where l = lower real limit of interval containing median N = total cases in distribution f_b = total number of cases below interval containing median f_w = number of cases *within* interval containing median i = score width of interval	1　Find $N/2$, or half of the cases in the distribution 2　Count cases up from the lowest score interval until you locate the interval containing the median 3　To determine how many cases you need out of the interval, subtract the frequency below the median interval f_b from half of the total cases $N/2$; this tells how many cases must come from the interval 4　Divide the number of cases needed (step 3) by the number of cases within the median interval f_w 5　Multiply the result of step 4 by the score width of the interval i 6　Add the result of step 5 to the lower real limits of the median interval l

---

Example　Given the following frequency distribution, find the median.

31–33	1
28–30	2
25–27	4
22–24	5
19–21	9
16–18	11
15.5	
13–15	10 ⎫
10–12	7 ⎪
7–9	6 ⎬ 28
4–6	4 ⎪
1–3	1 ⎭
	$N = 60$

$N/2 = 60/2 = 30$

l = lower real limit of interval containing thirtieth case
= 15.5

$f_b = 28$　　$f_w = 11$
$i = 3$ score points

$$\text{Mdn} = l + \left(\frac{N/2 - f_b}{f_w}\right)i = 15.5 + \left(\frac{\frac{60}{2} - 28}{11}\right)3$$

$$= 15.5 + \left(\frac{30 - 28}{11}\right)3 = 15.5 + .18(3)$$

$$= 15.5 + .54$$
$$= 16.04$$

The Mean

Most upper elementary school children can give you the average amount of money for three people who have $3, $5, and $2. A sixth-grade child would add up the amounts of money and divide by the number of persons who have that money. The result is the average

Table 3-6   Computing the Mean from Ungrouped Raw Scores

Formula (3-2)	*What it says to do*
$$\bar{X} = \frac{\Sigma X}{N}$$ where $\bar{X}$ = mean X = raw score of given individual N = number of individuals	Add together individual scores and divide by the total number of individuals: $$\bar{X} = \frac{X_1 + X_2 + X_3 + \cdots + X_N}{N}$$

Example   Find the $\bar{X}$ IQ for the five people whose individual IQs are 100, 105, 90, 115, 110.

$$\bar{X} = \frac{100 + 105 + 90 + 115 + 110}{5} = \frac{520}{5} = 104$$

known as the *arithmetic mean* or just the *mean*. It is found simply by adding together the values of the quantities and then dividing this sum by the number of quantities which were added.

This process can be simply written in a formula by using a few symbols to represent the quantities and the arithmetical processes involved. If we let X stand for the quantity of a given trait as measured for one individual, then $\bar{X}$ will represent the mean,* the typical quantity for the group. The letter N represents the number of individuals on whom we have the quantities designated X, and the Greek letter Σ (sigma) indicates that the quantities X are to be summed. Our shorthand designation for computing the mean is

$$\bar{X} = \frac{\Sigma X}{N} \qquad (3\text{-}2)$$

The calculations are given in detail in Table 3-6.

This procedure does not require raw data to be put into any kind of order, such as a frequency distribution. Computations can be made directly from the raw scores as they are observed. This procedure is especially recommended when a calculator or computer is available but is also handy without mechanical aids.

The mean, in contrast with the median and mode, represents a balance point in the distribution of scores. As an illustration of this fact,

---

* $\bar{X}$ represents the mean of a sample, but the Greek letter μ (mu) will be used to represent the mean of a population. In this book Greek letters are used to indicate population values and Latin letters to indicate sample values.

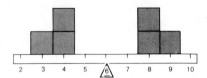

Figure 3-2

let us imagine individuals as blocks and their scores as inches on a 10-inch ruler. Suppose the score values are 3, 4, 4, 8, 8, and 9. The mean for this set of scores would be

$$\frac{3 + 4 + 4 + 8 + 8 + 9}{6} = \frac{36}{6} = 6$$

In Fig. 3-2 we discover an important idea about the mean. There are two blocks a distance of 2 inches below the mean and one block a distance of 3 inches below the mean. There are also two blocks 2 inches above the mean and one block 3 inches above it. Now we see that the total blocks-times-distance-from-the-mean arrangement below the balance point equals the total blocks-times-distance arrangement above it. This finding illustrates a fact about the mean as a balance point. Its position is influenced by the number of scores at any point in a distribution in relationship to the *location* of that point in the distribution. Since this frequency-times-distance-from-the-mean arrangement on one side of the mean will always be equal to the frequency-times-distance arrangement on the opposite side of the mean, the position of scores in a frequency distribution is important in computing and interpreting the mean.

Figure 3-3 also illustrates this point. Again we use blocks to represent individuals and the ruler to represent a continuum of scores for these individuals. Now suppose we have scores of 2, 4, 4, 6, and 9. The mean will be 5. Computing the frequency-times-distance values for scores on both sides of the mean we find:

Below the mean		Above the mean	
Distance	*Frequency*	*Distance*	*Frequency*
(4 − 5)(2 blocks) = −2		(6 − 5)(1 block) = 1	
(2 − 5)(1 block) = −3		(9 − 5)(1 block) = 4	
Total −5		Total +5	
Algebraic total = 0			

If the position of any block on the ruler is changed, the balance point will also be changed. Likewise, in a frequency distribution the

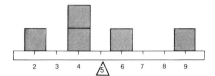

Figure 3-3

magnitude of each of the scores influences the location of the mean. In other words, the position of every score in a distribution plays a part in determining the location of the mean and does so to the extent to which a score deviates from the central area of the distribution. *This is not true of the median or mode.*

When the number of scores in a distribution is large, it is difficult to add them all up one at a time without a calculator, but we can shorten the adding process in several ways.* One way is illustrated by the data below. First, using the procedure described in Table 3-6, we find

X
10
8
8
10
8
6
6
$\Sigma X = 56$

$$\frac{\Sigma X}{N} = \frac{56}{7} = 8$$

If these data are arranged in a frequency distribution, we have

X	f	fX
10	2	20
9	0	0
8	3	24
7	0	0
6	2	12
		$\Sigma fX = 56$

$$\frac{\Sigma fX}{N} = \frac{56}{7} = 8$$

---

* Students uncertain of their basic algebra may wish to consult Appendix A before proceeding.

Multiplying each score by the number of times it appears gives fX; then adding these values gives ΣfX, the same result as ΣX above. Dividing ΣfX by N gives the same value for the mean as $\Sigma X/N$, and we added only three numbers instead of seven. The formula for dealing with data in such a frequency distribution, i.e., with intervals equal to 1 unit, is

$$\bar{X} = \frac{\Sigma fX}{N} \qquad (3\text{-}3)$$

where f is an interval frequency and X is the score that identifies the interval.

This procedure can also be generalized to grouped data in a frequency distribution like the one below. A useful formula in this case is

$$\bar{X} = \frac{\Sigma f \cdot \text{MP}}{N} \qquad (3\text{-}4)$$

where f is the interval frequency, MP is the midpoint of the interval, and the dot indicates multiplication. Here we let the midpoint of the interval represent all scores in the interval. Then we multiply the midpoint of the interval by the frequency $f \cdot$ MP and sum this product for all intervals, $\Sigma f \cdot$ MP. This gives us an approximation of ΣX. When divided by N the result is an approximation of the mean. In this procedure we assume that cases are equally spread across each interval; e.g., in the data below the three cases in the 20–22 interval really represent one case at score 20, one at 21, and one at 22. This in fact may not be true, but deviations from the equal spread tend to average out across many intervals. A worked example is shown below.

X	f	MP	f·MP		
47–49	4	48	192		
44–46	3	45	135		
41–43	7	42	294		
38–40	9	39	351		
35–37	13	36	468		
32–34	16	33	528	$\bar{X} = \dfrac{\Sigma f \cdot \text{MP}}{N} = \dfrac{2688}{79} = 34.03$	
29–31	10	30	300		
26–28	7	27	189		
23–25	7	24	168		
20–22	3	21	63		
	79		2688		

Exercises

3-7 Ten children ran the 50-yard dash in the following times in seconds:

$$8 \quad 11 \quad 14 \quad 10 \quad 9 \quad 7 \quad 11 \quad 13 \quad 10 \quad 12$$

What is their mean time?

3-8 Mary and George are studying typing. In six speed tests Mary typed the following number of words per minute:

$$32 \quad 76 \quad 54 \quad 60 \quad 45 \quad 51$$

while George typed

$$46 \quad 60 \quad 54 \quad 58 \quad 52 \quad 48$$

Compare the two students' mean performances.

3-9 I have the following data in a distribution. Calculate the mean.

Score	f
11	1
10	0
9	4
8	7
7	8
6	5
5	3
4	1
3	1

3-10 The following frequency distribution was developed by a newspaper research staff, but the mean was not included. Calculate the mean.

X	f
19–20	1
17–18	3
15–16	4
13–14	9
11–12	12
9–10	8
7–8	6
5–6	4
3–4	2
1–2	1

The Mean in Inferential Statistics

The mean is often used in making inferences about populations when we only have sample data. Typically samples will look much like the population from which they come; i.e., the mean of a sample, its range of scores, etc., will be in the vicinity of the population parameters. If we can decide how large our sampling error is, we can hypothesize what the population mean is, within a given range of error. This is very useful because often we cannot even begin to measure whole populations. Instead we can take a sample from the population, find its mean, and project this value onto the population—within the range of our sampling error.

This procedure can also be generalized to comparing two populations, as we do with experiments. We begin with two samples of people; the experimental group gets a special treatment and the controls do not. After the treatment are these two samples still from a common population, or do they represent different populations? We calculate a mean for each group on the measure of the dependent variable. Then we compare these means. If they are within the range of normal sampling error, we conclude that the samples are still from a common population, but if the difference between the means of the two samples is greater than normal sampling error, we conclude that the samples are no longer from a common population as far as the dependent variable is concerned. The actual procedure will be given in a later chapter.

The mean is a very useful statistic for inferential applications because of its stability from sample to sample. This reduces sampling error to a minimum, a feature lacking in the mode and the median.

We often refer to the mean as an *unbiased estimator*. By this we mean that if we have many samples of the same size, the average of their sample means is expected to equal the population mean. For example, if I have 1000 samples of 100 people each, measure their height, calculate a mean for each sample, and then average the means across all these samples, I expect to have a value equal (or nearly so) to the population height. It is this unbiased quality of the mean that allows us to make the inferences described above.

Linear Transformation of the Mean

For several reasons we may want to transform a distribution to a higher scale by adding a constant to all numbers. (A constant is a value that does not change across successive observations.) For example, if our numbers fall below zero, i.e., are negative values, we may decide to raise the scale so that all values will be positive. What does this do to the mean of the distribution?

Here is an example: If I have the following scores

$$
\begin{array}{r}
8 \\
6 \\
5 \\
3 \\
1 \\
0 \\
-4
\end{array}
$$

I may wish to add a constant $(+5)$ to all scores so that the lowest score will be 1. The mean of the above distribution is

$$
\bar{X} = \frac{\Sigma X}{N} = \frac{19}{7} = 2.71
$$

If I add 5 to each score, the new sum is 54 and the new mean is 7.71, or exactly 5 points (the value of our constant) larger than it was before. This happens because

$$
X + C = \text{any value of } X + \text{constant}
$$
$$
\Sigma(X + C) = \text{sum of all } X \text{ values} + \text{constant}
$$

which equals $\Sigma X + \Sigma C$, that is, the sum of the X values plus the sum of the constant we added to each X value. But $\Sigma C = NC$ because adding a constant is the same as multiplying it by the number of times we added it. For example, if we added our constant to five different values of X, the sum of those constants is

$$
\Sigma C = C + C + C + C + C
$$

We have five of the C terms or $5C$ or NC.

If $\Sigma(X + C) = \Sigma X + \Sigma C$, it must then be equal to

$$
\Sigma X + NC
$$

To calculate the mean we divide $\Sigma(X + C)$ by N or

$$
\frac{\Sigma X + NC}{N}
$$

which equals

$$
\frac{\Sigma X}{N} + \frac{NC}{N} = \frac{\Sigma X}{N} + C = \bar{X} + C
$$

So if we add a constant to all values in a distribution, we increase the mean by the value of the constant.

A similar result occurs from multiplying all values of X by a constant.

For example, multiplying all values in this distribution by 2 would give

$$
\begin{array}{rl}
8(2) = & 16 \\
6(2) = & 12 \\
5(2) = & 10 \\
5(2) = & 10 \\
4(2) = & 8 \\
\underline{1(2) =} & \underline{2} \\
29(2) & 58
\end{array}
$$

$$
\bar{X} = \frac{\Sigma X}{N} = 4.83 \qquad \frac{\Sigma XC}{N} = 9.66
$$

or the old mean 4.83 times the constant

$$
4.83(2) = 9.66
$$

Should we conclude that if we multiply all values by a constant the mean is also increased as $\bar{X}C$? The sum of the scores multiplied by the constant is

$$
X_1C + X_2C + X_3C + \cdots + X_NC = C(X_1 + X_2 + X_3 + \cdots X_N) = C\Sigma X
$$

Now if we divide by N to get the mean, we have

$$
\frac{C\Sigma X}{N} \qquad \text{or} \qquad C\bar{X}
$$

We can therefore conclude that multiplying all scores by a constant has the effect of multiplying the mean by that constant.

Averaging Means

Sometimes we have means on several samples from a population and would like to find the mean for the several samples combined into one large group. This grand mean can be calculated from the means of the samples, but first the data must be *weighted* in accordance with the number of cases in each sample. This is done with the formula

$$
\bar{X}_g = \frac{n_1\bar{X}_1 + n_2\bar{X}_2 + \cdots n_j\bar{X}_j}{n_1 + n_2 + \cdots + n_j} \tag{3-5}
$$

where $\bar{X}_g$ = grand mean

n_1, n_2, n_j = sample sizes

$\bar{X}_1, \bar{X}_2, \bar{X}_j$ = means for samples.

Formula (3-5) can be written more economically as

$$\bar{X}_g = \frac{\Sigma(n_j\bar{X}_j)}{\Sigma n_j}$$ (3-6)

where the subscript j can be replaced in turn by 1, 2, 3, etc.

Let us see what formula (3-5) does. The formula for the mean

$$\bar{X} = \frac{\Sigma X}{N}$$

can be rewritten as

$$N\bar{X} = \Sigma X$$

so that the numerator of formula (3-5) is simply adding up all the X values (ΣX) across all samples; i.e., it is equal to $\Sigma(\Sigma X)$, or the sum of the separate sample sums. Since the denominator is the sum of all the cases, formula (3-5) gives us the sum of all the scores divided by the sum of all the cases, which is how we would have calculated the mean if the data had not been divided into the samples beforehand.

If samples are of unequal sizes, formula (3-5) will give us the correct mean for the total across all samples, but merely adding sample means and dividing by the number of samples, i.e., finding an average mean, will *not* produce the correct answer. Suppose we have the following data:

Sample	A	B	C
$\bar{X}$	10.5	14.2	11.1
n	20	35	10

Formula (3-5) gives

$$\bar{X} = \frac{20(10.5) + 35(14.2) + 10(11.1)}{20 + 35 + 10} = 12.58$$

but simply averaging across the three means, i.e., $(10.5 + 14.2 + 11.1)/3$, gives us the wrong value 11.93. This is because simple averaging acts as if the same number of scores went into the calculation of each mean, and this is typically not the case. Formula (3-5), however, weights each sample according to the number of observations it is based upon.

Therefore, if we have several sample means, we cannot simply add up these means and divide by the number of them to find the grand mean for the combined groups. We must weight each sample mean in proportion to the number of cases involved in the computation of that mean.

Which Average to Use

We have discussed several characteristics of each of the indicators of central tendency. These characteristics should help us choose the indicator which best suits the data we are dealing with. The mean is tied to the positions of the scores and as such is dependent upon each observation in the distribution for its location. To this end it is the best indicator of the total group's performance unless the distribution is conspicuously skewed. Widely deviating scores in one direction from the center of the distribution tend to pull the mean toward them. This is not true, however, of the median, and for this reason the median may be more nearly central to the bulk of the scores in markedly skewed distributions than the mean. The following extreme example shows the mean, 7, entirely out of range of the majority of the scores, but the median, 5, is centrally located in the greater portion of the data:

$$2 \quad 3 \quad 3 \quad 4 \quad 5 \quad 6 \quad 6 \quad 6 \quad 28$$

Therefore, the fact that the median is not influenced by widely deviating scores makes it preferable to the mean when one wishes to describe typical behavior in a distribution that contains a few cases whose scores range in one direction far beyond the rest of the cases. The mode, like the median, is also insensitive to widely deviating scores, but its instability makes it the last choice even in markedly skewed data. Figure 3-4 indicates how a skewed distribution influences the relative position of the three indicators of central tendency.

Stability of the indicator is especially important when choosing among averages. From sample to sample in a population the means are expected to be more nearly alike than medians, and medians more alike than modes. Therefore, if we are using a sample as a basis for estimating the typical performance for the population, we would choose the mean because of its stability. This is also true if we are comparing the performance of two groups. However, if the distribution should be markedly skewed, the mean may not describe the group as precisely as the median.

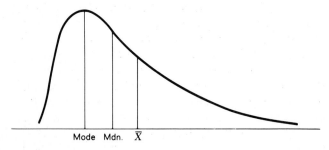

Mode   Mdn. $\overline{X}$

Figure 3-4

How we are applying the descriptive data may also be a consideration in choosing between the mean, the median, and the mode. It has often been said that "figures don't lie but liars can figure." It is therefore to our advantage to know the figuring that liars use in order to catch them in their ruses. One trick of deceit lies in the use of averages. How one's purpose influences the choice of an average can be illustrated by a devious example.

We have seen in Fig. 3-4 that it is entirely possible for the three indicators of central tendency to give completely different values in a distribution of scores. People who use statistics to promote their own objectives feel free to choose the average which best suits their ends rather than the one which best fits the data. For example, consider the following teachers' salaries. For simplicity suppose we have five teachers and a teaching principal, whom we shall include as a teacher also. Their salaries are

$$\$33,000$$
$$20,100$$
$$20,000$$
$$15,400$$
$$15,400$$

The chairman of the teacher salary committee declares to the PTA that the average teacher's salary is $15,400, the mode. The chairman of the school board later reports to the chamber of commerce that the teachers' salaries average $20,780, the mean. Neither person has really departed from the truth. Both have used a bona fide average, but that statistical tool has been employed which best suited the reporter's objective rather than the one that best described the data—in this case the median. So beware of people who argue with averages unless you know which average is being used and how well it fits the data it represents.

Summary

Central tendency is an important concept in statistics because it indicates what level is typical or "average" for a given group as assessed by a particular measuring device. It therefore helps us describe groups. But indicators of central tendency also help in inferential statistics. From samples we hypothesize what the average is for the population, within sampling error.

The three indicators of central tendency that are widely used are the mode, the median, and the mean. The mode is the most frequently appearing value in a distribution; it is a quickly found indicator of central tendency but tends to be less stable than other indicators. The median is that point on the scale which divides a distribution so that half of the cases are below it and half above it. It is especially useful when we deal with obviously skewed distributions.

The mean is the arithmetic average for the scores in a set of data. Since it is the most stable value from sample to sample, it is often used in inferential statistics to make hypotheses about the value of the population mean. The mean, however, is pulled toward widely deviating scores in a skewed distribution, and in these cases may not be the best indicator of the central accumulation of scores.

Since the units in measuring scales are often arbitrary, we can alter them without harming the relative status of individuals as shown by a given scale. We call these alterations transformations, and they are typically achieved by adding a constant to all individual scores or multiplying all scores by a constant. If we add a constant to all scores, the mean is increased (or decreased if the constant is negative) by the amount of the constant. If we multiply all scores by a constant, the new mean will be equal to the original mean multiplied by the constant.

Key Terms

central tendency	Σ
mode	constant
median	transformation
mean	unbiased estimator

Problems

3-1   A sociologist who surveyed 50 counties selected at random from five midwest states has recorded the following number of divorces per 1000 population recorded in the last 2 years:

No. of divorces	f
25	1
24	5
23	16
22	6
21	5
20	4
19	3
18	4
17	3
16	0
15	2
14	1
	50

(*a*) Locate the mode.

(*b*) Calculate the median.

(*c*) Calculate the mean.

(*d*) How do you account for the differences in the sizes of these three indicators of central tendency?

(*e*) The sociologist who collected the data reported that across these counties in the last 2 years the typical number of divorces was 23; you argue that it is more like 21. Who is right?

(*f*) Which figure, mode, median, or mean, is probably the best estimate for divorces across all counties of the five midwestern states? Why?

3-2 A school principals' association surveyed 30 school districts to see how many freshmen per 1000 enrolled (*X*) dropped out during the school year. The frequency distribution is

X	f
38–40	1
35–37	2
32–34	2
29–31	0
26–28	3
23–25	4
20–22	8
17–19	7
14–16	2
11–13	1
	30

(*a*) Calculate the mean number of dropouts.

(*b*) Calculate the median number of dropouts.

(*c*) Determine the mode.

(*d*) Which of the three indicators of central tendency best identifies the central grouping of scores? Why?

3-3 Ten boys shooting baskets from the free-throw line on a basketball court made the following number (*X*) out of 10 tries each:

Boy	A	B	C	D	E	F	G	H	I	J
X	6	9	5	6	7	5	3	8	4	7

(*a*) Calculate the mean.

(*b*) Arrange the scores from highest to lowest (record tied scores like 7s as separate scores) and subtract the mean from each score, that is $X - \bar{X}$ = ? Add the $X - \bar{X}$ values for scores above the mean and then add the $X - \bar{X}$ values for scores below the mean. Are absolute values of these two sums equal? What does this tell you about the mean as a balance point in a distribution?

3-4   An admissions officer at Eks University wants to compare high schools on the basis of the college entrance test scores. Test scores for Pine Crossing High School, Wolf Run High School, and Culdesac High School were as follows:

Scores	Pine Crossing	Wolf Run	Culdesac
65–69	1	1	2
60–64	1	2	1
55–59	2	8	0
50–54	11	10	2
45–49	9	0	6
40–44	16	9	8
35–39	12	6	4
30–34	1	2	4
25–29	1	0	3
20–24	1	2	0

(*a*) Compute a mean and median for each school.
(*b*) Are the means or the medians more alike for the three schools? How do you account for this?
(*c*) What is the mean for the combined groups from the three schools?

3-5   Take four coins and toss them up and let them settle on the desk. Count and record the number of heads. Repeat this procedure 10 times so that you have 10 figures, each representing the number of heads found in one toss of the four coins.
(*a*) Find the mean number of heads for the 10 tosses.
(*b*) Imagine each toss of the four coins as a separate "experiment." Will each experiment provide us with the same result as the other experiments? Will each experiment produce the same result as the average for the group? Are separate experiments good estimates of the average for the total for all experiments?

3-6   Suppose you are an instructor and you wish to increase the *median* performance for your class. Which group would you want to work with the most and why?
(*a*) Those who are now scoring the highest
(*b*) Those who are now just above the median
(*c*) Those who are now just below the median
(*d*) Those who are now at the lowest end of the scale

3-7   Suppose you wished to increase the *mean* performance of your class. Which of the alternatives in Prob. 3-6 would you wish to work with most? Why?

3-8 In the following groups of data, tell whether the mean or median best describes the data.

(a) 6 4 2 12 3   (b) 45 51 47 65 36
(c) 21 28 24 9 23

3-9 $\Sigma X/N$ is the mean of all X scores; that is, X's are our basic data. In your own words state what the basic data are for each of the following formulas:

(a) $\dfrac{\Sigma Y^2}{N}$ (b) $\dfrac{\Sigma (X - \bar{X})^2}{N}$

(c) $\dfrac{\Sigma XY}{N}$ (d) $\dfrac{\Sigma (X - \bar{X})(Y - \bar{Y})}{N}$

3-10 I have calculated the mean IQs for 30 first graders as 108. What will the mean be if I subtract a constant of 4 from all scores? If I multiply all scores by 10?

3-11 I gave a spelling test to 50 ten-year-old children. The mean was 21.4 for a list of 35 words. I discovered that everyone got the first two words correct but these were supposed to be practice words and not counted in the score. What will be the mean for the list without the practice words?

3-12 I have found data which I believe to be positively skewed. The skewness disappears if I multiply each score by .80. The original distribution had a mean of 12.6. What will be the mean of the transformed scores?

3-13 In three first-grade classes the mean sight vocabularies were as follows:

Class	A	B	C
Words $\bar{X}$	14.4	10.7	19.3
Children	12	16	10

What is the mean for the combined classes?

3-14 Four different groups of freshmen studying methods of concentration are tested on a stylus problem. The mean number of errors and group sizes were as follows:

Group	A	B	C	D
Size N	10	15	9	12
Mean no. of errors	4.7	6.2	3.6	9.1

What is the mean for the combined groups?

4

Indicators of Relative Position in a Distribution

Professor Wye has just administered a test-anxiety scale to a group of students who volunteered for group therapy. Dan Clutch has a score of 21. What does this tell me about Dan's test anxiety? Is it high? Low? The score alone is not helpful. We need to know how this score ranks among those made by various groups of people with different amounts of test anxiety. If it is near the middle of an unselected group of students who are still enrolled, we may decide that the score of 21 reflects no more than the typical amount of tension associated with taking tests. On the other hand, if we know that the score of 21 ranks among the highest in a group of students who dropped out of college because they "clutched" on examinations, we may conclude that the score indicates considerable test anxiety.

In general, scores and other types of observational data often have meaning only in terms of their status among other scores collected on a defined group of individuals. The arrangement of scores collected on the group gives us a benchmark against which to evaluate any single score. This is especially true when our scale does not have a true zero, i.e., zero on the test does not mean complete absence of the trait.

If we know that a child is 3 feet tall and his father is 6 feet tall, we can interpret the child's height as a fractional portion of the father's height. Here we know where true zero is, and our measurement begins at that point. But with many measuring devices used by behavioral scientists true zero is not known. The zero on their scale is an arbitrary one, and we do not know how far below this point the observed condition actually may extend. For example, intelligence test results conceivably could be zero. But even angleworms have been observed to learn. How far below test zero is the angleworm's intelligence? When we do not know where true zero is, we cannot describe one measurement as a fraction of another.

Although some problems arise from using scales that begin at an arbitrary zero, much information can be collected by using these devices. Even if we did know where true zero was, we would still want to apply many of the same techniques to data analysis that we use with scales which begin at an arbitrary point. We still get the greatest meaning from a measurement if we know how that measurement ranks with other similar marks, no matter whether we start from true zero or an arbitrary zero.

Methods of bringing order and meaning into a disarray of data must therefore include techniques for identifying the relative position of a given score among other scores. The purpose of this chapter is to present some common techniques for doing just this. Specifically, we shall deal with percentile ranks, quartiles, and deciles. They are by no means all the procedures available for showing relative position, but they are among the most common.

Percentile Ranks

A person's position in a group with reference to a given trait can be shown by pointing out what percent of the group has less of the trait than the person being considered. We have all sat tensely fumbling our test papers in our moist hands while the instructor puts the distribution of scores on the blackboard. Suppose that on your test you have 23 questions correct. We see from the instructor's distribution of scores that 30 of the 37 students in the class have scores lower than yours. Thus, 81 percent of the group did less well than you did. A statistician may say you were at the 81st *percentile rank*, or that your score of 23 is the 81st *percentile*. Out of the corner of your eye you notice that your neighbor's score is 26, and the class distribution shows that 91 percent of the group scored below 26. Your neighbor is at the 91st percentile rank, or his score of 26 is the 91st percentile.

The difference between percentile and percentile rank must be made clear. A percentile is always a point in the score distribution below which a given percentage of cases falls. In the above example since 23 is a point in the distribution below which 81 percent of the cases fall, 23 is the 81st percentile or the point that cuts the distribution so that 81 percent appears below the point.

On the other hand we may look at a set of scores and ask: What is the rank of a raw score of 23? It ranks above 81 percent of the group. The 81 tells us the rank of the score 23 in percentages, so that 23 has a percentile rank of 81. It lies at the point below which 81 percent of the cases fall; i.e., since it ranks 81st on the percentage-of-cases scale, 23 is at the 81st percentile rank. The number 23 identifies the scale point (the percentile) that corresponds with a percentage-of-cases scale (the percentile rank). The percentile is a point on the scale; the percentile rank says what percentage of the group falls below that scale point.

The first steps in computing percentiles have already been completed in building cumulative percentages. Let us look at an example. Table 4-1 shows a frequency distribution of heights of 30 girls. We created cumulative frequencies (cf) by adding the frequencies of each interval to the sum of all frequencies below that interval. Next, these cumulative frequencies were converted into cumulative percentages (cp) by dividing cf by N and multiplying by 100; for example, for the interval 55–57 the cf divided by N is 12/30 = .40; multiplied by 100 this becomes 40.0, the cumulative percentage. Now our table provides cumulative percentages that tell us what proportion of the group appears below a given interval. But the table gives us only a few percentile ranks, and we often want the whole range from 1 to 99. Moreover, this table yields fractional ranks, whereas we usually deal with percentile ranks in whole numbers.

We could plot these data on a cumulative percentage graph and read any point we wanted off the graph. The data given in Table 4-1 are therefore graphed in Fig. 4-1. (Why are the plots in this graph placed at the upper limits of each interval?) Now we can read the percentiles, at least to estimable points within an interval. The 50th percentile rank appears to be at about midway in the interval 58–60, or about equivalent to a score of 59. The 75th percentile rank is about equivalent to a score of 62, and so on.

Table 4-1   Heights of 30 Girls on Their Twelfth Birthdays

Height, inches	f	cf	cp
67–69	1	30	100.0
64–65	3	29	96.7
61–63	5	26	86.7
58–60	9	21	70.0
55–57	6	12	40.0
52–54	2	6	20.0
49–51	2	4	13.3
46–48	1	2	6.6
43–45	1	1	3.3

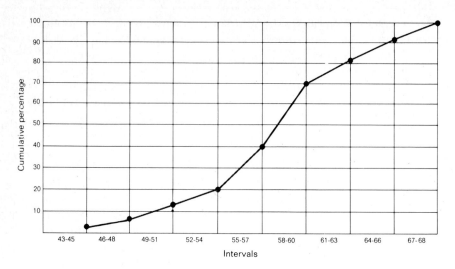

Figure 4-1

Although this procedure lacks the precision we would like to see in a method for finding percentiles, it is fast and especially efficient when we have only one score unit per interval, a condition which increases the accuracy of our estimations of points within various intervals.

Percentile points in a distribution can be computed directly, and indeed we have already learned how. You recall that the median was the point in the distribution that divided the group into an equal number of cases. This means that 50 percent of the cases are marked off below the median, placing it at the 50th percentile rank. The procedure applied to computing the median just needs to be generalized so that we can find not merely the scale point below which 50 percent of the cases fall but also points which correspond with any percentile rank we desire. What is the scale point that corresponds with the 30th percentile rank (below which 30 percent of the cases fall), what scale point corresponds with the 73d percentile rank, etc.? The procedure used to make these locations requires a slight modification of the one used to find the median, but the logic behind it is the same.

Our first step in computing the median was to find the number of cases that made up exactly half of the total group. We then proceeded to locate the point on the raw-score scale that corresponded to the point below which 50 percent of the group was cut off. Now suppose we want to find not the point below which there are 50 percent of the cases but the point below which there is 20 percent of the group (the 20th percentile). In locating the median (the 50th percentile) we divided N by 2, or found .50 of N. In locating the 20th percentile, we begin by taking .20 of N; otherwise, the basic steps are not really different. An example will bear this out.

Let us begin by finding the 50th percentile (the median, or P_{50}) in Table 4-2. First we find 50 percent of the group, or $.5N$, which is 45 cases. We count up to find the interval that contains the forty-fifth case, and we find it to be 28–30. We need one case from that interval (44 up to that interval plus 1 makes 45 cases). There are 19 cases in the interval, and so we need $\frac{1}{19}$ of those cases to add on to our 44. We also then take $\frac{1}{19}$ of the interval width to add on to the range of scores below it, or $\frac{1}{19}(3) = .16$. Now, .16 added to 27.5 (the real upper limit of the last interval below 28–30) gives us 27.66, which is the median and the 50th percentile.

Now suppose we want to find the score equivalent to the 20th percentile rank (P_{20}). We begin as before by finding how many cases are cut off by that figure (20 percent) and then proceed to find the point on the raw-score scale that matches with that number of cases. In our 90 cases, 20 percent is 18 cases. We find the interval in which the eighteenth case comes and proceed to find its corresponding point on the raw-score scale. The interval is 19–21, and we need 7 of its cases to make up the 18. This is $\frac{7}{9}$ of the width of the interval, or $\frac{7}{9}(3) = 2.33$ score points, which when added to the upper limit of the last interval, 18.5, becomes 20.83, the raw score which is the 20th percentile. This process is illustrated in Fig. 4-2. We have moved into the interval $\frac{7}{9}$ of the cases, so we calculate $\frac{7}{9}$ of the interval's score range. This portion of the score range is added onto the bottom of the interval to identify P_{20}. The generalized procedure for finding percentiles is given in Table 4-3 on page 69.

Table 4-2 The Number of Letter Cancellations Made in a 3-Minute Cancellation Test by 90 College Freshmen

Cancellations	*f*	*cf*
37–39	2	90
34–36	10	88
31–33	15	78
28–30	19	63
25–27	16	44
22–24	8	28
19–21	9	20
16–18	7	11
13–15	3	4
10–12	1	1

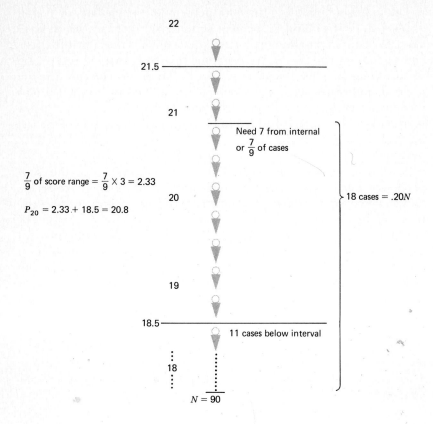

Figure 4-2

Now that we have mastered the logic of locating a score value that corresponds to a given percentile rank, let us review what we have done. First we multiplied P_X, the percentile rank in question (in decimal form) times the number of cases in our distribution P_XN. This indicates how many cases fall below the percentile rank with which we are dealing.* Next we calculated the number of cases f_b below the interval that contains P_X and subtracted this sum from P_XN to indicate how many cases we need from the interval.

Then we divide this number of cases $(P_XN - f_b)$ by the number of cases *within* the interval giving us the proportion of additional cases we need from within the interval to reach the percentile rank. We also add to the interval's lower limit this proportion of the interval's score width to find the score point that corresponds to the P_X value: $(P_XN - f_b)/f_w$ gives us that proportion. When multiplied by i, the interval width, this yields the score points to add onto the lower limits of the interval l. Our

---

* Recall that in calculating the median, P_XN was $.50N$, or $N/2$, as in formula (3-1).

Table 4-3 Procedure for Finding Values Corresponding to Given Percentile Ranks

1 Where P_X is the desired percentile, multiply the corresponding proportion X times N (for example, for P_{38} and 150 cases, multiply .38 times 150); this tells how many cases will fall below the desired score

2 Then find the point on the raw-score scale that corresponds to that portion of the cases as follows:

    (*a*) Locate the interval that contains the desired number of cases (X times N)

    (*b*) Besides the number of cases below this interval, how many additional cases are needed from the interval itself to equal XN? Divide the cases taken from the interval by the total number within that interval. This is the proportion of cases needed from the interval, $(P_XN - f_b)/f_w$.

    (*c*) Now find the same proportion of the interval width by multiplying the proportion found in *b*) by the interval score width, *i*.

    (*d*) Add this portion of the interval to the lower real limits of the interval with which you are working; this gives the raw-score point (the percentile) which corresponds with the desired percentile rank.

Example For the data below locate the 80th percentile.

X	f	cf
27–29	1	57
24–26	3	56
21–23	6	53
18–20	10	47
15–17	9	37
12–14	11	28
9–11	10	17
6–8	3	7
3–5	3	4
0–2	1	1
	N = 57	

1 .80(57) = 45.6 cases; these make up 80 percent of N

2 (*a*) The 45.6th case falls in the interval 18–20

    (*b*) 45.6 − 37 = 8.6 cases needed from the interval; the corresponding portion of the interval width is 8.6/10 of 3 score points (the interval width) or 2.58

    (*c*) 2.58 added to the upper limits of the interval below gives 17.5 + 2.58 = 20.08, the score corresponding to the 80th percentile rank

formula for finding a score value that corresponds with a given percentile rank is

$$X_p = l + \left(\frac{P_XN - f_b}{f_w} \right) i \qquad (4\text{-}1)$$

Table 4-4   Procedure for Finding Score Values Corresponding to Given Percentile Ranks

Formula (4-1)	*What it says to do*
$$X_P = l + \left(\frac{P_X N - f_b}{f_w} \right) i$$ where X_P = score point that corresponds with percentile rank we are interested in l = lower limit of interval containing $P_X N$th case P_X = percentile rank in decimal form N = total number of cases f_b = cumulative frequency below interval containing $P_X N$th case f_w = frequency within interval i = score width of interval	1 Multiply P_X, the desired percentile rank in decimal form, times the total number of cases in the distribution 2 (*a*) Counting from the lowest score interval, locate the interval that contains the $P_X N$th case (*b*) Subtract from $P_X N$ the total number of cases up to the bottom of the interval containing the $P_X N$th case 3 Divide this difference by the frequency within the interval containing the $P_X N$th case 4 Multiply this fraction by the width of the interval 5 Add the result to the lower limit of the interval containing the $P_X N$th case; this yields the score value corresponding to the desired percentile rank

where l = lower real limit of interval containing percentile
$\quad\quad P_X$ = desired percentile rank in decimal form (.80, .63, etc.)
$\quad\quad f_b$ = sum of frequencies below interval
$\quad\quad f_w$ = frequency within interval
$\quad\quad i$ = width of interval in score points

The application of formula (4-1) is shown in Table 4-4.

As we have seen to this point, the basic procedure used for computing the median can be generalized to locate the score that corresponds any given percentile rank. However, we often wish to reverse this procedure and calculate the percentile rank that corresponds to a given score. For example, in Table 4-3, what percentile rank is equal to a raw score of 12?

To find the percentile rank which corresponds with a given score value we begin by locating the score within its interval. What portion of

Example Locate the 80th percentile, i.e., the score value that corresponds to a percentile rank of 80.

X	f	
27–29	1	
24–26	3	
21–23	6	
18–20	10	
15–17	9	
12–14	11	
9–11	10	37
6–8	3	cases
3–5	3	
0–2	1	
	N = 57	

$$X_P = 17.5 + \left(\frac{45.6 - 37}{10}\right)3$$

$$= 20.08$$

1 .80(57) = 45.6 cases; these make up 80 percent of the total cases

2 (*a*) The 45.6th case is in the interval 18–20

   (*b*) Up to that interval we have 37 cases; so 45.6 – 37 = 8.6 cases are needed from the 10 in the interval

3 8.6 of the 10 cases in the interval is 8.6/10 = .86; since .86 of the cases in the interval is needed to reach 45.6, we take .86 of the score range by step 4

4 .86 (3 score points) = 2.58 score points

5 Add these to the lower limit of the interval, 17.5

     17.5 + 2.58 = 20.08

   This is the score value corresponding to the 80th percentile rank

the interval is below the score? We shall need the same portion of the people in the interval as the portion of the interval below the score.

For example, suppose in Table 4-3 we want to know the percentile rank for a score of 13. That score is in the interval 11.5 to 14.5. The interval is 3 score points wide, but only 1.5 points (13.0 − 11.5 = 1.5) of the interval fall below the score of 13. If that portion of the interval comes below 13, we shall also find out how many people in the interval fall below a score of 13 by taking 1.5/3 times 11, since there are 11 people in the interval. This value is 5.5. This says that 5.5 people in the interval rank below a score of 13.

But we want to know what portion of all the people in our distribution rank below a score of 13. Therefore we add the people we located in the interval below 13 (5.5 people) to the cumulative frequency of all intervals below the interval 12–15. In Table 4-3 this is 17

Table 4-5   Procedure for Finding the Percentile Rank for Any Given Score Point

Formula (4-2)	What it says to do

$$P_X = \left\{ \frac{[(X - l)/i]\,f_w + f_b}{N} \right\} 100$$

where P_X = percentile rank for score X

X = score of interest

l = lower limit of interval in which score X falls

i = width of interval

f_w = frequency within interval

f_b = sum of all frequencies below interval

N = total number of cases

1. Locate the interval in which the raw score of interest falls; subtract the lower limit of the interval from the score

2. Divide this difference by the width of the interval; this indicates what portion of the interval our score is above

3. Multiply this figure by the frequency within the interval. If the score is above a given portion of the interval, it is also above the same portion of the cases in the interval; multiplying by f_w gives this portion of cases

4. This portion of cases within the interval is added to all cases below the interval f_b to give the total number of cases below X

5. To convert this total to a percentage divide by N and multiply by 100, which produces the percentile rank for score X

---

people. The total number of people who fall below a score of 13, then, is $17 + 5.5 = 22.5$ people.

Our next job is to find what percentage 22.5 is of the total sample of 57 people. We divide 22.5 by 57 and get 39.47 percent. This tells us now that 39+ percent of the total sample fell below a score of 13. A score of 13 is at the 39th percentile rank.

Now that we have made a logical analysis of locating the percentile rank for a given score, a generalized formula for what we did is useful:

$$P_X = \left\{ \frac{[(X - l)/i]f_w + f_b}{N} \right\} 100 \qquad (4\text{-}2)$$

where X = score for which percentile rank is wanted

Example Given the following distribution of scores, locate the percentile rank for a score of 52.

Scores	f	cf
70–74	3	105
65–69	4	102
60–64	7	98
55–59	12	91
50–54	14	79
45–49	21	65
40–44	17	44
35–39	15	27
30–34	10	12
25–29	2	2
	$N = 105$	

$$P_{52} = \left\{ \frac{[(52-49.5)/5]14 + 65}{105} \right\} 100$$

$$= 69 \text{ percentile rank}$$

1 The score of 52 is in the interval 50–54 or (49.5–54.5), the portion of the interval below 52 is $52 - 49.5 = 2.5$ points, or $(X - l)$

2 The interval is 5 points wide; 2.5 points is .5 above the bottom of the interval, i.e., $2.5/5 = .5$, or $(X - l)/i$

3 If 52 is above half of the score points in the interval, it is also above half of the people in the interval; to locate this portion of the people we multiply $.5(14) = 7$ cases, or $[(X - l)/i]f_w$

4 We next add 7 to the total cases below the interval to determine the total cases above which a score of 52 ranks, or $7 + 65 = 72$ cases, 72 cases below a raw score of 52

5 Dividing 72 by the total of all cases, we get .69; multiplying it by 100 to convert this proportion to a percentage, we get 69; so a score of 52 is at the 69th percentile rank

$l =$ lower limit of interval in which score appears
$i =$ score width of interval
$f_w =$ frequency within interval
$f_b =$ sum of all frequencies below interval
$N =$ total number of cases

The procedure for locating the percentile rank that corresponds to a given score is shown in Table 4-5.

Exercises

4-1 Fill in the following blanks:

(a) In a distribution of 1000 IQs of six-year-old children, 16 percent fall below an IQ of 85. An IQ of 85 is the _____, and 16 is the _____.

(b) In a survey of high school girls, 32 percent had had their first formal date by fourteen years of age. Age fourteen is the _____, and 32 is the _____.

4-2 From the data in Table 4-2 compute the raw score (percentile) that corresponds to

(a) The 81st percentile rank

(b) The 36th percentile rank

4-3 The freshmen in Centertown High School took a reading comprehension test, and the school counselor wishes to develop local norms for the test by computing percentile ranks. The scores are as follows:

$$
\begin{array}{llllll}
10 & 11 & 32 & 26 & 18 & \qquad 24 & 27 & 22 & 26 & 25 \\
\end{array}
$$

10 11 32 26 18	24 27 22 26 25
31 28 27 31 30	13 16 26 18 24
23 26 19 21 26	18 26 22 24 34
15 20 30 24 24	34 23 26 20 27
19 23 20 26 25	19 22 28 19 29

(a) Begin with the score of 10 and make a distribution using intervals two scores wide and find the percentile rank for a score of 31.

(b) Find the percentile rank for a raw score of 24.

(c) Find the score equivalent to the 75th percentile rank.

(d) Joe Smith, a freshman at Centertown High School, had a raw score on the reading test of 28. What is his percentile rank?

Occasionally a zero frequency appears in an interval of a distribution of scores and complicates the calculation of percentiles. For example, for the data in Table 4-6 the interval 140–149 contains no frequency. What is the weight below which 20 percent of the boys (six cases) appeared? Any weight in the range from 139.5 to 149.5 could be considered as the answer. However, the fact that each score of this range is the same percentile seems to be leading us into problems that complicate rather than simplify our description of the data. Therefore, when we have intervals with zero frequencies, we agree to follow a standard practice, i.e., to define the midpoint of the interval as the point we are seeking. Thus, in Table 4-6 a weight of 144.5 is the 20th percentile.

Frequency distributions with zero frequency in an interval usually occur when N's are small. In these distributions the computations of percentiles may not be worth the effort. In fact, the rank with which we identify a given score begins to be most stable only when the number of cases involved is several hundred. In practice we probably would not bother to compute percentiles on fewer than 100 cases in a given distribution, although we have done so for illustrative purposes in this chapter. Although as N's get larger the intervals with zero frequency

Table 4-6 Weight of 30 Freshmen Boys in High School

X	f	cf
210–219	1	30
200–209	1	29
190–199	2	28
180–189	6	26
170–179	5	20
160–169	6	15
150–159	3	9
140–149	0	6
130–139	3	6
120–129	2	3
110–119	1	1

become rare, they do not become extinct, and so some plan for dealing with them is necessary.

Another important characteristic of percentile ranks is the shape of the histogram plotted with percentile ranks on the base line. Suppose we tabulate our frequency distributions in intervals of percentile ranks instead of raw scores and our intervals are 10 ranks wide. What percent of the cases will fall within the first interval 1–10? What percent will fall in the second interval 11–20? The fifth interval 41–50? The answer to each of these questions is 10 percent. We have seen that most raw-score distributions have few frequencies in the extremes and more frequencies near the center of the score range. This means that in a typical distribution it must take more raw-score points to achieve the desired 10 percent of the cases at the extremes of a distribution than it would near the center of the distribution. This fact is illustrated in Fig. 4-3. The raw-score scale is presented in equal-sized units, but the corresponding percentile ranks are broad at the ends of the distribution compared with those at the center. For example, between scores 10 and 11, a one-point range, we have compressed 14 percentile ranks, while between scores 3 and 4 we have only 3 percentile ranks. Percentile ranks are very close together at the center of a distribution and far apart at the extremes. In other words, the percentile ranks are not equal-sized units along our base line. This fact is very important; it means that we cannot add or subtract percentile ranks. If we want to perform arithmetic computations with percentile ranks, we must first convert them into their raw-score equivalents, perform the computations, and then convert the results back into percentile ranks.

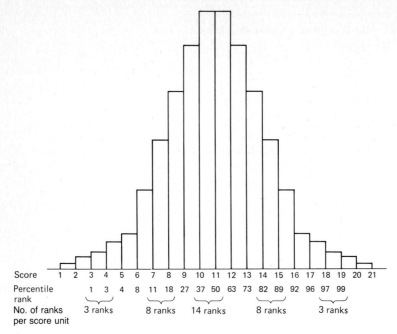

Figure 4-3

Application of Percentile Ranks

When a psychologist, counselor, or other test user calculates a score on a test, the next step is to get some meaning out of the score. Since most measures in the social sciences, including tests, are not ratio scales, we have no absolute zero as a starting point for interpretation. Therefore, we derive meaning out of scores and other observations by noting how each ranks within a sample from some defined population.

Here are some examples. A military personnel officer classifying job-skill potential for a new recruit notes that the recruit has a mechanical aptitude test score of 38 out of 50 items on the test. Is this a "good" score? It all depends on how all other recruits have scored on this test. If 38 is above 95 percent of scores made by previous recruits, *then* it may be a high score, but if it ranks only above 46 percent of other recruits, it is only average. The personnel officer needs a percentile rank for the score of 38 to decide what this recruit's comparative potential is.

Percentile ranks are often useful in other ways. Suppose you have the amount of money spent on schools in thousands of dollars and the number of juvenile arrests per thousand population under age 21. You have collected these data across 150 middle-sized towns in 25 states and 4 Canadian provinces. Aspirinville spent 1750 thousand dollars on schools last year and had 6 arrests per thousand juveniles. Is Aspirin-ville's rank higher in expenditures or arrests? We need a systematic ranking system that shows position within a defined group before we can answer this question. Percentile ranks will serve this purpose quite well.

Deciles and Quartiles

Two kinds of figures besides percentiles are also frequently used to show relative standing in a group. These are *deciles* and *quartiles*, both of which are similar to, and indeed can be read from, percentile tables. Deciles are points that divide the distribution of raw scores into segments of 10 percent each. Thus, the 1st decile D_1 would be that point on the distribution below which 10 percent of the cases fall, D_2 the point below which 20 percent of the cases fall, etc. Deciles can be computed in the same manner as percentiles, since D_1 is P_{10}, D_2 is P_{20}, etc.

Deciles, like percentiles, are points on a scale. Therefore, a score can be between the 3d and the 4th deciles, i.e., in the fourth lowest 10 percent of the group, but it cannot be *in* the 3d decile, since that decile is only a point on the scale.

Quartiles divide the distribution of raw scores into segments of 25 percent each. Thus, the 1st quartile Q_1 is the point that cuts off the lowest 25 percent, Q_2 the lowest 50 percent of the group (what is another name for this point?), and Q_3 the lowest 75 percent of the distribution.

It should be emphasized, however, that deciles and quartiles, like percentiles, are *points* along the scale. They are not *segments* of that scale. It is wrong to say that case X *is in the 3d quartile* or something similar. This is an error because the 3d quartile is only a point on the scale.

Summary

One way to bring meaning into a collection of measurements is to show where given individuals rank among the group. Thus, several procedures have been devised to locate relative position of various raw-score values. Principal among the devices is the percentile rank which identifies the percentage of the group falling below a given score point. For example, if 30 percent of the group falls below a score of 28 on a given test, 28 is at the 30th percentile rank, or the 30th percentile is 28.

Percentile points are much wider in score points at the extremes of a normal distribution than near the center. For this reason the use of percentile ranks in computations which assume equal interval units is often not advisable.

Deciles and quartiles are like percentiles in that they show relative status in a group by indicating portions of the group that come below given points. The decile divides the distribution by segments of 10 percent, D_1 cutting off the lowest ranking 10 percent, D_2 the lowest ranking 20 percent, etc. Quartiles divide the distribution by 25 percent segments, Q_1 cutting off the lowest 25 percent, Q_2 the lowest 50 percent, etc.

Percentiles, deciles, and quartiles are dividing points in the distribution of scores; they are not the segments of scores between these points. Thus it is *not* correct to say that a student scored within X decile.

Key Terms

percentile rank	P_X
percentile	X_P
decile	Q_i
quartile	D_i

Problems

4-1 Fifty psychology students took a clerical aptitude test. Their scores were as follows:

31 38 26 50 42	47 39 28 61 33
27 48 60 58 53	41 36 57 51 66
45 56 54 41 35	43 49 32 58 53
47 43 68 43 31	48 34 47 59 36
32 52 55 59 41	65 71 62 60 45

Construct a frequency distribution beginning with the score of 26, with an interval width of three score points.

(a) What is the cumulative frequency through the interval 35–37? 50–52?

(b) Locate the score values that correspond with the following percentile ranks: P_{20}, P_{67}, P_{90}.

(c) Locate the percentile rank for each of the following score values: 36, 55, 41.

(d) Locate the 3d decile.

(e) Locate the 2d quartile. What is another name for this point?

4-2 You are a high school counselor who has given a hand-eye coordination test to Helen Jackson and found her score to be at the 78th percentile rank (based on test norms for 8126 high school seniors). What does Helen's score tell us about her? Above what percent of the norm group does Helen rank?

4-3   In Medianville 120 students took an academic aptitude test. The test had two parts, one verbal, one numerical. The distributions are below.

Verbal X	f	Numerical	f
37–39	3	45–47	1
34–36	15	42–44	7
31–33	19	39–41	8
28–30	28	36–38	14
25–27	19	33–35	23
22–24	13	30–32	25
19–21	10	27–29	17
16–18	8	24–26	12
13–15	4	21–23	10
10–12	1	18–20	3
	120		120

(*a*) Kim has a Verbal score of 34; what is her percentile rank?

(*b*) Jose has a Numerical score of 32; what is his percentile rank?

(*c*) Tuff Tech requires a percentile rank of 87 on the Numerical test to be considered for admission. How many items must a student get to reach the percentile rank of 87?

(*d*) Laidbach College requires a Verbal percentile rank of 24 for admission. What score on the Verbal test is the minimum for admission to Laidbach?

(*e*) Tyrone has a score of 28 on the Verbal test and 32 on the Numerical test. On which test does he rank higher? (Calculate the percentile rank for each score.)

(*f*) Ralph has a Verbal score of 32. What Numerical score will he have to make to rank at the same point on the Numerical test as on the Verbal test?

5

Variability

Indicators of central tendency, such as the mean or median, tell us what typical or average performance is for a specified group. Although they define the middle point around which the observations cluster, this point is not enough to describe a distribution of observations. We also need to know how widely spread out the observations are above and below this central point.

Suppose Commonville has a mean income for all workers of $28,000. The highest income there is $32,000; the lowest is $24,000. The neighboring town of Spreadsville also has a mean income of $28,000, but its highest income is $60,000 and its lowest is $1000, with incomes spread across the range between these extremes. Although both towns have the same mean incomes, $28,000, the economic circumstances of the residents are obviously quite different from one town to the next. The point is, before we make comparisons or generalizations about sets of data, we need not only an indicator of central tendency but also an indicator of how widely the scores are dispersed around that central point. Figure 5-1 illustrates this idea by showing the distribution of IQs in two elementary schools. Obviously, the array of "talent" in one school is not like that in the other, and instructional programs in school A would not be equally effective in school B. Knowing that children in both schools have the same mean IQ does not provide us with a sufficient description of the "talent" context within which program planning should take place. We need also an indicator of dispersion.

A slightly different situation similarly points up our need for information about spread of scores. Joe and Bill are throwing darts at a

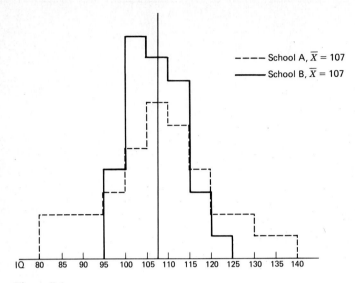

Figure 5-1

target. They throw 20 darts each and record the scores noted in Table 5-1. Their mean scores are identical, but the quality of Bill's perform-ance is clearly different from Joe's. We can see that the mean alone does not adequately describe the characteristic being observed; we must also know the extent to which scores are spread out around that average.

The following situation represents a third example in which more

Table 5-1   Scores Made by Two Boys on 20 Trials Each

| | f | |
Score	Joe	Bill
10	1	
9	2	
8	4	5
7	6	8
6	2	5
5	3	2
4	1	
3	1	
	$\bar{X} = 6.8$	$\bar{X} = 6.8$

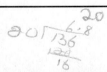

than an average is needed to describe the observed condition. Suppose that the towns of Centerville and Spanville each have 15 third-grade classes and that each class has 25 students (unusual, but pretend it is true). The *mean* reading speed for the classes in each town are computed. If we combine the classes in Centerville, the mean will be equal to 100, and so will the grand mean for Spanville. Are the third grades essentially comparable groups in reading speed? A plot (Table 5-2) of the distribution of class averages may help us decide.

Clearly, the programs in Centerville can be quite different from those in Spanville. Spanville must have a wider array of materials to accommodate the conspicuous differences in abilities; more individualization may seem advisable; and more demands will be placed on diagnostic specialists than in Centerville.

Several terms have been used to describe this tendency for data to be dispersed around the average. Of course, *dispersion* is one name often given this property, *spread* is another, and so is *scatter*. But probably the term most widely used by statisticians when they are referring to this condition is *variability*. The data in Fig. 5-1 represent variability among individuals in a group. The range of scores shown there is due to the fact that some individuals got low scores, some were in the middle range, and some high. The subjects differed in the scores they received. This is sometimes referred to as *within-group variability*.

The data in Table 5-1 represent differences between successive scores made by single individuals on several occasions. The performance of any single person on a given task is likely to vary from day to day or hour to hour. These differences are reflected in *within-subject variability*. But samples from a common population, like those in Table 5-2, also may differ from each other in typical performance. These

Table 5-2   Mean IQs for 15 Classes in Two Towns

IQ means	Spanville	Centerville
130–139	1	
120–129	2	
110–119	3	4
100–109	3	7
90–99	3	4
80–89	2	
70–79	1	

differences between samples are referred to as *between-groups* variability.

The computation of within-individual variability does not differ from the computations of between-individual variability, but the decisions to which the two types of dispersion apply are different. Should the manager of the Yankees send "Slugger" Slayton in as a pinch hitter? His average is .320, but in some games he gets a hit every time he bats; in some games he strikes out every time. "Fingers" Fletcher has an average of .300, but he got this average by hitting twice in every game he played. Here within-individual variability is important to know in deciding who should be selected as a pinch hitter.

On the other hand, how should the Yankees arrange their batting order, and how should the opponents arrange their fielding strategy? These decisions are tied primarily to the variability within the group of Yankee batters. Some are good hitters, some not so good. The essential differences are between the members of the team, i.e., within-group variability.

Occasionally we deal with several samples from a common population. No matter how carefully selected, the means of these samples will not be the same from sample to sample. How is this variation between group means useful in decision making? Suppose Sudsy Soap Company is making a survey of sales potential in Machineville. Since they first wish to know how much soap the typical Machinevillian uses a month, they randomly select from the telephone book five samples of 25 people each. Next they find out what the average soap use is for each of the five samples. The five group means are not identical but do cluster somewhat together. It is likely that the means of other samples of this town's citizens, as well as the mean for the entire population of Machineville, will fall in the vicinity of the range of the sample means already observed, and Sudsy can proceed with fair confidence in their estimate of soap use typical for Machinevillians without having to survey the whole population in the town.

Exercises

5-1   List five situations where an indication of variability within a group of individuals will help us describe the data.

5-2   List five situations where within-individual variability will be important in making decisions.

5-3   List three situations where variability between groups may be an important characteristic to know.

If we agree that variability is an important property of a group of observations because it helps us (1) to describe those observations and (2) to make better decisions based upon the available data, we are ready to look at statistical ways to determine variability.

Range

The simplest and most readily calculated approach to establishing the amount of dispersion in a set of scores is to note how many scale points are included from the lowest value to the highest, inclusive. This statistic, called the *range*, is simply computed by subtracting the smallest score from the largest and adding one point. For example, in Table 5-1 Joe's scores begin at 3 and run through 10. The range is $10 - 3 + 1$, which is 8; that is, from 3 through 10 there are 8 scale points.

As a measure of variability the range is less reliable than some other indicators. Its numerical value rests upon only two scores in a distribution, the most extreme scores. These two values may or may not be representative of the range of the bulk of the scores. A wide gap may exist between the lowest score and the next lowest one, in which case that lowest score would not reflect the spread of the greater portion of the data. An example is given in Table 5-3, where although the scores progress continuously from 16 to 21, a range of 6, the actual range for the distribution is 10. The one person who got a score of 12 determines this broader range compared with the narrower range within which the bulk of the data is dispersed.

For this reason the range as an indicator of variability falls into the

Table 5-3 Data Illustrating Problems in Use of the Range

X	f
21	3
20	4
19	7
18	9
17	8
16	5
15	0
14	0
13	0
12	1

same category as the mode as an indicator of central tendency. Although they are both simple to compute and are quickly obtained descriptive data, they are unreliable from sample to sample, especially when the number of observations in each sample is small. For this reason we shall look for other methods of describing data whenever possible.

Quartile Deviation

One problem with the range is that it depends only on the capricious location of the extreme observations in a distribution, a condition which lends instability to the size of the range. If we could choose our upper and lower points a given distance from the upper and lower ends of the distribution, we could avoid the problem illustrated in Table 5-3. The inset points usually chosen are those points which cut off the highest scoring 25 percent and the lowest scoring 25 percent of the observations. These points are called the *1st quartile* (cutting off the lower 25 percent of the group) and *3d quartile* (cutting off the upper 25 percent of the group). Table 5-4 illustrates these points. The median becomes the 2d quartile point since it cuts off two quarters, or 50 percent of the number of cases in the distribution.

Variability of a group of scores can now be indicated by using the two points, Q_1 and Q_3. In a perfectly symmetrical distribution, the distance in scores from the median to Q_1 will be the same as the score distance from the median to Q_3. Therefore, taking the median as a central point and thinking of the scores being dispersed above and below the median, we can get an indication of the extent of this dispersion by noting the deviation of the quartile points from the median. The average of these median-to-Q_j points is called the *quartile deviation* because it shows how far the quartiles deviate from the center

Table 5-4   Data Illustrating Quartile Points

X	f	
15	1	
14	2	
		Q_3
13	3	
		Mdn
12	3	
		Q_1
11	2	
10	1	

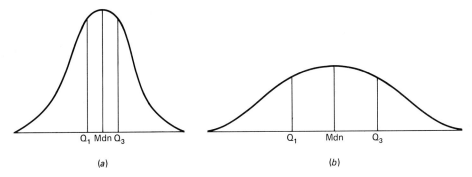

Figure 5-2

of the distribution. If the score range is relatively wide, the distance from the median to the quartiles will be large; if small, the median-to-quartile range will be much less (Fig. 5-2). The data in Fig. 5-2*a* show a distribution with very slight scatter, whereas Fig. 5-2*b* shows a considerable spread of scores. Similarly, the distance from the median to either quartile point in Fig. 5-2*a* is much less than it is in Fig. 5-2*b*.

Not all distributions of scores or other observations are symmetrical. Some are skewed. When we have skewness, the median-to-Q_1 distance will not equal the median-to-Q_3 distance, as shown in Fig. 5-3. Therefore, we average the Q_1-to-median and Q_3-to-median distances to find the quartile deviation,

$$QD = \frac{(Q_3 - Mdn) + (Mdn - Q_1)}{2}$$

If we take the parentheses off, we have

$$QD = \frac{Q_3 - Mdn + Mdn - Q_1}{2}$$

which equals

$$QD = \frac{Q_3 - Q_1}{2} \qquad (5\text{-}1)$$

Some writers refer to QD as the *semi-interquartile range*, an impressive but cumbersome term. In this book we use simply quartile deviation.

How do we find the quartile points? The basic procedure is the one used to find the median with one small variation. The first step in computing the median was to find $N/2$, since half of the scores come below our median. The first step in obtaining Q_1 is to find $N/4$ since only one-fourth of the scores fall below the 1st quartile. Similarly, to

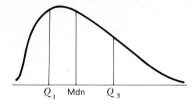

Figure 5-3

find Q_3 we begin by finding $\frac{3}{4}N$, since three-fourths of the group will fall below the 3d quartile. With this exception, the procedure for computing quartiles is the same as that for finding the median.

The quartile deviation is a more stable statistic from sample to sample than the range, but, like the range, it is not an algebraic function of all the scores in the distribution; nevertheless, because it depends upon a larger portion of scores than the range, it tends to be a more reliable indicator of spread.

Standard Deviation

The most desirable indicator of dispersion of observations would be one tied to the deviation of each and every score from some origin. Possibly the average of these score deviations would be a good indicator of "typical" deviation of scores around a central point. What origin or central point should we use from which to calculate the deviation of a given score? We could find how far scores range above the lowest one in the distribution, but lowest scores tend to be unstable from sample to sample. We need a stable origin from which to calculate our score deviations, and the mean is that stable point. We shall calculate our score deviations from the mean, that is, $X - \bar{X}$, find the average of these deviations, and use it as an indicator of dispersion. But wait! Figure 5-4 points out a flaw in this process. Here are children standing at points on a foot scale representing distances broadjumped from a standing position. The mean is 5 feet. How widely dispersed are the marks around the mean? We could compute the distances from each observation to the mean, a value called a *deviation*, and average these values.

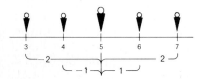

Figure 5-4

But, as we noted in Chap. 3, the sum of the frequencies times distances above the mean equals the sum of the frequencies times distances below the mean. That is how the mean gets its "balance" characteristic in a distribution. These values for the five cases in Fig. 5-4 are

Below the mean	Above the mean
$3 - 5 = -2$	$7 - 5 = +2$
$4 - 5 = -1$	$6 - 5 = +1$
Total -3	Total $+3$

The sum of all five $X - \bar{X}$ values adds up to zero, and a value of zero as an indicator tells us very little about spread of scores around the mean. Therefore, a method of eliminating the minus signs appears to be appropriate here. We find it in squaring the deviations so that all values will be positive numbers, averaging them, and then "unsquaring" them. The resulting value is called a *standard deviation*. Specifically, it is the *root-mean-squared deviation*. This definition can be simplified by looking at its separate parts.

First, we are dealing with deviations. A deviation is merely the distance, in score points, that a given observation is from the mean of the distribution; that is, $X_1 - \bar{X}$ is the deviation for X_1; $X_2 - \bar{X}$ is the deviation for X_2; etc. As we saw above, the deviations must be squared to avoid ending up with a total of zero. Now we have the deviation squared portion of our definition.

Next we add another term from the above definition—*mean-squared deviation*. The mean of anything is simply the sum of the values (here squared deviations) divided by the number of values that were summed. Thus, we compute the $X_1 - \bar{X}$ value and square it, $(X_1 - \bar{X})^2$, then find $(X_2 - \bar{X})^2$ and add this value to $(X_1 - \bar{X})^2$, etc., until we have summed all the squared deviations of our scores from the mean. Then we divide this sum by the number of deviations we have, that is,

$$\frac{(X_1 - \bar{X})^2 + (X_2 - \bar{X})^2 + \cdots + (X_N - \bar{X})^2}{N}$$

The result is the mean-squared deviation. This value is often referred to as the *variance*, σ_X^2, and we shall deal with it more later.

The last term in our definition completes the picture—*root-mean-square deviation*. It merely says to take the square root of the mean-squared deviation just computed. The result is the *standard deviation*,

Table 5-5   Computation of the Standard Deviation

Formula	What it says to do
$$\sigma_X = \sqrt{\dfrac{\Sigma(X - \bar{X})^2}{N}}$$	1 Subtract the mean from each score and square each of these differences, $(X_1 - \bar{X})^2, (X_2 - \bar{X})^2, \ldots, (X_N - \bar{X})^2$ 2 Add together all these squared deviations 3 Divide by the number of deviations to get the mean-squared deviation, or variance 4 Take the square root of this value

Example   Given scores of 6, 10, 5, 8, 6, and 7, where $\bar{X} = 7$, compute the standard deviation.

Step 1: $(6 - 7)^2 = 1$ $(10 - 7)^2 = 9$ $(5 - 7)^2 = 4$
$\qquad\qquad\;\;(8 - 7)^2 = 1$ $\;\;(6 - 7)^2 = 1$ $(7 - 7)^2 = 0$

Step 2: $\Sigma(X - \bar{X})^2 = 16$

Step 3: $\dfrac{\Sigma(X - \bar{X})^2}{N} = \dfrac{16}{6} = 2.67$

Step 4: $\sqrt{2.67} = 1.63$

σ_X. The steps we have just performed are given by

$$\sigma_X = \sqrt{\frac{\Sigma(X - \bar{X})^2}{N}} \tag{5-2}$$

This definitional formula is helpful in illustrating what the standard deviation is. It is also useful in writing a computer program for calculating with large quantities of data. For these reasons attention is especially directed to formula (5-2) and what it tells us to do, as spelled out in Table 5-5.

Exercises

5-4   Ten children taking a test in maze tracing have made the following number of errors: 10, 7, 9, 5, 6, 13, 8, 4, 7, 9. Compute the standard deviation of errors for this group of children.

5-5 A runner has posted the following times in the 100-yard dash during time trials in the past week: 10, 12, 11.2, 11, 10.7, 10.5, 11.6 seconds. What is the standard deviation of his times?

5-6 Compare the effort in computing the standard deviation in Prob. 5-4 with that in Prob. 5-5. Which is easier and why?

Formula (5-2) functions well in computing the standard deviation if the mean is a whole number or we are using a computer. Since many samples we shall study are not large enough to warrant the time and effort in putting our data into a computer-usable mode, i.e., punching into cards, recording on tape, or setting up a computer file via a terminal, we need a handy computational formula for these situations. That formula is found by performing some algebraic manipulations on the definitional formula, as illustrated in Table 5-6.

Our definitional formula (5-2) is shown in Table 5-6 to equal

$$\sigma_X = \sqrt{\frac{\Sigma X^2 - (\Sigma X)^2/N}{N}} \tag{5-3}$$

This will be our *computational formula* for the standard deviation. Like our definitional formula it yields the square root of the average of all squared deviations of scores from the mean. This standard deviation is in score points and is not only a useful indication of dispersion but also (as we will see in Chap. 6) extremely useful in identifying the location of scores in a distribution. The application of formula (5-3) is illustrated in Table 5-7 on page 93.

Now that we have computed a standard deviation, let us see what it looks like applied to a frequency distribution. Remember that σ begins at the mean and measures deviations above (+) and below (−) that point. The resulting application will look like Fig. 5-5. The normal range of a distribution of scores rarely runs beyond ±3 standard deviations, although theoretically scores in a given population may range beyond these values. In the classical bell-shaped curve approximately two-thirds (68 percent) of the observations will fall between +1 and −1 standard deviation.

How does this information apply to a distribution of data? Suppose we have weight data on 100 women who work on the mail routes of a midwestern city. The distribution appears quite bell-shaped. The mean is 140 pounds; the standard deviation is 15 pounds. Since we begin at the mean and add or subtract standard deviations, +1σ would give us a weight of 140 + 15 = 155 pounds. Women at 1 standard deviation above the mean would be at 155 pounds, and 1 standard deviation below the mean would be 140 − 15 = 125 pounds.

Table 5-6 Development of a Computational Formula for Standard Deviation from the Definition Formula

Algebraic operation	Explanation
$\sigma_X = \sqrt{\dfrac{\Sigma(X - \bar{X})^2}{N}}$	Definitional formula
$\sigma_X^2 = \dfrac{\Sigma(X - \bar{X})^2}{N}$	Squaring both sides to simplify the algebra
$\sigma_X^2 = \dfrac{\Sigma(X^2 - 2X\bar{X} + \bar{X}^2)}{N}$	Squaring a binomial of the form $(a - b)^2 = a^2 - 2ab + b^2$ (see sec. A-6)
$\sigma_X^2 = \dfrac{\Sigma X^2 - \Sigma 2X\bar{X} + \Sigma\bar{X}^2}{N}$	The sum of several terms (like those in parentheses in step 3) is equal to the sum of the terms separately (see Sec. A-2)
$\sigma_X^2 = \dfrac{\Sigma X^2 - 2\bar{X}\Sigma X + \Sigma\bar{X}^2}{N}$	The change is in the middle term; $2\bar{X}\Sigma X$, the sum of a constant (like $2\bar{X}$) times a variable (like X) is equal to the sum of the variable (ΣX) times the constant ($2\bar{X}$), or $2\bar{X}\Sigma X$
$\sigma_X^2 = \dfrac{\Sigma X^2 - 2\bar{X}\Sigma X + N\bar{X}^2}{N}$	Note the last term; summing a constant (like $\bar{X}$) is the same as multiplying it by the number of cases: if $\bar{X} = 6$, then $6 + 6 + 6 = 18$ and $3(6) = 18$
$\sigma_X^2 = \dfrac{\Sigma X^2 - \dfrac{2\Sigma X\Sigma X}{N} + \dfrac{\not{N}\Sigma X}{\not{N}}\dfrac{\Sigma X}{N}}{N}$	Since $\Sigma X/N = \bar{X}$, we substitute this value for $\bar{X}$; then cancel Ns
$\sigma_X^2 = \dfrac{\Sigma X^2 - \dfrac{2(\Sigma X)^2}{N} + \dfrac{(\Sigma X)^2}{N}}{N}$	Multiplying a term by itself $(\Sigma X)(\Sigma X)$ produces the term squared $(\Sigma X)^2$
$\sigma_X^2 = \dfrac{\Sigma X^2 - (\Sigma X)^2/N}{N}$	Subtraction; two negative terms reduced by one positive term yield one negative term
$\sigma_X = \sqrt{\dfrac{\Sigma X^2 - (\Sigma X)^2/N}{N}}$	Take the square root of both sides to get back to the standard deviation

Table 5-7   *The Computational Formula for Standard Deviation*

Formula (5-3)	*What it says to do*

$$\sigma_X = \sqrt{\dfrac{\Sigma X^2 - (\Sigma X)^2/N}{N}}$$ where σ_X = standard deviation of observations on variable X
 	
 X = individual observation on variable X
 	
 N = number of observations made
	1 Using raw scores, we
	  (*a*) Sum all scores ΣX
	  (*b*) Square each of the scores and sum the squared values ΣX^2
	  (*c*) Square the ΣX value, which becomes $(\Sigma X)^2$
	  (*d*) Divide $(\Sigma X)^2$ by N, the number of cases
	2 Subtract $\Sigma X^2 - (\Sigma X)^2/N$
	3 Divide the result by N
	4 Find the square root of the quotient in step 3

Example (follows steps above)

Step 1 (*a*)	Step 1 (*b*)
X	X^2
6	36
10	100
5	25
8	64
6	36
7	49
$\Sigma X = 42$	$\Sigma X^2 = 310$

Step 1 (*c*): $(\Sigma X)^2 = 42^2 = 1764$

Step 1 (*d*): $\dfrac{(\Sigma X)^2}{N} = \dfrac{1764}{6} = 294$

Step 2: $\Sigma X^2 - \dfrac{(\Sigma X)^2}{N} = 310 - 294 = 16$

Step 3: $\dfrac{16}{6} = 2.67$

Step 4: $\sqrt{2.67} = 1.63 = \sigma_X$

Within the range of $\pm 1\sigma$ we have about 68 percent of the distribution, so that the middle two-thirds (roughly) of women postal route workers range between 125 and 155 pounds. Since scores rarely range beyond ± 3 standard deviations from the mean, in our sample of postal women we would expect not to find weights below 95 pounds or above 185 pounds. Intuitively, these figures verify the fact that rarely do we find cases below -3σ and above $+3\sigma$. Only a few adult females in the American population are outside the limits of 95 to 185 pounds.

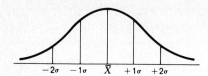

Figure 5-5

A second collection of data will further illustrate the point that the standard deviation is an indicator of dispersion or spread of data. Suppose we go into office buildings along the above postal routes and take a sample of 100 women clerical workers. They also have a mean weight of 140 pounds, but the standard deviation of their distribution of weights is only 10 pounds. The middle 68 percent will lie between 130 and 150 pounds, a range of 20 pounds for clerical workers but 30 pounds for postal workers. The ±3σ range for clerical workers will be 110 to 170, much less than the 95 to 185 range for postal workers. The spread of weights for the two groups is drawn in Fig. 5-6. The distribution of the group with the smaller standard deviation clearly has the smaller spread.

So far we have computed variability of scores by means of the standard-deviation procedures which apply to data based on the entire population as limited by definition. But often we work with samples

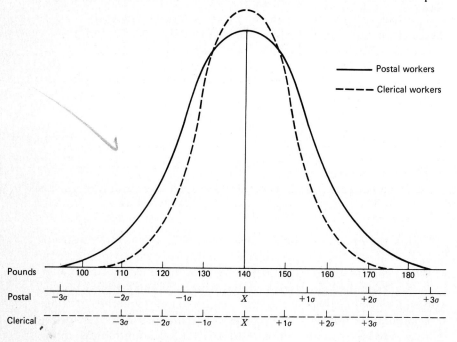

Figure 5-6

from a population, estimating the population properties (parameters) from the sample properties (statistics). When we compute a standard deviation on a sample from a population and wish to use this statistic as an estimate of the population value, a slight change in our procedure must be made. We change our formulas and symbols as follows:

$$s_X = \sqrt{\frac{\Sigma(X - \bar{X})^2}{N - 1}}$$ (5-2a)

$$s_X = \sqrt{\frac{\Sigma X^2 - (\Sigma X)^2/N}{N - 1}}$$ (5-3a)

where we have labeled the sample estimate of the population value s to distinguish it from the population value σ. The use of $N - 1$ in the denominator is especially important when a sample is small, say 30 or less, but should be used every time we are estimating a population standard deviation from a sample from that population.

This kind of situation occurs often. For example, I want to know what the distribution of height looks like for all ten-year-old children in Chicago. Since I cannot measure them all, I take a carefully planned sample and from it estimate the characteristics of the population. We frequently do this kind of thing in studying all phases of human development. Also, in conducting experiments we may wish to draw conclusions about the effect of a treatment on all adults in a given category or all children of a defined group or all rats of a given strain. Since we find it impossible to test them all, we resort to estimating population characteristics from carefully selected samples of the population. When we make such estimates, we should use $N - 1$ in the denominator of the formula for standard deviation.

Exercises

5-7 We have a sample of 10 high school senior girls, selected at random from Localville High School. We have given them a clerical test which involves cancelling as many as possible of the letters *a* and *o* in script in 1 minute's time. The numbers of letters canceled by the 10 girls are 9, 11, 8, 18, 12, 15, 7, 16, 14, 10.

(*a*) Suppose we define our population as all the girls who were selected to take the test. Compute the standard deviation for this population.

(*b*) Suppose we define the population as all girls in Localville's senior class. Now compute the standard deviation for our sample as an estimate of the population.

(*c*) Compare the two figures you have computed. What kind of error would we make if we used a sample standard deviation as an estimate of the population parameter without using $N - 1$ in the denominator? Can you think of a possible reason why this would be true? Would this error be greater for samples when $N = 100$ or when $N = 15$?

5-8   A group of 15 college freshmen was involved in a clairvoyance experiment. As the experimenter held a card in his hand, the students wrote down the symbol (star, circle, square) they believed was on the card. This exercise was repeated through 20 cards. The number of correct responses for each of the students is given below. Compute the standard deviation.

Student	A	B	C	D	E	F	G	H	I	J	K	L	M	N	O
No. correct	4	9	6	10	13	3	8	11	14	10	8	11	7	10	15

Table 5-8   Calculating a Standard Deviation from Grouped Data

Formula (5-4)	*What it says to do*
$$\sigma_X = \sqrt{\dfrac{\Sigma f{\cdot}MP^2 - (\Sigma f{\cdot}MP)^2/N}{N}}$$ where MP = midpoint of interval f = frequency for interval N = total number of cases	1 Locate the midpoint of each interval, multiply each midpoint by the frequency for its interval ($f{\cdot}MP$), sum these $f{\cdot}MP$ values, and square the sum, $(\Sigma f{\cdot}MP)^2$
	2 Divide the sum found in step 1 by N
	3 Square the MP values and multiply each MP^2 value by the frequency for its interval ($f{\cdot}MP^2$); sum these $f{\cdot}MP^2$ values
	4 Subtract the value found in step 2 from the value found in step 3
	5 Divide the result by N; if this is an estimate of the population σ, divide by $N - 1$
	6 Take the square root of the value found in step 5

$$\sqrt{\frac{\Sigma f \cdot MP^2 \cdot (\Sigma \times MP)^2}{N}}$$

5-9 A sociologist was studying teenage communications. As part of this study high school students were asked to tabulate the number of telephone calls they made in a week's time. The data were as follows. Compute the standard deviation of the sample as an estimate of the population.

Student	A	B	C	D	E	F	G	H	I	J
No. of calls	22	14	16	9	12	15	8	11	18	10

Standard Deviation with Grouped Data

Sometimes we have data in a frequency distribution and wish to calculate a standard deviation on the data as tabulated. An *estimate* of the standard deviation can be arrived at as follows. Given the frequency distribution in Table 5-8, we assume that all frequencies are at the

Example Given the following scores on a paper maze test, compute the standard deviation

X	f	MP	f·MP	f·MP²
30–32	1	31	31	961
27–29	2	28	56	1,568
24–26	4	25	100	2,500
21–23	7	22	154	3,388
18–20	12	19	228	4,332
15–17	10	16	160	2,560
12–14	8	13	104	1,352
9–11	4	10	40	400
6–8	0	7	0	0
3–5	2	4	8	32
$N = $	50		881	17,093

Step 1: $(\Sigma f \cdot MP)^2 = 881^2 = 776,161$

Step 2: $\dfrac{(\Sigma f \cdot MP)^2}{N} = \dfrac{776,161}{50} = 15,523.22$

Step 3: $\Sigma f \cdot MP^2 = 17,093$

Step 4: $\Sigma f \cdot MP^2 - \dfrac{(\Sigma f \cdot MP)^2}{N} = 17,093 - 15,523.22 = 1,569.78$

Step 5: $\dfrac{1,569.78}{50} = 31.40$

Step 6: $\sqrt{31.40} = 5.60 = \sigma_X$

midpoint of the interval. We then square each midpoint (MP) value
and multiply the result by the frequency f of the interval. Summing
these $f \cdot MP^2$ values gives us an approximation of ΣX^2. Next we
calculate the value of frequency times MP for each interval. Summing
the $f \cdot MP$ values gives an approximation of ΣX. When we substitute
these terms into formula (5-3), we get

$$\sigma_X = \sqrt{\frac{\Sigma f \cdot MP^2 - (\Sigma f \cdot MP)^2/N}{N}} \qquad (5\text{-}4)$$

The application of this procedure is given in Table 5-8.

The calculation of σ in the grouped-data situation is rather tedious,
but if raw scores cannot readily be resurrected, the procedure provides
a close estimate of the value of the standard deviation calculated
directly from raw scores. If raw scores are available, formula (5-3) is
recommended. But if the data are already in a frequency distribution,
formula (5-4) can be used to estimate the standard deviation.

Exercises

5-10 After teaching a given method of dart throwing to 30 students, I allow
each student to throw 25 darts at a target and record the number of hits.
The scores are shown below in a frequency distribution. Compute the
standard deviation using the grouped-data procedure.

Hits	f
21–22	1
19–20	0
17–18	2
15–16	2
13–14	5
11–12	9
9–10	4
7–8	3
5–6	2
3–4	1
1–2	1

5-11 The data below represent raw scores on a statistics test taken by a class of
30 college students.

```
32 36 38 40 42    52 52 53 54 51
44 45 46 47 47    62 62 63 65 66
48 49 50 51 51    68 69 71 73 77
```

(*a*) Arrange the scores into a frequency distribution with intervals 5 points wide, beginning with 30–34, and compute a standard deviation for the group [formula (5-4)].

(*b*) Now, using the ungrouped-data method [formula (5-3)], compute the standard deviation again.

(*c*) Which procedure, (*a*) or (*b*) is easier in terms of the arithmetic involved?

Variance

A value closely related to the standard deviation is the *variance*, symbolized s^2 for sample estimates of the variance and σ^2 for population variance. Indeed we calculate it when we calculate the standard deviation because it is the square of the standard deviation σ^2. We define the variance as the average squared deviation of scores around the mean. It is calculated as follows:

$$s_X^2 = \frac{\Sigma(X - \bar{X})^2}{N - 1} \tag{5-5}$$

$$= \frac{\Sigma X^2 - (\Sigma X)^2/N}{N - 1} \tag{5-6}$$

Since the sample estimate of the population is about the only variance calculated, only that formula, with $N - 1$ in the denominator, is given.

The variance has additive qualities that make it especially useful in more advanced topics, like analysis of variance, covered in later chapters. In these processes the variance is an essential statistic. However, as an indicator of dispersion describing a set of measurements on a defined group of people, the standard deviation is much more useful. The standard deviation can be stated in terms of score points along the base line of a distribution. The variance cannot. This alone makes the standard deviation preferable in describing data. But the variance also has its unique functions, as we shall see later.

Which Indicator to Use

The most stable indicator from sample to sample is the standard deviation. It also has some relevance to portions of people marked off by various points on the base line of a distribution. It is used with the mean and cuts off segments of scores above and below the mean. Where the data allow we shall choose the standard deviation.

The quartile deviation, which is not as stable as is the standard deviation, is used with the median. Where data are clearly skewed, the median may be the preferred indicator of central tendency; in this case the quartile deviation will also be chosen.

The range has only one good feature in its favor. It is quick and easy to calculate. If this is relevant, then clearly you will choose the range. It is used with the mode but can be used with either the mean or median. When knowledge of the full range of observations is relevant, the range will be used, and indeed it can be cited along with other indicators of dispersion because it alone cites the highest and lowest points on our continuum.

The selection of an indicator of dispersion depends on the function we want it to serve. The first thing to ask is: What do our data look like and what feature do we want to illustrate? At this point we can turn to selection of the indicator that best fits the data and best serves our purpose.

Altering All Scores by a Constant: Effect on σ

Adding a constant to all scores changes the mean an amount equal to the constant. This is not true of the standard deviation. Note that by adding a constant, $\sigma^2 = \Sigma(X - \bar{X})^2/N$ would become

$$\sigma^2 = \frac{\Sigma[(X + C) - (\bar{X} + C)]^2}{N}$$

$$= \frac{\Sigma(X + C - \bar{X} - C)^2}{N}$$

$$= \Sigma(X - \bar{X})^2/N$$

The effect on the standard deviation of adding a constant to all scores is zero. The standard deviation remains the same as before the constant was added. Here is an example. Suppose I have the following data:

X	C	X + C
10	2	12
8	2	10
7	2	9
6	2	8
4	2	6

I have added a constant of 2 to each score. If we calculate the standard deviation for the scores before adding C, we get

$$\bar{X} = 7.0$$

$$\sigma_X^2 = \frac{\Sigma(X - \bar{X})^2}{N} = \frac{20}{5} = 4.0$$

$$\sigma_X = 2.0$$

If we calculate the standard deviation $X + C$, we find

$$\sigma_{X+C}^2 = \frac{\Sigma[(X + C) - (\bar{X} + C)]^2}{N} = \frac{20}{5} = 4.0$$

$$\sigma_{X+C} = 2.0$$

which equals exactly the standard deviation of the scores before the constant was added. However, *multiplying* each score by a constant alters the standard deviation:

$$\sigma_{XC} = \sqrt{\frac{\Sigma(XC - \bar{X}C)^2}{N}}$$

$$= \sqrt{\frac{C^2\Sigma(X - \bar{X})^2}{N}} = \sigma_X C$$

If we multiply each score by a constant, in effect we multiply the standard deviation by that constant. For example, if the X data above are multiplied by a constant of 2, the resulting distribution will be

X	XC
10	$10(2) = 20$
8	$8(2) = 16$
7	$7(2) = 14$
6	$6(2) = 12$
4	$4(2) = 8$

and the standard deviation will be

$$\sigma_{XC}^2 = \frac{\Sigma(XC - \bar{X}C)}{N} = \frac{80}{5} = 16.0$$

$$\sigma_{XC} = 4.0$$

Previously we noted that the standard deviation of the X's was 2. If we multiply this value by the constant (2 × 2) the product is 4, which equals σ_{XC}. Our conclusion is that if we multiply all scores by a constant, the effect on the standard deviation is to multiply it by that constant.

Standard Deviation in Statistical Inference

One further note on standard deviation is of interest here. So far we have looked at the computation of the standard deviation to help us describe a set of raw observations, height, IQ, etc. But statisticians also calculate standard deviations on other kinds of data including all types of statistics. For example, if we take many samples from a population and calculate the mean on each sample, we shall notice that each sample mean is not exactly equal to every other mean in the group. Indeed we can make a distribution of these sample means and calculate a standard deviation for the means in this distribution. In this case our basic data are sample means, not raw scores.

What will such a standard deviation tell us? It will provide a range within which we can expect the mean of a sample from that population to fall. If I have a sample whose mean is somewhat out of range for this population, I must wonder whether this sample is indeed representative of the population. This is a very useful application of the concept of standard deviation with which we shall deal in some detail later.

Summary

In describing a set of scores it is not enough to know the central point around which scores pile up. We must also know how spread out, or how dispersed, the scores are. Therefore, we need an indicator of variability. Several such indicators are readily computed. The range is merely the number of points from the highest to the lowest score, inclusive. It depends only on the position of these two extreme points and as such is not a reliable indicator of variability although it is an easily determined one.

The quartile deviation is the difference between Q_3 and Q_1 divided by 2, and in a symmetrical distribution the median plus and minus the quartile deviation indicates the spread of the middle 50 percent of the scores. The quartile deviation is more stable than the range, but it is not a function of the position of all the scores and lacks the reliability of the standard deviation.

The standard deviation is the square root of the arithmetic average of the squared deviations of scores from the mean of the distribution. It divides the score range (not the frequencies) into about three equal units above the mean

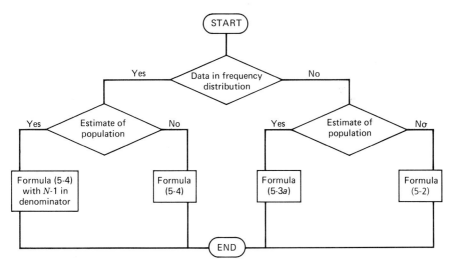

Figure 5-7

and about three below the mean. Thus in a distribution with a narrow scatter of scores these units cannot be large, whereas in a distribution with wide scatter these units must be large. This fact reveals how the standard deviation is an indicator of variability.

The variance is the square of the standard deviation. It has its greatest use in more advanced statistical procedures, where it is a valuable indicator of dispersion. As a strictly descriptive indicator, the standard deviation is more useful than the variance.

The selection of which indicator of dispersion to use depends on the nature of the data and what purpose the indicator is intended to serve. These questions must be dealt with on a situation-by-situation basis, but the standard deviation is often preferred because of its stability and relevance to portions of the bell-shaped curve.

In sorting out the procedures used with the standard deviation the flow chart in Fig. 5-7 provides a useful summary of the various alternatives.

Key Terms

range	between-group variability
quartile deviation	within-subject variability
semi-interquartile range	deviation
standard deviation	σ
root-mean-squared deviation	s
variance	s^2
within-group variability	

Problems

5-1 The counselor in Wolf Crossing High School wishes to analyze aptitude-test scores. The following results on 40 senior boys were found:

31	38	26	50	42	42	39	28	61	33
27	48	60	58	53	41	36	55	51	66
43	56	57	40	35	43	48	32	58	53
47	43	68	43	31	44	34	47	57	36

(a) Compute a standard deviation using formula (5-3).

(b) Compute a standard deviation using grouped-data procedures. (Begin your bottom interval at 26 with an interval width of three score points.) Compare your answer with that to part (a). How do you account for the difference?

(c) Compute a quartile deviation.

5-2 A researcher studying teaching styles of elementary school teachers sits in each of 20 classrooms for exactly 1 hour, during which time she tabulates the number of "prompts" the teacher gives to students:

(a) Calculate the standard deviation using formula (5-2).

(b) Calculate the standard deviation using formula (5-3).

(c) Compare the complexity of arithmetic involved.

(d) Calculate the standard deviation as an estimate of standard deviation of the population from which these 20 teachers are a sample.

Teacher	A	B	C	D	E	F	G	H	I	J	K	L	M	N	O	P	Q	R	S	T
Prompts	21	33	14	18	26	42	36	34	12	17	27	21	29	24	19	25	23	26	29	22

5-3 Would you typically expect a wider range of scores between 1 standard deviation below the mean and 1 standard deviation above the mean or between Q_1 and Q_3?

5-4 The data below represent IQs for two sixth-grade classes. Scores in each class were quite evenly distributed.

Class	$\bar{X}$	σ
I	101	12
II	105	6

(a) In which class would you find the larger number of bright children (IQ of 115 or more)? Why?

(*b*) In which class would you find the larger number of dull children (IQ of 85 or less)? Why?

5-5 Two groups of ten-year-old children had the following scores on a spatial-relations test:

Group A	Group B
10 6 9 7 6	6 9 15 12 1
3 8 5 6 7	8 7 6 5 7

(*a*) Using formula (5-2), compute a standard deviation for each group (round calculations to two decimal places).

(*b*) Compute the quartile deviation for each group (use intervals of one score width).

(*c*) Compare the dispersion of the groups based first on the quartile deviation and then on the standard deviation. How do you account for the differences between the two indicators of dispersion?

5-6 In a physical education class two types of data were recorded: the number of pushups each student could do and the time (in seconds) it took each student to run 50 yards. These data are below. Calculate the standard deviation for each set of data, first as a descriptive statistic for this set of data and then as an estimate of the population of similar physical education classes.

Student	A	B	C	D	E	F	G	H	I	J	K	L	M	N	O	P	Q	R	S	T
Pushups	10	14	18	15	24	9	17	28	21	32	23	26	19	22	38	25	26	23	20	25
Time	10	6	6	7	5	9	8	7	9	5	6	7	9	5	6	5	7	6	8	6

5-7 The middle two-thirds (approximately) of the scores in Prob. 5-6 fall within +1 and −1 standard deviation of the mean. Calculate this range for pushups and runs using the standard deviation you calculated as a descriptive statistic for the data. Now calculate the percentage of scores that fall between +1 and −1 standard deviation from the mean. Does it approximate 68 percent?

5-8 A researcher interviewed 25 adolescents about their leisure time activities. For one question dealing with hours spent watching television in the preceding week the times to the nearest hour are given below. Calculate:

(*a*) The range

(*b*) The quartile deviation

(*c*) The standard deviation for this group of students

(*d*) The estimated standard deviation for the population of adolescents from which this sample came

Subject	time	Subject	time	Subject	time	Subject	time	Subject	time
A	21	F	9	K	28	P	26	U	24
B	20	G	22	L	27	Q	25	V	30
C	31	H	24	M	19	R	29	W	25
D	26	I	36	N	23	S	17	X	23
E	14	J	12	O	16	T	33	Y	18

6

The Normal Curve
and Its Applications

On many of the measurements we make of natural phenomena we find most individual scores not far from average. A few, however, will deviate noticeably from average, and a very few will be markedly different from average. In fact, when large numbers of data are collected through physical and psychological measurements, their frequency polygon often resembles a vertical cross section of a bell. For this reason we often refer to these graphs as *bell-shaped curves*.

An excellent example of such a distribution of data is presented in Fig. 6-1. These data were collected from children on whom the 1937 Stanford revision of the Binet tests was based. The familiar shape of the bell appears clearly in this figure. These data represent a rather large group of people. Smaller samples of data often do not conform so closely to the shape of the bell, although its general characteristics do appear; i.e., most of the cases pile up near the center of the distribution, and fewer cases are found at either extreme.

Usually we cannot measure an entire population; instead we must work with samples from a given population. Although data from a sample may not graph into an obvious bell shape, we believe that for many variables the population data would provide the familiar bell configuration and indeed the sample data approach that pattern closely enough to warrant our assumption about the population. Therefore, we proceed to treat sample data as though they represent the typical form of the population even though the frequency polygon of the sample data may deviate slightly from the form of the bell.

Why are we interested in knowing the shape of the frequency polygon anyway? Suppose we have plotted 1000 IQ scores into a

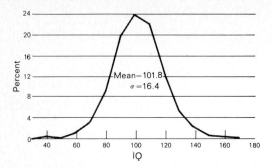

Figure 6-1 Distribution of composite IQs of 1937 Stanford-Binet standardization group. *(From Louis M. Terman and Maud A. Merrill, "Stanford-Binet Intelligence Scale," Manual for the Third Revision, Form L-M, p. 18, Houghton Mifflin, Boston, 1960, reprinted by permission of Houghton Mifflin.)*

frequency polygon and it looks like Fig. 6-1. IQ scores are along the base line (the abscissa), and the number of people are tabulated along the vertical axis (ordinate). Now suppose I have a child whose IQ is 90. I can locate that IQ on the base line. How many people from this group have IQs higher than 90? How many below 90? If I know the shape of the curve, I can figure out the answer to these questions with just a little mathematics.

In fact the problem is even simpler than that. For a given polygon, like the bell-shaped curve, if someone has already figured out areas under the curve that fall below given points along the base line, we need do only a little simple arithmetic to answer the question: How many people have scores below and above point X? Indeed, someone has already figured out these areas for the bell-shaped curve.

Before we can tie scores or other observations to areas of the curve, we must translate them to a common base-line metric, the standard deviation unit. If we know how many standard deviations (or fractions thereof) a score is above or below the mean and if the distribution is bell-shaped, we can read from a previously devised table the area under the curve that is expected to be above or below any given score.

First we must be more specific about the mathematical nature of the bell curve. The frequency polygon that has this distinctive shape is referred to as the normal curve.*

---

* The normal curve has often been called the *gaussian curve* after K. F. Gauss (1777–1855), believed to have developed the formula for it. Later it was discovered that Abraham Demoivre (1667–1754) had arrived at the formula before Gauss. The application of the normal curve to the study of human behavior received great impetus from the work of Adolphe Quételet (1796–1874).

The Normal Curve

The *normal curve* is precisely defined by a mathematical formula from which the curve can be constructed:

$$Y = \frac{N}{\sigma\sqrt{2\pi}} e^{-(X-\bar{X})^2/2\sigma^2} \qquad (6\text{-}1)$$

where Y = ordinate* for given value of X
π = constant = 3.1416
e = constant = 2.7183 the base of natural logarithms

We must first determine the values for $\bar{X}$ and σ for a given set of data. These figures are then put into the formula along with the constants, and values of X are repeatedly substituted in for the computation of successive ordinates in the construction of the normal curve. For example, suppose that we have for a group of 1000 cases a mean IQ of 100 and a standard deviation of 15. We can determine the heights of our ordinates in plotting the normal curve by the formula

$$Y = \frac{1000}{15\sqrt{2(3.1416)}} \, 2.7183^{-(X-100)^2/2(15^2)}$$

Substituting IQ scores for X along a scale from low to high and computing the values of successive ordinates gives the necessary points for constructing the normal curve. The amount of calculation in this operation is considerable and fortunately not necessary for the typical user of statistics.

The normal curve is what statisticians call a mathematical model. It is a mathematical description of many social and physical variables in which we are interested. It has been tested in a large number of circumstances and found to be useful in describing natural phenomena. Although statisticians also use other mathematical models, probably none of them is so well known or so widely applied.

The normal curve is not actually a single curve but a whole family of curves. This idea becomes clear when we study the formula in more detail. In the example, suppose we retain the mean of 100 and N of 1000 but change the standard deviation to 5, to 10, to 20. Applying these data to the formula produces three normal curves, but the three would not look exactly like each other. They would probably look like Fig. 6-2. Each of these curves fits the requirements of normality, yet their shapes are clearly different. We could similarly vary N and $\bar{X}$ and produce a still greater variety of curves.

---

* Students may wish to refer to Sec. A-4 for a review of graphing nomenclature.

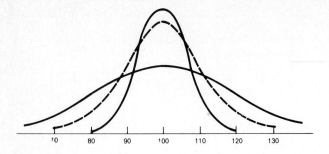

70   80   90   100   110   120   130

Figure 6-2

Normal curves may vary from rather narrow distributions with relatively long ordinates to rather wide distributions with relatively short ordinates. Curves which deviate from normality tend to vary in two ways: either they have too few cases in the central area of the curve and are therefore flattened in the middle of the distribution, or they have too many cases in the central area and as a result are too peaked. When samples of data plotted into frequency polygons are too peaked in the central area, i.e., too many cases fall near the mean, we say that the distribution is *leptokurtic,* from Greek words meaning "thin" and "curvature." Indeed these curves do appear thin because their central areas project upward at the expense of areas just above and below the mean. When the distribution is rather flattened in the central area, i.e., too few cases fall in the middle of the curve, the distribution is said to be *platykurtic,* from the Greek *platys,* meaning "flat." These curves distribute cases somewhat widely above and below the mean, with too few cases in the central area to match the normal curve constructed by formula (6-1).

The fact that we can assume normality in the populations of data we work with allows us to perform several useful procedures based on the relations between the mean, the standard deviation, and the various areas under the curve.

In working with the normal curve, we typically use the mean as our starting point and work up or down from there for two reasons: (1) we often do not know where true zero is for the characteristic we are dealing with and so we cannot begin with zero; (2) the curve theoretically does not touch the base line at the extremes because of the possibility (though rare) of locating in the population a case which scores still higher than our highest score or lower than our lowest score. For these reasons the mean is the most logical benchmark from which to begin our operations.

We also need a unit of measurement to apply to distances above and below our point of departure. This unit is the standard deviation, which gives us a yardstick for siting points along the base line of the curve.

Now we are ready to apply these elements—the mean as a starting point and the standard deviation as a unit of measurement above and below that point on the base line of the normal curve. Some useful facts emerge. We find (Fig. 6-3) that if we erect an ordinate at 1 unit (1 standard deviation) above the mean, the portion of the area under the curve enclosed by this ordinate, the mean, and the base line is 34.13 percent of the total area under the curve. The same portion of the area is cut off by 1 unit below the mean. Similarly, 2 units above or below the mean cut off an additional 13.59 percent of the area under the curve, and 3 units in either direction cut off another 2.28 percent. *These portions pertain to all curves which fit the criteria of normality,* even though the distribution may be rather widely dispersed (Fig. 6-3A), rather narrowly dispersed (Fig. 6-3B), or between these extremes.

Since portions of the area under the curve represent portions of the total number of individuals we have observed we can apply the information from Fig. 6-3 to solving many problems. Suppose we know that the mean IQ for sixth-grade children in Pinetown is 100 and the standard deviation is 15 IQ points. What portion of the sixth-grade pupils will have IQs below 115? This score is 1 standard deviation above the mean. From Fig. 6-3 we see that below $+1\sigma$ we have 34.13

Figure 6-3

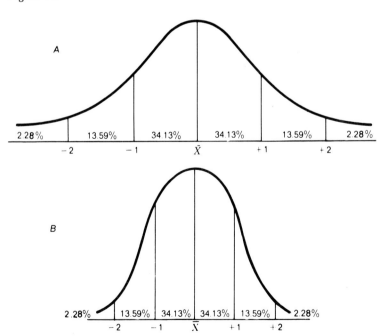

percent, 34.13 percent, 13.59 percent, and 2.28 percent, which add up to 84.13 percent. We therefore expect that about 84 percent of the sixth-graders will have IQs under 115.

How many of Pinetown's sixth-grade children had IQs below 85? As before, we find that 85 is 1 standard deviation below the mean. Adding up the portions of the curve that fall below that point (13.59 and 2.28 percent), we find our answer to be 15.87 percent.

Now let us make a third attack on the problem of bringing additional meaning to score distributions. Suppose Pinetown wants to provide a special program for slow learners in the sixth grade. They have defined slow learners as children with IQs from 70 to 85. What portion of their children will come within this range?

Using the procedures illustrated in the previous paragraphs, we find that 85 is 1 standard deviation unit below the mean, 70 is 2 units below, and between these points 13.59 percent of the cases are expected to appear.

These brief examples show how the normal curve can be applied to solving various problems. The basic procedure tells us to begin at the mean of a distribution, move up or down the scale the necessary number of standard deviations to locate the point in question, and then determine the portion of the curve below that point or above it, as the problem may indicate.

Exercises

6-1 A high school counselor has given a general-anxiety questionnaire to all entering freshmen in her high school. The mean for all freshmen was 70; the standard deviation was 10.
(*a*) Herman has a score of 80. What portion of the class is expected to have lower anxiety scores than Herman's?
(*b*) Janette has a score of 50. What percent of the class is expected to have anxiety scores higher than Janette's?

6-2 A manufacturer of schoolroom furniture was interested in making chair desks in as few sizes as possible to fit the varying heights of senior high school students. He decided to build one model for the middle 68 percent of the high school population. He found that the mean height for the students was 68 inches and the standard deviation was 4 inches. What will be the range of height for the middle 68 percent of this high school group?

---

z Scores

By now you realize that there are essentially two basic steps in solving problems with the normal curve. There is a point in the distribution about which we have a question. Call it X. We find how many standard deviation units point X is above or below the mean $\bar{X}$. Then we find the portion of the curve that lies above or below the point. The procedure translates into the formula

$$z = \frac{X - \bar{X}}{\sigma}$$

(6-2)

So far all our z values have been whole numbers; i.e., all our X values have deviated from the mean $(X - \bar{X})$ in units of the standard deviation. Since in real problems we must deal with points that deviate from the mean in fractional parts of the standard-deviation unit, a more precise approach to calculating areas under the curve will be required. Fortunately, areas under the curve for all fractional values of z are already calculated for us.

The z score is nothing more than a raw score converted to standard-deviation units. Since standard deviations are measured from the mean, z scores begin at that point and range up and down the scale. A score which is 1 standard deviation above the mean would have a z-score equivalent of $+1$; if a raw score is ½ standard deviation below the mean, the z score will be $-.5$, etc.

We convert a raw score to z scores in order to determine how many standard-deviation units that raw score is above or below the mean. This allows us then to use the characteristics of the normal curve in solving problems. The procedure for computing a z score is given in Table 6-1.

Since the z score is reported in *standard-deviation units*, it allows us to locate points on the base line of the normal curve. We can then determine portions of cases that come above or below these points by calculating areas under the normal curve. We now know what portions of the group are cut off by whole-unit z scores, but what if we have fractional parts of units? If we are adept at the calculus, we can figure out areas under curved portions of graphs, but fortunately we do not have to do this, since these areas have already been figured out for us and have been put into table form.* Appendix Table C-2 is such a table. A portion of it is repeated as Table 6-2 so that we can see how it works.

Suppose we have a set of data on heights of children in a given

---

* Suppose we wish to know the area under the normal curve and have no reference to go to. We could slice the area under the curve into many very narrow rectangles, so narrow that their width approaches zero as a limit. (Note figure at right.) Then we could find the area of each of these rectangles and add together as many of these areas as we need to find the magnitude of the portion of the curve we are interested in. This is essentially the procedure used in the calculus.

Table 6-1   Computation of z Scores

Formula (6-2)	What it says to do
$$z = \dfrac{X - \bar{X}}{\sigma}$$ where X = any given raw score for which we want a z equivalent $\bar{X}$ = mean of distribution of X scores σ = standard deviation of X scores	1 Determine the mean and standard deviation 2 To find a z equivalent for X first subtract the mean from X to find how many raw-score points X is from the mean 3 Divide this difference by σ to determine how many standard-deviation units this difference is equal to

Example For a college freshman class at Hillside University the mean college aptitude-test score is 48 and the standard deviation is 8. What is the z-score equivalent of a score of 43 at Hillside?

$$z = \frac{X - \bar{X}}{\sigma} = \frac{43 - 48}{8} = -.625$$

That is, a raw score of 43 is .625 standard-deviation unit below the mean.*

---

* The student who needs a review of negative numbers should consult Appendix Sec. A-3. Negative numbers regularly appear in z-score computations.

inner-city school. The mean is 41 inches and the standard deviation 4 inches. Jay's height is 42 inches. We calculate z

$$z = \frac{42 - 41}{4} = .25$$

Now this value can be located in the left-hand column of Table 6-2 and in Table C-2. The next column to the right says that there is .0987 or 9.87 percent of the area of the normal curve between the mean and z. This is illustrated in Fig. 6-4. The shaded area is the portion of the curve from the mean to a z value of .25. It is the area in our case between the mean of 41 inches and a height of 42 inches.

What portion of our students is expected to be beyond a height of 42 inches? The last column on the right of Table 6-2 tells us that .4013 or 40.13 percent of the group will be beyond 42 inches tall. This is also illustrated in Fig. 6-4.

All z values in Tables 6-2 and C-2 are positive. These data then apply only to the upper half of the curve, the area above the mean. Since the normal curve is symmetrical, the solutions applied to z values above the mean also apply to z values below the mean. Suppose our example

Table 6-2   Portion of Table C-2

$z = \dfrac{X - \bar{X}}{\sigma}$	Area between mean and z	Area beyond z
.25	.0987	.4013
.26	.1026	.3974
.27	.1064	.3936
.28	.1103	.3897
.29	.1141	.3859

above had dealt with a z score of $-.25$, instead of .25. The solution then would look like Fig. 6-5. If one keeps clearly in mind that z scores begin at the mean and move up and down from that point, using Table C-2 should cause no problem.

Some applications of z scores and the corresponding areas of the normal curve will help stabilize what we know about the topic. Suppose a high school counselor is faced with a senior boy who is considering enrollment at Midstate University. The boy has an academic aptitude-test score of 650. Is this a high score? We cannot tell from the score alone; we need to know how this score ranks in relation to the average performance of college applicants, i.e., the mean. In order to show this relationship we need a yardstick to determine how far 650 is above or below the mean. This yardstick is the standard deviation, and the distance 650 is from $\bar{X}$ in standard-deviation units is our z score for 650.

We find that the mean of the academic aptitude-score distribution is 500 and the standard deviation is 100. Applying the z-score formula, we find that 650 is 1.5 standard deviations above the mean. What portion of the students is expected to fall below this point? Table C-2

Figure 6-4

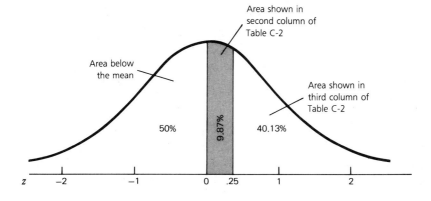

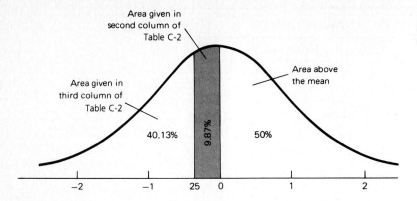

Figure 6-5

tells us that .4332 or 43.32 percent of the group lies between the mean and 650, our z of 1.5. Since 50 percent of the students are below the mean, we add 50 to 43.32 and find that a score of 650 lies above 93.32 percent of the scores of all test takers. Is a score of 650 a high score? Certainly it is. Only 6.68 percent of the group who took the test scored higher.

Sometimes we have records of a person's performance on two or more different kinds of measurements and would like to compare the performances. Unless the scales of two measuring tools are the same, we cannot make a direct comparison. For example, Janie has spelled 28 out of 50 words on her spelling test and has solved 11 out of 18 problems in arithmetic. Direct comparison of her status in arithmetic and spelling cannot be made from these data because they are clearly not in the same scale of measurement. If we convert the two scores to z scores, we put them on a common scale since a z-score unit in one normally distributed set of data represents the same thing as a z-score unit in any other normally distributed variable.

Did Janie do better in spelling or in arithmetic? The spelling-test scores had a mean of 30 and a standard deviation of 5; the arithmetic mean was 10 with a standard deviation of 3. Janie's comparison looks like this:

Spelling	**Arithmetic**
$z = \dfrac{28 - 30}{5} = -.40$	$z = \dfrac{11 - 10}{3} = .33$

and she appears to have done comparatively better in arithmetic than in spelling.

Exercises

6-3 In a learning experiment students were asked to learn as many as possible of a list of 30 nonsense words in 5 minutes. They were tested at the end of the 5-minute period to see how many words they had learned. The mean number of words learned was computed to be 12, and the standard deviation was 4.

(*a*) Student A had 15 correct words on the test. What is A's z score?

(*b*) Sketch a normal curve, locate A's z score on its base line, and erect an ordinate at that point. What percent of the group is expected to be below A? Above A?

6-4 In the experiment described in Exercise 6-3 student B had 10 correct words.

(*a*) What is B's z score?

(*b*) What percent of the group is expected to score below B? Above B?

6-5 The subjects in the experiment in Exercise 6-3 also took a spatial-relations test. The score was the number of various-shaped blocks placed in correct holes in a form board in 3 minutes. The mean score for students in the experiment was 24; standard deviation was 6. Student B in Exercise 6-4 placed 22 blocks correctly.

(*a*) What was B's z score on this test?

(*b*) Did this student do better on the learning test or the spatial-relations test?

(*c*) What percentage of the group is expected to score below student B on the spatial-relations test?

6-6 Student A in Exercise 6-3 has a spatial-relations score of 28.

(*a*) Was A better on the learning task or the spatial-relations test?

(*b*) What percentage of the group should be between A and B on the spatial-relations test?

(*c*) What percentage of the group is expected to score above A on the spatial-relations test?

T Scores

The use of the z score has illustrated the frequent need for changing raw-measurement data to some type of standard scale. Since the z-score scale is often used for this purpose, z scores are sometimes referred to as *standard scores*. The z score, however, is not the only standard-scoring method. In fact a wide variety of standard-scoring methods could be devised. Any time we express a score in terms of its variation from the mean with the standard deviation as the unit along the base of the curve we have a standard score.

Another commonly used standard-score method is the *T* score. Some people find it inconvenient working with a scale that has zero in

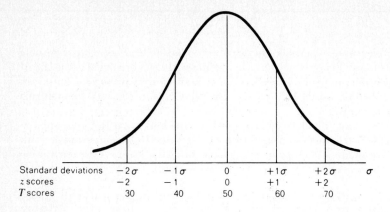

Standard deviations	-2σ	-1σ	0	$+1\sigma$	$+2\sigma$	σ
z scores	-2	-1	0	$+1$	$+2$	
T scores	30	40	50	60	70	

Figure 6-6

the middle and half the scores with minus values. The *T*-score scale has solved this problem by placing the mean raw score equal to a *T* score of 50 and equating the raw-score standard deviation to 10 *T*-score points. Thus a score 1 standard deviation above the mean (*z* score of 1) would have a *T* score of 60, that is, 50 plus 10. A score 1 standard deviation below the mean (a *z* score of -1) would have a *T* score of 40. Comparable standard-score points for *z* and *T* scales are shown in Fig. 6-6.

Table 6-3

Formula (6-3)	*What it says to do*
$T = 50 + 10z = 50 + 10\,\dfrac{X - \bar{X}}{\sigma}$	To convert a raw score to the *T*-score scale:
where $z = \dfrac{(X - \bar{X})}{\sigma}$	1 Compute the *z* score for that raw score
X = given raw score $\bar{X}$ = mean of distribution of X scores σ = standard deviation of X scores	2 Multiply the *z* value by 10 and add 50
	Example Given a mean of 42 and a standard deviation of 6, what is the *T*-score equivalent of a raw score of 30?
	Step 1: $\quad z = \dfrac{30 - 42}{6} = -2$
	Step 2: $\quad T = 50 + [-2(10)] = 30$

The computation of a T score is essentially a matter of manipulating the comparable z score. You have seen that the z scale has a mean of zero and a standard deviation of 1.0. Our objective in T scores is to convert z values into a scale with a mean of 50 and a standard deviation of 10. Since z scores are already in standard-deviation units, we can simply multiply the z value by 10 and have a result which tells us how far the score in question is from the mean in terms of a 10-unit standard deviation. This $10z$ value we add algebraically onto our T-score mean of 50. The result will be a T-score value for the raw score under consideration. The formula is

$$T = 50 + 10z \tag{6-3}$$

and its use is demonstrated in Table 6-3.

Exercises

6-7 Vanguard High School has developed its own scholastic aptitude test. The raw-score mean is 110; the standard deviation is 20. The school's counselor wishes to report scores in a T scale. In the table below convert the raw scores first to z, and then to T and find the portion of the group that will score below each point.

Raw score	$X - \bar{X}$	z	T	% below
132				
126				
121				
90				
86				
70				

6-8 If you were a counselor explaining to students the results of their aptitude-test scores, would you prefer to use raw scores, z scores, or T scores? Why?

Other Types of Standard Scores

Besides z scores and T scores several other widely used score conversions are tied to standard-deviation units. Actually we can create a system with any mean we like and with any standard deviation by manipulating the z-score procedure. If we use a subscript m to stand for "new," the procedure is

$$\text{Conversion} = (\bar{z} + \bar{X}_m) + z(\sigma_m)$$

but since $\bar{z}$ is always zero, this becomes

$$\text{Conversion} = \bar{X}_m + z(\sigma_m)$$

where $\bar{X}_m$ is the mean of the new distribution and σ_m is the standard deviation of that new distribution.* For example, the formula for the T scores we just computed above would be

$$T = 50 + z(10)$$

that is, 50 is to be the mean of the new scale, and its standard deviation is to be 10.

Suppose we want a scale with a mean of 100 (let us call it a C scale) and a standard deviation of 20. The C scale would be developed as

$$C = 100 + 20z$$

and our raw scores could be converted to a scale with a mean of 100 and a standard deviation of 20.

The z value in our conversion tells us how many standard deviations a score is above or below the mean. Multiplying z by a given value, say 20, says that there are z standard deviations each 20 units long; so $20z$ converts my scale to one with a standard deviation of 20 points.

Some widely used tests began with a converted mean of 500 and a standard deviation of 100. Here the raw scores were changed to z values and then

$$V = 500 + 100z$$

converted the scores to a scale whose mean is 500 and whose standard deviation is 100.

Exercises

6-9 I have built a test to measure quantitative reasoning ability and have given it to 384 entering college freshmen. The 50-item test has a mean of 32 and a standard deviation of 8 points.

(a) Convert a raw score of 39 into a scale with a mean of 100 and standard deviation of 20.

(b) Convert this same raw score, 39, into a scale with a mean of 500 and standard deviation of 100.

(c) Now convert 39 into a scale with a mean of 250 and standard deviation of 50.

(d) Convert 39 to a T score.

---

* Recall that adding a constant to each score (here $\bar{X}_m$) is the same as adding the constant to the mean: multiplying each score by a constant is the same as multiplying the standard deviation by the constant.

6-10 I have a raw score of 24 on the above test. Convert this to a scale with a mean of 100 and standard deviation of 20.

Standard Scores, Samples, and Populations

We have seen how standard scores can be applied to locating one's position in the group. A similar operation can be applied to samples from a population to see how a given sample ranks in reference to other samples from the same population.

Suppose we have 480 boys who report to the city park in June to participate in organized baseball. We randomly assign 12 boys to each of 40 teams, and the boys play all summer. At the end of the summer we compute the average number of hits per game for the entire group of 480 boys as 4 hits per game. Now we also compute the mean number of hits per game for *each of the 40 teams* separately. Not every team will have the same mean number of hits, but if we plot these 40 means on a graph, we probably will see something like a normal distribution appearing around a point which will be the population (480 boys) mean (Fig. 6-7). A standard deviation can also be computed for this distribution of means. We call this standard deviation the standard error of the mean to distinguish it from a raw-score standard deviation. But the use of the standard error is just like the standard deviation: it is a unit of measurement which tells us how far a given point is from the mean.

Now we can apply something like a z score to solving some problems. Team 7 had an average of 3.5 hits per game for the season. What percent of the teams ranked higher than team 7? We could solve this problem if we actually computed the grand mean and the standard error.

Now suppose I am going to be a coach. What are my chances of being assigned a team which will have an average of 5 hits per game for the season? Suppose I hear about a team with only 2.3 hits per game. Is it likely this team comes from our population of boys? Scanning the distribution of Fig. 6-7, we would have to conclude that such a team could only rarely come from random selection from our population of

Figure 6-7

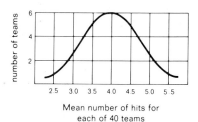

Mean number of hits for
each of 40 teams

boys. This application of the basic idea of the standard-score principle, as well as its relation to portions of the normal curve, actually fits into the topic of statistical inference, dealt with later. However, we solve these problems dealing with many means of samples the same way we solved problems dealing with raw scores except that here we calculate the standard deviation in a slightly different way. We shall deal with this procedure in a later chapter.

Summary

The magnitude of a given measurement or similar observation is not readily apparent when only a raw score is available, but when a raw score can be shown in terms of its deviation from the mean, its magnitude begins to be evident. The yardstick we need to evaluate the extent of the deviation is found in the standard deviation. A score expressed as a deviation from the mean in terms of multiples of the standard deviation is known as a standard score. Two types of standard scores are commonly used: z scores and T scores. With z we equate raw scores to a scale with a mean of 0 and a standard deviation of 1. The T scale has a mean of 50 and a standard deviation of 10. We can create a standard scoring scheme with any mean and standard deviation we wish by adding the desired mean to the product of z and the desired standard deviation.

One of the most useful applications of standard scores, used especially with z, is based on the assumption of normality in the population out of which the measured sample has come. Since areas under the normal curve, in relation to the standard deviation and fractional parts of it, are known, once we obtain a person's position in the group, as shown by the standard score, we can determine the portion of people likely to perform better or worse.

Standard scores are also useful for comparing two measures with each other when the two scales are not in the same units. If we wish to compare intelligence test results with achievement scores, we can hardly do so until the results of both tests are translated into a common scale. Such a scale is found in z, T, or similar standard scores.

The idea of portions of the normal curve falling above or below various standard-score points can be applied not only to data on individuals but also to a collection of means of samples from a population. Thus we can determine the likelihood of getting a sample whose mean is above or below various magnitudes and consequently decide whether a given sample is indeed a random selection of cases from a given population.

Key Terms

bell-shaped curve	z score
normal curve	T score
leptokurtic	standard score
platykurtic	

Problems

6-1 Eks University gave college admissions examination A and found their raw-score mean to be 97 and the standard deviation to be 12. Compute a z score and a T score for each of the following raw scores:

X	z	T
91		
119		
110		
86		
97		

6-2 On an admissions test B, Eks University officials found a mean raw score of 78 and a standard deviation of 15. For the following students state whether their performance relative to all the students at the university is better on test A of Prob. 6-1 or on test B.

Student	Raw score A	Raw score B
Joe	95	74
Mary	110	95
Bill	88	73

6-3 Convert the test A score of each student in Prob. 6-2 to a standard score with a mean of 500 and a standard deviation of 100.

6-4 Convert test scores in B in Prob. 6-2 to standard scores with a mean of 10 a standard deviation of 3.

6-5 For each of the following z scores state the area under the normal curve between the mean and the z score, the area in the larger portion of curve, and the area in the smaller portion of the curve.

z	Area $\bar{X}$ to Z	Area in larger	Area in smaller
1.20			
−.65			
.75			
−.10			

6-6 Given a distribution of raw scores with a mean of 28 and standard deviation of 6, complete the following table for each raw score

Raw score	z	Area to $\bar{X}$	Total area above z
32			
36			
21			
16			
24			

6-7 Given a distribution with a mean of 50 and a standard deviation of 15:

(a) What percent of the group is expected to have scores greater than 68?

(b) What percent will lie between scores of 47 and 60?

(c) What percent will lie between scores of 40 and 47?

6-8 I have a set of arithmetic scores on 50 children in grade 6. The mean is 31 and standard deviation 5. I also have spelling-test scores. The spelling mean is 18.4 and the standard deviation is 4.

(a) Ian got 38 on his arithmetic test. How many spelling words will he have to get right to rank at the same point in the distribution on both tests?

(b) Kim has an arithmetic score of 25. If she takes the test again and moves up to a score of 34, what percent of the group will she have passed between her first and second scores?

(c) Jess has a spelling score of 12 words right. To pass the next 20 percent of the students just above him, what score will he have to make on a retest?

6-9 Across counties in three midwestern states the mean number of divorces per 1000 households was 73. The standard deviation was 5. Assuming that the data are normally distributed, find:

(a) The percentage of counties with more divorces than Casey County, which had 81 per thousand households

(b) The percentage of counties that had fewer than 83 but more than 70

6-10 Given the following test data for students in Possum Trot High School

Test	$\bar{X}$	σ	Smith's score
Arithmetic	45.8	5.2	53
Reading	33.4	6.1	35
Science	74.1	8.4	73

(a) What is Smith's z score for arithmetic? For reading? For science?

(b) What percent of the students in the school are expected to fall below Smith in arithmetic? In reading? In science?

(c) Does Smith rank lowest among his peers in reading, arithmetic, or science?

(d) Rivera has an arithmetic score of 48. What percent of the group is expected to score between Smith and Rivera?

(e) Volsky has a score of 44 in arithmetic. What percent of the group is expected to score between Rivera and Volsky?

(f) Convert Smith's three scores into a common scale with a mean of 200 and standard deviation of 50.

7

Correlation

Looking over the records of last fall's entering freshmen, the admissions officer at Midstate University notices that students who had high grade-point averages in high school tend to do better in college than those who had mediocre high school averages. Across town a young mother who just took a course in child psychology decides to keep a monthly growth record of her child. After 6 months she sees a fairly clear picture of increases in height accompanied with increases in weight. Back on campus an experimental psychologist notices that the more lists of nonsense syllables a student learns the fewer trials it takes to learn successive lists.

Each of these is a situation in which change in one condition is associated with a change in a second condition; i.e., two conditions are seen to change together. In none of these situations would we expect one condition to change a definite prescribed amount with each unit change in the other condition. The kinds of behaviors described are not that stable or closely related. Nevertheless, clear trends of change do appear in pairs of variables in many situations in the world about us.

In the behavioral sciences we occasionally see two variables which change together almost unit for unit; in other conditions we observe only a tendency for one to change as the other changes a given amount; and in still other cases we see no relationship between two variables at all. Because relationships between pairs of variables may range from very close to none, we need a technique to indicate the extent to which changes in X are indeed reflected by changes in Y. This technique is found in what are known as *correlational methods*.

The Basic Concept

Correlation in nonstatistical use means that there is a correspondence between two conditions. In statistical usage correlation refers to techniques which indicate in numerical terms the extent of the relationship between two variables. It is a measure of relationship with specified limits, stated in coefficients ranging from 0 to ±1.00.

Suppose I ask you to guess the income of 30 adult males from their shoe sizes. Chances are that shoe size alone will tell us nothing about income; i.e., a large shoe size is not necessarily associated with a large income nor a small shoe size with a small income. Such a condition is indicated by a correlation coefficient of zero. A zero correlation coefficient means that knowledge of one variable does not allow us to speculate, beyond a guess, about the value of another variable. If I know that your shoe size is $8\frac{1}{2}$, this does not allow me to say, beyond a guess, what your income is. With a zero correlation coefficient knowledge about the value of A tells me nothing about the value of B.

But suppose I give you 30 pairs of shoes and ask you to guess the size of each left shoe from the size of the corresponding right one. You are almost certain to guess the size of every left shoe correctly because as the size of right shoes changes, the size of the left shoe changes an equal amount. In this case the correlation coefficient would be +1.00, indicating that the relationship is perfect (as one condition changes 1 unit, the other also changes a specific amount). Moreover, the *direction* of change is the *same* for both conditions, indicating a *positive* (or plus) relation; i.e., as one variable increases in magnitude, so does the other variable. We have a positive relationship when variables change together in the same direction: as X gets larger Y gets larger; as people grow taller they tend to grow heavier; the larger the right shoe the larger the left. Height and weight and left and right shoe size are examples of *positively correlated* variables. As one variable increases in magnitude, the other tends to increase also.

Sometimes we find that as one condition increases, the other decreases. The higher their high school grade-point averages the fewer failures we expect among college freshmen. The longer parents attended school the fewer the high school dropouts among their children. If one condition increases as the other decreases, there still is a definite relationship between these two conditions; but since one variable is diminishing as the other expands, we say that the relationship is *negative* (or minus). If for every unit increase in X we have a prescribed decrease in Y, we have a perfect negative correlation, or −1.00.

Negative correlations are just as useful as positive correlations. The uninitiated in statistics often want to interpret negative correlations as meaning a lack of relationship between the two variables. This is *not* true; the positive and negative signs merely tell us the direction of change expected to occur in one variable when the correlated variable changes.

For example, the number of cubic feet of natural gas used for heating per day is negatively correlated with the temperature. During warm weather there is little need for natural gas, but when the temperature goes down, there is increased demand for gas. Cubic feet of gas is *negatively correlated* with temperature, but being negatively correlated does not mean that there is no relationship; it merely means that as the amount of one variable increases, the amount of the other decreases. We can predict one variable from the other just as well when they are negatively related as when they are positively related.

Briefly, then, if two variables change together in magnitude in the same direction, the relationship is positive and the coefficient of correlation between them will be a positive number between 0 and +1.00; if two variables change together in magnitude in opposite directions (one increases while the other decreases), the relationship is negative and the coefficient of correlation between these variables will be a negative number between 0 and −1.00. If a change in magnitude of one variable tells us nothing about the likely condition of the other variable, there is no relationship and the coefficient of correlation will be zero.

Most pairs of conditions observed in the behavioral sciences do not change perfectly together, nor do many even approach a perfect relationship. As a result we see many more correlation coefficients between .50 and .60 than between .90 and 1.00. The same is true for correlation coefficients on the negative side. Outside the physical sciences we simply do not expect perfect negative or perfect positive relationships between pairs of variables.

The Scatter Diagram

We have noted that correlation coefficients are figures ranging from −1.00 to 0 to +1.00, which indicate the direction of the relationship and how closely two variables change together. Our job now is to see where these figures come from and how to calculate them. We begin by plotting data on a graph to illustrate the basic concepts. Then we shall interrelate data from two variables to show how the values between −1.00 and +1.00 are found and what they tell us about the relationship of the two variables.

The relationship between two variables can be illustrated by plotting them on a bivariate graph as we did in Chap. 2. For example, suppose we have the following data, where an X value and its corresponding Y value represent two measurements on the same person:

X	1	2	3	4
Y	5.0	5.5	6.0	6.5

Graphed on a bivariate format, these data would look like Fig. 7-1. The graph shows that as X increases 1 unit Y also increases a specific increment. In other words, as X increases a given amount, Y increases in a proportionate amount, although the increments are of different size. The relationship is therefore a perfect positive one, and the plotted points lie in a straight line.

But not all pairs of variables progress together unit for unit. Their relationship is less than perfect, although there may be a clear increasing trend in Y for each unit increase in X. Here is an example of this type of relationship, where again corresponding X and Y values are two measurements of the same person:

X	1	2	3	4
Y	4.9	5.7	5.9	6.6

Graphed, the data would look like Fig. 7-2. Here the trend of an increase in Y for each unit increase in X is clear, but the data do not progress unit for unit. Thus, the plotted points fall near, but not directly on, the straight line. Although it is positive, the relationship is not perfect, and the correlation coefficient would be less than 1.00.

Similar examples can be used to illustrate the different relationships. Figure 7-3*a* shows a perfect negative relationship: as X increases 1 unit, Y decreases a specific increment; Fig. 7-3*b* shows a negative relationship which is less than perfect: as X increases 1 unit, Y has a decreasing trend but it is not an increment-for-increment change; and Fig. 7-3*c* shows a zero relationship; change in X tells us nothing at all about change in Y. We call bivariate graphs such as these *scatter diagrams* or *scattergrams*.

We now put some real data for X and Y into our scatter diagrams,

Figure 7-1

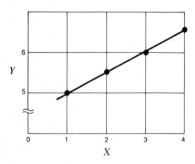

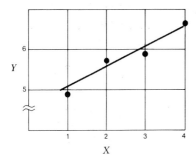

Figure 7-2

using grouped data showing a relation between academic aptitude and grade-point average (GPA). These data are selected because of their wide familiarity. Figure 7-4*a* and *b* shows two different levels of relationship between GPA and aptitude. For each interval of aptitude we can see how many students achieved various GPA levels.

In Fig. 7-4*a* and *b*, as aptitude goes up, so does the mean GPA. However, with Fig. 7-4*a* our ability to predict GPA from aptitude is limited. For example, suppose I have an aptitude score in the interval 110–119. Most of the cases within that aptitude range fall between a GPA of 1.00 and 3.99. This is a broad range in which to predict GPA. However, in Fig. 7-4*b*, if my aptitude score is in the same 110–119 range, it appears most likely (although not with 100 percent accuracy of prediction) that my GPA will be in the 2.00–2.99 range.

In Fig. 7-4*a* we have data that represent only a moderate correlation, possibly midway between 0 and 1.00. In Fig. 7-4*b*, however, we have a higher correlation between the two variables, possibly .80 to .90. These data illustrate the point that as correlation coefficients get larger, i.e., approach +1.00 or −1.00, the change in one variable becomes increasingly predictable from knowledge of the second variable.

Figure 7-3

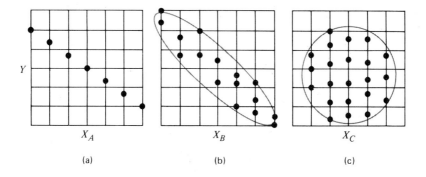

(a)   (b)   (c)

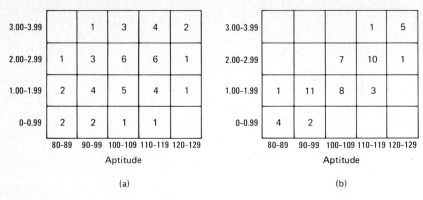

Figure 7-4

Let us summarize what we have learned so far.

1 We collect two measures on the same individuals (or events or agencies).

2 The relationship between these two measures is portrayed by the correlation coefficient, which ranges from −1.00 to +1.00.

3 A *positive* relation means that as we find larger and larger scores on one measure, we also find larger and larger scores on the other. If a unit increase in one variable is associated with specified increase in the other variable, the correlation coefficient will be +1.00. Here individuals ranking high on one variable will rank high on the other variable.

4 A *negative* relation exists if as one measure gets larger and larger the other gets smaller and smaller. If a unit increase in one variable is associated with a specified decrease in the other, the correlation coefficient is −1.00. In this case individuals who rank high on one variable will be low on the other variable.

Exercises

7-1 For each of the pairs of conditions below, do you think the correlation will be positive (high, moderate, or low), zero, or negative (high, moderate, or low)?

(*a*) The number of automobiles on the highway and the number of accidents

(*b*) The height and age of elementary school children

(*c*) Mental age and shoe size for adults

(*d*) The birthrate and socioeconomic level

(*e*) IQ and the number of failures in college

(*f*) The length of the base of a square and the length of its diagonal

The Product-Moment Approach

As one variable changes, another variable may also change in a predictable way. The more highly correlated the two variables the more precisely the change in Y can be predicted from the change in X. When two variables show common changes, we say they *covary*. If Joe varies 1 standard deviation above the mean, that is, $X - \bar{X} = 1\sigma_X$, to what extent does he rank above the mean on the other variable $Y - \bar{Y}$? The extent to which the two amounts $X - \bar{X}$ and $Y - \bar{Y}$ change together tells us to what extent the variables covary.

If we multiply an $X_i - \bar{X}$ value by its corresponding $Y_i - \bar{Y}$ value, where i represents individuals 1 to N, add up these cross products, and divide by the number of pairs of scores N, we have a mean cross product of the deviations. Any time we add up a set of values and divide by N, we get the mean. This time since we added the $(X_i - \bar{X})(Y_i - \bar{Y})$ cross products, dividing by N gives the mean cross product.

In the nineteenth century British statisticians noted that this mean cross product gets larger when the change in X is associated with a definite change in Y but gets smaller when X changes are not closely associated with changes in Y. This point is illustrated in parts I and II of Table 7-1. In part I the change of 1 point in X is associated with a 2-point change in Y, and the mean cross product is 4. However, in part II the 1-point change in X is not so precisely tied to changes in Y, although as X gets larger Y tends to also. In this case the mean cross product is 2.4. In other words, the larger the mean cross product of deviations

$$\frac{\Sigma(X_i - \bar{X})(Y_i - \bar{Y})}{N}$$

the closer the association between X and Y.

This conclusion, however, depends on the magnitude of the scores involved in the distribution. If we double the scores in the Y distribution in part II of Table 7-1, we get part III. By doubling the scores we have not changed the order in which cases A, B, etc., rank themselves, but we have extended the range of the distribution. The mean product of deviations in part III is 4.8, larger than the mean product of deviations in part I, where the X and Y scores covaried perfectly. It appears that although the deviation product means do reflect covariation between two sets of scores, they are not good indicators of this relationship because they are too closely tied to magnitudes of scores and changing range of distribution, as demonstrated in Table 7-1. What we need is a product mean based on scores that are converted to a standard-scoring procedure so that we can change all sets of raw scores to a common mean and deviations from that mean can be expressed in terms of a common standard deviation. A z-score conversion will do this.

The above discussion leads to a formula for correlation r_{XY} based on

Table 7-1   *Mean Cross Products as an Indicator of Covariation*

Case	X	Y	$X - \bar{X}$	$Y - \bar{Y}$	$(X - \bar{X})(Y - \bar{Y})$

Part I

Case	X	Y	$X - \bar{X}$	$Y - \bar{Y}$	$(X - \bar{X})(Y - \bar{Y})$
A	1	2	−2	−4	8
B	2	4	−1	−2	2
C	3	6	0	0	0
D	4	8	+1	+2	2
E	5	10	+2	+4	8
Means	3	6			20/5 = 4

Part II

Case	X	Y	$X - \bar{X}$	$Y - \bar{Y}$	$(X - \bar{X})(Y - \bar{Y})$
A	1	4	−2	−2	4
B	2	6	−1	0	0
C	3	2	0	−4	0
D	4	10	+1	+4	4
E	5	8	+2	+2	4
Means	3	6			12/5 = 2.4

Part III

Case	X	Y	$X - \bar{X}$	$Y - \bar{Y}$	$(X - \bar{X})(Y - \bar{Y})$
A	1	8	−2	−4	8
B	2	12	−1	0	0
C	3	4	0	−8	0
D	4	20	+1	+8	8
E	5	16	+2	+4	8
Means	3	12			24/5 = 4.8

covariation between two sets of scores as shown by the mean product of pairs of z scores. This procedure is known as the *Pearson product-moment* correlation method, after the English statistician Karl Pearson, who developed it. The fundamental procedure is

$$r_{XY} = \frac{\Sigma z_X z_Y}{N}$$

(7-1)

Since the sum of anything divided by the number of items summed is the mean of those items, we can now see that the correlation coefficient is a mean, the mean of the cross multiplications of corresponding z scores for X and Y values.

The basic method for computing a correlation coefficient is shown in Table 7-2, where the raw scores are first converted to z scores for both the X and Y variables. Then for each person the z_X value is multiplied by the z_Y value. These cross products are then summed and divided by the number of persons on whom the X and Y values were observed. This summing and dividing by N produces the mean cross product of the z scores, or the correlation coefficient. In Table 7-2 this value is .91, a value which approaches 1.00, indicating a fairly close

Table 7-2 Computing a Correlation Coefficient

Formula (7-1)	What it says to do*
$r_{XY} = \dfrac{\Sigma z_X z_Y}{N}$	1 Compute a z value for each X and each Y value
where $z_X = \dfrac{X - \bar{X}}{\sigma_X}$	2 Multiply z_{X_1} by z_{Y_1}, z_{X_2} by z_{Y_2}, ..., z_{X_N} by z_{Y_N}
$z_Y = \dfrac{Y - \bar{Y}}{\sigma_Y}$	3 Add $z_{X_1}z_{Y_1} + z_{X_2}z_{Y_2} + \cdots + z_{X_N}z_{Y_N}$
N = number of pairs of X, Y scores	4 Divide this sum by the number of pairs of X, Y values

Example Given the following pairs of z scores representing seven children's positions on the IQ scale and on a reading test, what is the correlation between IQ and reading comprehension R?

			Step 1		Step 2
Child	IQ (X)	R (Y)	z_{IQ}	z_R	$z_{IQ}z_R$
A	122	25	1.4	1.0	1.40
B	116	27	1.0	1.4	1.40
C	111	23	.7	.7	.49
D	100	20	.0	.0	.00
E	91	14	−.6	−1.2	.72
F	81	14	−1.2	−1.3	1.56
G	80	17	−1.3	−.6	.78
				Step 3: $\Sigma z_{IQ}z_R =$	6.35

Step 4: $r_{XY} = \dfrac{6.35}{7} = .91$

* Students who need a review of subscripts should turn to Appendix Rule A2-3 before proceeding.

relationship between the reading scores and IQ. When we see a high IQ, we also tend to see a high reading score; when we see a low IQ, we expect to see it associated with a low reading score. But the relationship is not perfect since it is not 1.00, so our generalization about IQ and reading scores will have some exceptions.

This seems like a good point to explain why correlation coefficients never exceed 1.00. The highest $\Sigma z_X z_Y$ value we can get occurs when each z_X equals its corresponding z_Y. In this case $\Sigma z_X z_Y$ will be equal to Σz_X^2, and since z scores are deviation scores, $\Sigma z_X^2/N$ will equal σ_z^2, the variance for the z scale, which is always 1. If z_X values are not equal to their corresponding z_Y scores, $\Sigma z_X z_Y$ must be less than Σz_X^2 and hence less than 1.00; therefore correlation coefficients never exceed ± 1.00.

The procedure of Table 7-2 has a disadvantage in that our data are usually in raw-score form rather than z-score form. We therefore need a raw-score procedure to circumvent the intermediate step of computing z scores. We get the desired process by simply substituting the z-score formula in Eq. (7-1):

$$ r = \frac{\Sigma \left[\left(\dfrac{X - \bar{X}}{\sigma_X} \right) \left(\dfrac{Y - \bar{Y}}{\sigma_Y} \right) \right]}{N} \tag{7-2} $$

If we multiply $X - \bar{X}$ and $Y - \bar{Y}$ values together and then substitute $\Sigma X/N$ for $\bar{X}$ and $\Sigma Y/N$ for $\bar{Y}$, it can be shown that the numerator in formula (7-2) becomes the numerator in (7-3). We can similarly deal with the denominator of (7-2).

$$ r_{XY} = \frac{\Sigma XY - \Sigma X \Sigma Y/N}{\sqrt{\Sigma X^2 - (\Sigma X)^2/N} \; \sqrt{\Sigma Y^2 - (\Sigma Y)^2/N}} \tag{7-3} $$

The result is a handy raw-score formula (7-3) for computing r_{XY}; it is developed in Appendix B. Formula (7-3) can be rearranged algebraically to give

$$ r_{XY} = \frac{N\Sigma XY - \Sigma X \Sigma Y}{\sqrt{N\Sigma X^2 - (\Sigma X)^2} \; \sqrt{N\Sigma Y^2 - (\Sigma Y)^2}} \tag{7-4} $$

which works more efficiently with some calculators.

Formulas (7-3) and (7-4) provide the same results as formula (7-1), but the intermediate step of converting all scores to z scores is avoided. The result is the mean cross product of z values in either case, and the interpretation of r_{XY} is identical in both procedures.*

The use of formula (7-3) is illustrated in Table 7-3. We begin with

---

* The curious student may follow this development in Appendix B.

Table 7-3   Computation of r_{XY} from Raw Scores

Formula (7-3)	What it says to do
$$r_{XY} = \frac{\Sigma XY - \Sigma X \Sigma Y/N}{\sqrt{\Sigma X^2 - (\Sigma X)^2/N}\ \sqrt{\Sigma Y^2 - (\Sigma Y)^2/N}}$$ where X = first observation on an individual Y = second observation on an individual	1 (a) Sum the X scores and sum the Y scores (b) Square the X scores and add up the X^2 values; square the Y scores and add up the Y^2 values (c) For each person multiply the X score times the Y score and add up these products 2 Square the values ΣX to get $(\Sigma X)^2$ and ΣY to get $(\Sigma Y)^2$ 3 N is the number of persons on whom we have pairs of scores 4 Proceed with solving for r_{XY}

Example 1   Given the following scores on an arithmetic text X and a reading comprehension test Y, correlate the two sets of scores.

Child	X	Y	X²	Y²	XY
A	3	6	9	36	18
B	2	4	4	16	8
C	4	4	16	16	16
D	6	7	36	49	42
E	5	5	25	25	25
F	1	3	1	9	3
	$\Sigma X = 21$	$\Sigma Y = 29$	$\Sigma X^2 = 91$	$\Sigma Y^2 = 151$	$\Sigma XY = 112$

140.16

$$r_{XY} = \frac{112 - 21(29)/6}{\sqrt{91 - 21^2/6}\ \sqrt{151 - 29^2/6}} = \frac{10.50}{13.77} = .76$$

Example 2   Given scores on an arithmetic test X for 8 ten-year-old children and speed in seconds Y for running 40 meters, find the correlation.

Child	X	Y	X²	Y²	XY
A	12	6	144	36	72
B	10	9	100	81	90
C	9	7	81	49	63
D	7	11	49	121	77
E	6	10	36	100	60
F	5	14	25	196	70
G	4	8	16	64	32
H	2	12	4	144	24
	55	77	455	791	488

-4.18

$$r_{XY} = \frac{488 - 55(77)/8}{\sqrt{455 - 55^2/8}\ \sqrt{791 - 77^2/8}} = -.67$$

our pairs of observations, two observations on each person, in this case test data for arithmetic X and reading Y. Our procedure requires us to perform two operations on each variable (sum the scores and sum the squared scores) and one operation on the scores together (multiply for each person the X value times the Y value). From this point on it is a matter of solving the equation for r_{XY}. Table 7-3 carries out the procedure step by step.

In Example 1 in Table 7-3 the positive correlation value of .76 shows that reading scores and arithmetic tend to increase together, although by no means does this generalization apply to every case. Although the coefficient is less than 1.00, the common trend is clear. In Example 2 the negative value of $-.67$ shows that as arithmetic scores decrease, speed of running tends to increase. Again this is a trend, not true of every case. The coefficient did not reach -1.00.

Exercises

7-2   When I gave a spelling test of 10 words arranged in increasing order of difficulty, I read the words and my class wrote them on a prepared sheet of paper. In a second test students were to check all incorrectly spelled words in a list of 20 words, some of which were misspelled. I wish to correlate the written spelling test scores with the recognition test scores:

Student	A	B	C	D	E	F
Written	2	5	9	6	7	3
Recognition	5	8	15	12	11	9

(a) Compute the correlation between the two spelling methods by formula (7-3).

(b) The mean for writing is 5.33, and the standard deviation 2.36; the mean for recognition is 10.00, and the standard deviation 3.16. Compute the correlation coefficient by formula (7-1). How does your result compare with your result in part (a)? Which method is simpler?

7-3   I am studying the relationship between age (to nearest birth date) and the number of telephone calls made during 1 week's time for a sample of 20 children. Correlate the two variables.

Child	A	B	C	D	E	F	G	H	I	J	K	L	M	N	O	P	Q	R	S	T
Age	7	14	9	10	8	17	11	13	16	12	8	18	14	9	10	8	12	9	15	8
Calls	1	8	4	3	3	7	5	6	8	4	3	5	3	3	2	1	5	1	6	2

7-4   I gave a pretest on knowledge of labor unions to college freshmen and then played a recorded speech about collective bargaining. I asked the students to complete a scale that reflected how well the speech was delivered; low scores mean the student perceived the presentation as poor. Correlate the pretest on knowledge with scale that reflected delivery.

Student	A	B	C	D	E	F	G	H	I	J
Presentation	15	18	12	21	17	27	11	14	19	20
Delivery	6	5	2	10	4	9	5	4	8	7

Interpretation of r_{XY}

Now that we have computed a correlation coefficient, what does it tell us? There are several ways of interpreting r_{XY}. They deal with the common variance between X and Y and with the *coefficient of alienation*. Other procedures also elaborate r_{XY}, but we postpone discussing them until the next chapter, where we look at accuracy of prediction.

First, r_{XY} can be used to illustrate the portion of variance in Y which is associated with X. We observe that as X deviates from its mean, Y also deviates from its mean in a predictable direction. If so, a portion of the variance in Y can be predicted from variance in X. In fact, the larger the correlation coefficient the more closely X and Y deviations correspond, i.e., the more closely X deviations predict Y deviations. How much of that variance in Y is associated with variance in X? We find this out by squaring r_{XY}. This gives us the portion of variance σ^2 in one condition which is associated with variance in the other. If r_{XY} is .80, then .64 or 64 percent of variance in Y is associated with X.

The word "associated" is important. Correlation does not show that X *causes* a given change in Y; it merely shows that the two variables are showing simultaneous alterations. For example, the amount of juvenile delinquency between 1955 and 1965 is correlated with the number of jet aircraft used by commercial airlines. Can we say that the increase in jet aircraft caused adolescents to resort to antisocial behavior? Even more absurd, can we say that the increase in the number of delinquents caused airlines to turn from piston to jet aircraft? Probably both conditions were at least in part a result of increasing industrialization, but they have no logical causative relationship. And so when we look at r_{XY}^2, we see the portion of X variance associated with Y variance, but this tells us nothing about causative relationships. Certainly some variables do have a direct influence on others, but correlation techniques do not identify this condition.

A second way we can interpret a correlation coefficient is in terms of the lack of relationship between the two variables. To do this, we compute $\sqrt{1 - r_{XY}^2}$; this index of lack of relationship is called the *coefficient of alienation*. The term under the radical tells us the portion of Y variance which is *not* associated with X. In other words, the coefficient of alienation is essentially the opposite of the coefficient of correlation.

Exercises

7-5 (a) Compute r_{XY}^2 for each of the following values of r_{XY}:

.20   .40   .50   .60   .70   .80   .90   1.00

(b) Plot the data from part (a) on a graph like the one below. Is the variance in X associated with Y a straight-line function, or is it curvilinear? What does this tell us about units of common variance between X and Y as r_{XY} gets larger? What does this say about adding (or subtracting) correlation coefficients?

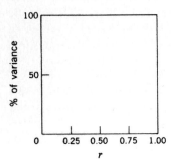

Exercise 7-5

7-6 (a) For the correlation coefficients below compute r_{XY}^2 and the coefficient of alienation; in your own words state what each of these figures tells us about Y once we know X.

r_{XY}	r_{XY}^2	$\sqrt{1 - r_{XY}^2}$
.20		
.40		
.60		
.80		
1.00		

(b) As r_{XY} increases, what is the direction of change for r_{XY}^2? Is it a rectilinear relationship?

(c) As r_{XY} increases, what is the direction of change for $\sqrt{1 - r_{XY}^2}$? Is it rectilinear?

Doing Arithmetic with r_{XY}: The z Transformation

Part (b) of Exercise 7-5 illustrates the fact that in terms of variance in X associated with Y the correlation coefficient is not an equal-unit index at all points between 0 and ± 1.00. As r_{XY} gets larger, the amount of variance in X associated with Y gets disproportionately greater. For this reason, simple addition or subtraction of two r_{XY} values is probably not advisable. We can solve this problem by transforming r_{XY} into a z value. The British statistician R. A. Fisher introduced the procedure for accomplishing this transformation

$$z = \tfrac{1}{2} \ln \frac{1 + r_{XY}}{1 - r_{XY}} \qquad (7\text{-}5)$$

where ln stands for the natural logarithm. Fortunately we do not have to calculate the value of z for each r_{XY} we wish transformed. It has already been done in Table C-10. We simply run down the column headed r_{XY} until we come to the value we want transformed and then read the z value to the right.

How shall we apply these z data once we have them? Here is an example. I have found the correlation of height and weight on a sample of 25 senior citizens to be .78. I locate a second sample of 80 additional cases. The correlation of height and weight here is .86. What is the average correlation across the two samples? In Chap. 3 weighting means for unequal-sized groups was demonstrated. We shall use a similar idea, plus our skill in converting r_{XY} into z, to solve the problem.

First the converted values for r_{XY} are

r_{XY}	.78	.86
z	1.045	1.293

To find the average value of z we use

$$\bar{z} = \frac{(n_1 - 3)z_1 + (n_2 - 3)z_2}{(n_1 - 3) + (n_2 - 3)} \qquad (7\text{-}6)$$

which gives

$$\bar{z} = \frac{22(1.045) + 77(1.293)}{22 + 77} = 1.237$$

We now convert the $\bar{z}$ value 1.237 back into r_{XY} by using Table C-10. Here we find that the nearest z value to 1.237 is 1.221*, which converts into an r_{XY} of .84. The weighted average of the above two correlation coefficients is .84.

Although at this point we use z transformations only to do simple arithmetic, as above, in later chapters, where we ask more profound questions about r_{XY}, z transformations will be invaluable.

* When working with two-place correlation coefficients, interpolation is usually not profitable.

Exercises

7-7 (*a*)   Convert the following correlations into *z* values:

r_{XY}	z
.11	
.78	
.52	
.24	
.98	

(*b*)   Convert the following *z* values back into r_{XY}:

z	r_{XY}
.060	
.245	
.604	
.281	
1.832	

7-8   I have three samples of kindergarten students. On each I correlate reading readiness and strength of dominant-hand grip (a developmental index). Find the average correlation coefficient for these data:

Group	A	B	C
n	20	45	31
r_{XY}	.40	.21	.65

7-9   In the *Psychological Bulletin* a reviewer reported four studies of length of therapy and gains in adjustment as measured by a personality inventory. Find the average correlation across these studies if their data were as follows:

Study	A	B	C	D
n	36	28	56	10
r_{XY}	.10	.21	.73	.60

Assumptions in Computing r_{XY}

The nature of the scatter diagram is a very important consideration in formulating correlation procedures. The scatter diagrams shown in Figs. 7-1 to 7-3 are simple because so few cases are shown. In typical studies we expect to have many more cases, and the data are more likely to distribute themselves somewhat as in Fig. 7-5 for a correlation

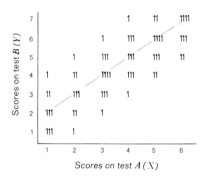

Figure 7-5

of about .70. Here we can see that as X increases, Y tends to increase, too, but the relationship is certainly not perfect.

Each row and each column in Fig. 7-5 is called an *array*. Let us now compute the means for each of the column arrays and see what these points look like. The small cross in an array locates its approximate mean. As we look at these points, we notice that they lie in quite a straight line. The means of the array do not always look this way. Sometimes we have curvilinear relationships like those shown in Fig. 7-6.

Here is the first assumption underlying the product-moment procedure for computing a correlation coefficient. The relationship between X and Y is assumed to be rectilinear; i.e., the means of the arrays are expected to lie in a straight line. Slight departure from this arrangement can be tolerated, but if a curvilinear relationship exists, techniques other than the product-moment method must be used.*

Since the distribution of points in any array is similar to that in any other array, in Fig. 7-5 we could compute a standard deviation for each of the arrays, giving an indication of dispersion for each of them.

---

* The interested students will find the appropriate procedure for handling curvilinear relationships listed in other sources under the topic of correlation ratio.

Figure 7-6

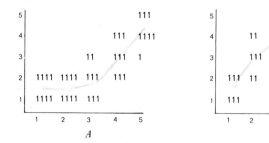

The second assumption of the product-moment procedure is tied to these array dispersions. We assume that all array dispersions are equal. If they are, we say that they have *homoscedasticity*, from Greek roots meaning "same" and "tendency to scatter." This is a necessary condition for the correlation method described above.

Typically, we do not apply exact tests to discover whether our data meet these two basic assumptions. A careful visual study of the scatter diagram usually is satisfactory. More precise tests of the assumptions are available but beyond the scope of this chapter.

Exercise

7-10   For each pair of given conditions indicate whether you believe the assumptions for the product-moment correlation are met and if not, which assumption is violated.

(*a*) Socioeconomic status and income

(*b*) Age and IQ between ages six and eighteen

(*c*) Age of females from ten to 40 years and number of dates per week

(*d*) IQ and GPA for high school freshmen

(*e*) Automobile speed and rate of gasoline consumption between 20 and 80 miles per hour

---

Special Correlation Methods

We do not always have two interval or ratio-scale measures on the persons we wish to study. Sometimes we deal with ordinal data and sometimes with dichotomous data. For example, a school psychologist has ranked 10 children on the severity of their learning disabilities. A classroom teacher has ranked these same children on the same problems. How closely do these two sets of ranks agree? In other circumstances we have data in which one or both variables are a dichotomy, e.g., sex or answering yes or no or preference for option A or B. We may want to know whether the categories in the dichotomy are related to a second variable which may be continuous or dichotomous.

For example, I want to correlate a test item (scored right = 1 and wrong = 0) with total score on a test-anxiety scale to see whether item scores are related to test anxiety. Or I correlate attitude-toward-art scale scores with sex to see whether high scores are related to one sex and low scores to the other. We may also correlate two dichotomies with each other. For example, in a yes/no situation is one sex more likely to say yes, the other no? If so, yes/no is correlated with sex. We shall look at a few of the most useful special methods designed for each of these procedures.

Correlation of Ranked Data

Sometimes the data to be correlated cannot reasonably be supposed to form a scale made up of successive equal units. For example, suppose we ask a teacher to rank the students in her class from the most capable to the least capable. The extent to which child A exceeds child B in ability is not expected to be equal to the extent to which child B exceeds child C, etc. Thus, the ranks are not believed to represent equal amounts of ability.

Occasionally we have rating-scale data that probably cannot be considered as equal-unit scales. Questionnaire data may also be of this type. In any case, we find occasions when the product-moment method is inappropriate because the data are not from a measuring device with a common unit at all points along the scale. The auxiliary correlation procedure needed to handle these situations is the *Spearman rank-order method*.

Charles Spearman, an English statistician, devised the statistic ρ, called *rho*, which represents the correlation of data when the subjects have first been ranked in order of magnitude of the trait in question. The correlation then represents the relationship of the ranks for individuals on two characteristics. For example, when teachers rank students on ability, we correlate this with IQ scores to see whether teachers estimate ability about the same as intelligence tests. The teachers' reports would already be in ranks, but we would have to rank IQ data, giving a rank of 1 to the highest IQ, 2 to the next highest, etc. In this manner we would have two ranks for each child, the teacher's ranking of the child and the rank of the child's IQ among his peers; these parallel sets of ranks can be correlated to show the relationship of teachers' estimates of relative ability and test results.

The basic procedure for Spearman's ρ is found in the product-moment method. In calculating r_{XY}, we needed ΣX, ΣY, etc. Comparable values can be determined from ranks. For example, the sum of a set of ranks is $N(N + 1)/2$. To devise the correlation procedure for ranked data, we merely substitute into the product-moment formula (7-1) comparable forms for ranked data, for example, $N(N + 1)/2$ for ΣX, $N(N + 1)(2N + 1)/6$ for ΣX^2, etc.; the result is formula (7-7).

Occasionally two or more persons have the same score, which complicates the ranking of these individuals. Suppose in the example in Table 7-5 that children B and C were both given a scale value of 15 by observer 1. Which child will get a rank of 2, and which 3? We resolve this question by averaging ranks and assigning each child the average of the ranks in question. Therefore, we would add ranks 2 and 3, divide by 2, and give each child a rank of 2.5. The same procedure is also used if we have more than two tied ranks. If B, C, and D were all given scale values of 15 by observer 1, we would determine their ranks by $(2 + 3 + 4)/3$, and each would have a rank of 3.

Tied ranks present a problem in the use of ρ. Actually ρ is a good estimate of the product-moment procedure to the extent that we have relatively few tied ranks. The more tied ranks there are the more ρ departs from the product-moment correlation coefficient. An example will illustrate the deviation of ρ from r_{XY} when tied ranks are evident.

Suppose we have 10 more nursery school children rated by two observers using a sociability scale (see Table 7-4). In these data we have two children (B and C) tied at rating of 16 by observer 1 and two others (E and F) tied at 11. Similarly, observer 2 had ties for E and F and for G and H. In this procedure we first rank each person on the first variable and then rank them on the second variable. (The person with the highest score gets a rank of 1, the second highest gets a rank of 2, etc.) Then for each person we subtract the second-variable rank from the first-variable rank, giving the difference d; when this is squared and summed for all subjects, we have the Σd^2 value. N is the number of persons ranked.

Now we are ready to calculate the value of ρ from formula (7-7).

$$\rho = 1 - \frac{6(\Sigma d^2)}{N(N^2 - 1)} \qquad (7\text{-}7)$$

we find

$$\rho = 1 - \frac{6(6)}{10(100 - 1)} = .96$$

The entire procedure is illustrated in Table 7-5.

Table 7-4   Nursery School Children Ranked by Two Observers

Child	Observer 1	Observer 2	Rank 1	Rank 2	d	d^2
A	19	16	1	2	-1	1.00
B	16	17	2.5	1	1.5	2.25
C	16	14	2.5	3	$-.5$	.25
D	12	13	4	4	0	0
E	11	12	5.5	5.5	0	0
F	11	12	5.5	5.5	0	0
G	10	9	7	7.5	$-.5$	.25
H	9	9	8	7.5	.5	.25
I	8	6	9	10	-1	1.00
J	7	7	10	9	1	1.00
					0	6.00

Table 7-5 Calculation of the Correlation between Two Sets of Ranked Data

Formula (7-7)	*What it says to do*
$$\rho = 1 - \frac{6(\Sigma d^2)}{N(N^2 - 1)}$$ where ρ = correlation between ranked data	1 Where X is a set of data to be correlated with Y, rank all individuals on the X scale beginning with rank of 1 for the highest X value, 2 for the next highest, etc.; similarly rank Y data
d = difference between the two ranks for given subject N = number of subjects ranked	2 Subtract Y ranks from X ranks to get d (the sum of all d values should be zero)
	3 Square each d value and add squared values
	4 Multiply Σd^2 by 6 and divide by $N(N^2 - 1)$, where N is the number of subjects ranked
	5 Subtract the result of step 4 from 1; the result is the correlation between X ranks and Y ranks

Example The following data represent sociability ratings submitted by two graduate psychology students after observing a group of 10 nursery school children for 1 hour. Correlate the ratings of the two observers.

Child	Observer 1	Observer 2	Rank 1	Rank 2	d	d^2
A	18	15	1	2	−1	1
B	14	16	4	1	3	9
C	15	14	3	3	0	0
D	17	13	2	4	−2	4
E	12	9	6	7	−1	1
F	13	10	5	6	−1	1
G	10	8	7	8	−1	1
H	9	7	8	9	−1	1
I	7	11	9	5	4	16
J	6	6	10	10	0	0
					0	34

$$\rho = 1 - \frac{6(34)}{10(100 - 1)} = 1 - .21 = .79$$

If we apply the product-moment correlation to the actual raw data on page 146 assuming that they each represent an equal-unit scale, we get

$$r_{XY} = \frac{1489 - 119(115)/10}{\sqrt{1553 - 119^2/10}\ \sqrt{1445 - 115^2/10}} = .93$$

The difference between ρ and r_{XY} in this case is not dramatic, but it does illustrate the fact that the two indicators of relationship will not be identical if ranks are tied. The product-moment correlation for the data in Table 7-4, however, will produce $r_{XY} = .79$, just like the rank-order method, because there are no tied ranks in these data. (Interested students may wish to prove this statement by computing r_{XY} for the data in Table 7-4.)

Exercise

7-11   Two English teachers have each ranked eight essays on overall quality. Correlate the two teachers' rankings.

Paper	Teacher 1	Teacher 2
A	4	4.5
B	7	6
C	5	4.5
D	1	3
E	8	7
F	3	1
G	6	8
H	2	2

Correlating a Dichotomous Variable with a Continuous Variable

Two methods are available for correlating a dichotomous variable with a continuous one, *biserial correlation* and *point-biserial correlation*. In biserial correlation we have two continuous variables, one of which we have artificially dichotomized. For example, we wish to calculate the correlation of a test score with GPA, and so we have designated grades A, B, and C as "pass" and D and F as "fail." We then correlate the test data with the dichotomy pass/fail.

Biserial correlation has relatively few applications in modern data analysis. It also assumes that the distribution of data underlying the dichotomy is normal. Since data, like GPAs cited above, often are not normally distributed, biserial correlation will not be discussed in detail. The interested student may wish to look in books on correlational procedures for computational methods. On the other hand, since point-biserial correlation has a number of applications in the social sciences and is especially useful in psychometrics, this procedure will be worked through.

In point-biserial correlation we have a natural dichotomy (like correct/incorrect, female/male, voted/did not vote) which is correlated with a continuous variable (like height, IQ, anxiety). The procedure is widely used to see how well specific test questions correlate with total scores on the test. Does item 7 on my test measure the same thing as the total test? If it does, it will correlate with the total score on the test. Although item 7 is scored 1 for correct, 0 for incorrect, i.e., scored in a dichotomy, this poses no problem. Point-biserial correlation is designed exactly for this type of situation, the correlation of a dichotomous variable with a continuous one. The formula is

$$r_{\text{pb}} = \frac{\bar{X}_p - \bar{X}_q}{\sigma_X} \sqrt{pq} \tag{7-8}$$

where $\bar{X}_p$ is the mean of the continuous variable for all persons in one of the dichotomous categories and $\bar{X}_q$ is the mean of the continuous variable for all persons in the other category of the dichotomous variable. For example, if we were correlating test item 7 with the score on the total test, $\bar{X}_p$ would be the mean score on the total test for all persons who passed item 7 and $\bar{X}_q$ would be the mean score on the total test for persons who failed item 7. Of course, σ_X is the standard deviation of the continuous variable calculated across scores for all persons. The other items in formula (7-8) are p (the proportion of persons in one category, e.g., pass) and q (the proportion of persons in the other category, e.g., fail).

In formula (7-8) the difference between the means for the two dichotomous groups may be observed to give us an intuitive grasp of how point-biserial correlation reflects the relationship between the categorical variable and the continuous one. The scatter diagrams at the beginning of this chapter showed that if the data lie in long, narrow ellipses, the correlation is high; if the ellipses approach roundness, the correlation is low. Now let us dichotomize variable X and look at the mean scores on the other variable (Y) within each of the two categories of the dichotomy (Fig. 7-7). In Fig. 7-7*a* we have a narrow ellipse, suggesting a high correlation. The mean of the Y scores for passers will

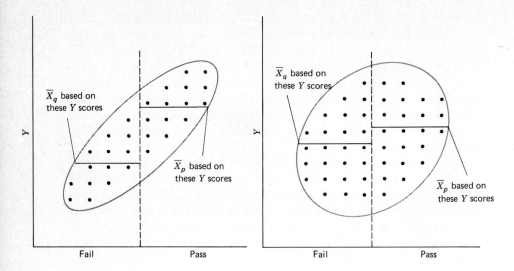

Figure 7-7

be somewhat greater than the mean of the Y scores for failers. On the other hand, in Fig. 7-7b the ellipse is broad, suggesting a low correlation. If we calculate the Y mean for passers, we shall find it only slightly different from the Y mean for failers. Hence, the narrower the ellipse (and the higher the resulting correlation) the greater the $\bar{X}_p - \bar{X}_q$ difference. Observing this relationship between the Y means for the two categories of the dichotomy should give us some appreciation for how the point-biserial procedure reflects correlation.

We are now ready to solve a problem with point-biserial correlation.

Table 7-6 provides a worked example in which voting behavior in campus elections (the dichotomous variable) is correlated with hours of study each week (the continuous variable). To simplify the arithmetic the table includes too few subjects (only 10). Voters were separated from nonvoters and the mean hours of study were calculated for each group, but the standard deviation is based on the entire sample of 10 cases. The resulting r_{pb} of .45 shows that there is a modest positive relationship between voting/not voting and number of hours of study, i.e., voters tend to study more than nonvoters, but the size of the relationship is not large enough to allow us to make this generalization with great confidence.

Point-biserial correlation is widely used in test analysis. Psychometricians often wish to know whether each item in a test is assessing the same dimension of behavior as the test as a whole. To this end a correlation of each test item, the dichotomous variable (right/wrong), with the whole test score, the continuous variable, is instructive. Items tied closely to the domain of the total test should correlate quite highly with total test scores; items not tied closely to that domain will not

Table 7-6 Application of Point-Biserial Correlation

Formula (7-8)	What it says to do
$$r_{pb} = \frac{\bar{X}_p - \bar{X}_q}{\sigma_X}\sqrt{pq}$$ where $\bar{X}_p$ = mean of continuous variable for persons in p $\bar{X}_q$ = mean of continuous variable for persons in q σ_X = standard deviation of all scores on continuous variable p = proportion of persons in one category of dichotomous variable q = proportion of persons in other category of dichotomous variable	1 Sort persons into dichotomous categories p and q 2 For all persons in p calculate a mean on the continuous variable; do the same for all persons in q 3 Calculate the standard deviation across the entire sample (p and q) 4 Calculate p by tabulating the number of cases in p divided by total number of cases; repeat for q 5 Enter the above data in formula (7-8)

Example When 10 students were interviewed in a survey, they were asked (1) if they voted in the last campus election and (2) how many hours on the average they studied per day. What is the correlation between voting behavior and hours of study?

Student	Voted	Hours		
A	Yes	3.00	$\bar{X}_p = 3.06$	
B	Yes	4.50		
C	Yes	2.00	$p = .400$	
D	Yes	2.75		$\sigma_X = 1.01$
E	No	1.50		
F	No	2.00	$\bar{X}_q = 2.13$	
G	No	1.00		
H	No	3.50		
I	No	2.25	$q = .600$	
J	No	2.50		

$$r_{pb} = \frac{3.06 - 2.13}{1.01}\sqrt{.40(.60)} = .45$$

correlate well with the total. Test analysis, however, is only one of many applications of point-biserial procedures.

The point-biserial correlation of .45 is interpreted like any other correlation coefficient, but one caution needs to be cited. The r_{pb} is an underestimate of the Pearson r_{XY}, and this underestimate increases as p deviates from .50. Nevertheless, point-biserial correlation has a number of applications in the social sciences where it produces useful information.

Correlating One Dichotomous Variable with Another

Sometimes both variables, X and Y, are in dichotomous categories. If X and Y are continuous variables which were artificially dichotomized, the proper procedure is tetrachoric correlation. As with biserial correlation, tetrachoric assumes normality of each variable, a condition which may not exist. Also there are few situations where the Pearson correlation cannot be applied to determine the relationship between two continuous variables. Therefore, we shall pass on to a more useful procedure, in which we correlate X and Y, both of which are natural dichotomies. This procedure is called *phi*, and uses the Greek symbol ϕ.

Suppose we want to correlate marital status with passing or not passing a test question. Here married/single is one dichotomy and pass/fail is the other. First we lay out our data in a 2×2 *contingency table* like this.

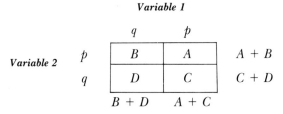

The value of ϕ is then found simply from

$$\phi = \frac{AD - BC}{\sqrt{(A + B)(A + C)(B + D)(C + D)}} \qquad (7\text{-}9)$$

The procedure for applying formula (7-9) to a real problem is given in Table 7-7, which deals with the relationship between getting a test item correct and marital status. The 2×2 table is first constructed and row and column totals calculated. Then appropriate cross products are found. The rest is solving formula (7-9). The value of ϕ is .24, showing a rather low relationship between marital status and getting the test item correct.

Table 7-7 Solving a φ Problem in Correlation

Formula (7-9)	What it says to do

$$\phi = \frac{AD - BC}{\sqrt{(A + B)(A + C)(B + D)(C + D)}}$$

1 Calculate the cross products AD and BC

2 Calculate the row and column sums

3 Apply these data in formula (7-9) and solve for ϕ

Variable 1

	q	P	
p	B	A	A + B
q	D	C	C + D
	B + D	A + C	

Variable 2

Example A test item dealing with how home budgets are·managed is to be correlated with marital status. The sample contains 40 people, 18 married and 22 unmarried. Of the 18 married 8 got the answer wrong; of the 22 who were not married 15 got it wrong. Calculate the correlation between marital status and getting the item right.

Married

		No	Yes	
Item	Correct	7	10	17
	Incorrect	15	8	23
		22	18	40

When these data are put into formula (7-9), we find

$$\phi = \frac{10(15) - 7(8)}{\sqrt{(17)(18)(22)(23)}} = .24$$

An intuitive grasp of the procedure may emerge from again recalling the scatter diagram at the beginning of this chapter. If our data distribute themselves into an elongated ellipse, the correlation is higher than if the ellipse begins to approach roundness. If we divide our scatter plot by dichotomizing each variable, as in Fig. 7-8, we can see that with higher correlations (the narrow ellipse, Fig. 7-8a) the proportion of people in quadrants A and D is much larger than in B and C. This is much less true when correlations are lower (the broad ellipse, Fig. 7-8b). By starting with the value of AD-BC our formula reflects this arrangement of scores in the scatter diagram. If AD is large

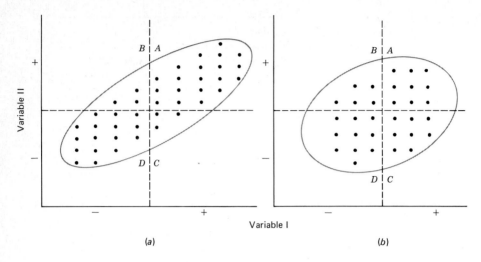

Figure 7-8

relative to *BC*, the ellipse is long and narrow and suggests a high correlation (Fig. 8*a*), but if *AD* approaches *BC*, the ellipse is becoming less narrow, suggesting lower correlations (Fig. 7-8*b*).

Phi is based on Pearson correlation procedures, and like Pearson correlation may range from -1.00 to $+1.00$. However, it is only under somewhat unusual arrangements of data that phi ever reaches these limits. Nevertheless, for completely dichotomous data phi is a useful procedure that should be exploited in reducing the many kinds of dichotomous data to appropriate generalizations.

Exercises

7-12 A group of students is asked whether they prefer open-housing dormitories, and their responses are to be correlated with GPA. Complete the correlation.

Student	A	B	C	D	E	F	G	H
Choice	No	Yes	No	No	Yes	Yes	No	Yes
GPA	3.2	2.6	2.8	3.0	2.1	2.3	2.6	1.8

7-13 Correlate sex with preference in the same study as Exercise 7-12:

Student	A	B	C	D	E	F	G	H	I	J	K	L	M	N	O	P
Choice	No	Yes	No	No	Yes	Yes	No	Yes	No	Yes	No	No	Yes	No	Yes	Yes
Sex	F	N	N	F	F	M	F	F	F	M	M	F	F	M	M	M

Summary

Correlation is a procedure for indicating in quantitative terms the relationship between two variables. The relationship is stated in terms of a coefficient of correlation which runs between 0 and ±1.00. If as X increases 1 unit Y also changes by a constant number of units, the coefficient will be 1.00; if as X changes 1 unit we can say nothing at all about Y, the correlation is 0; and if as X increases 1 unit Y decreases a constant number of units, the coefficient will be -1.00. Most correlations in the behavioral sciences are somewhat less than ±1.00, indicating a less than perfect relationship between variables being observed.

There are several ways to interpret a correlation coefficient. It can be used to show the portion of variance in X which is associated with Y; it can be used to show the lack of relationship between X and Y; it can be used to show the precision of prediction of Y from X.

The standard procedure for computing a correlation coefficient, the product-moment method, assumes that the relationship between X and Y is rectilinear and that the distribution within a given array has a dispersion equal to that of all other arrays. We call this characteristic homoscedasticity.

Sometimes data are taken from a scale which does not have a common unit at all points on that scale. This situation suggests rank ordering of the individuals. In this case the product-moment method is not appropriate, and Spearman's rank-order method (rho) is used to indicate relationships between pairs of ranked data.

Occasionally one or both variables in a correlation situation are dichotomized. Here the most useful procedures are the point-biserial correlation if one variable is a dichotomy and the other variable is continuous, and phi if both variables are a dichotomy. Both procedures yield coefficients similar to the Pearson product-moment correlation coefficient and for the most part can be interpreted like Pearson coefficients.

Correlation procedures indicate relationship but do not reveal causality. If X is related to Y, we cannot state that X is causing Y or vice versa.

The flowchart in Fig. 7-9 may help in ordering the correlation procedures processes discussed in this chapter.

Key Terms

correlation coefficient	homoscedasticity
positive correlation	coefficient of alienation
negative correlation	ρ
scatter diagram	rank-order correlation
cross product	biserial correlation
Pearson product-moment	tetrachoric correlation
correlation	point-biserial correlation
linearity	ϕ

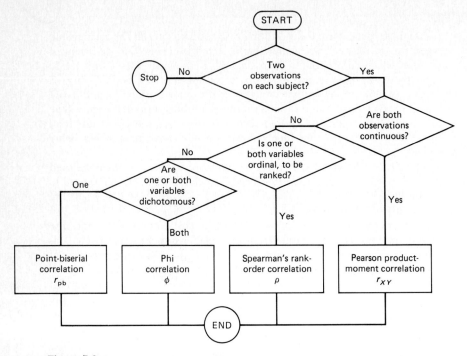

Figure 7-9

Problems

7-1 After a teacher had assessed the sight vocabulary (in words) for eight-year-old educable mentally retarded children, a psychometrist gave the same children a simple social-knowledge questionnaire. Correlate sight vocabulary with questionnaire score:

Child	A	B	C	D	E	F	G	H	I	J
Vocabulary	18	10	7	14	19	6	9	10	13	4
Questionnaire	31	22	10	17	16	12	17	13	18	9

7-2 I have measured test anxiety with a standardized test and have also collected social maturity scale scores on the same students. How does test anxiety correlate with social maturity?

Person	A	B	C	D	E	F	G	H	I	J
Social maturity	13	19	17	9	14	10	15	19	21	18
Test anxiety	12	10	16	18	15	21	15	10	8	11

7-3 I have collected data on teenage arrests per thousand teen population in 12 counties in my state. For each county I also have data on teen pregnancies per thousand girls in the county. (Note that the county is the unit of observation with two measures on each county.) Correlate arrests with pregnancies.

County	A	B	C	D	E	F	G	H	I	J	K	L
Arrests	4	14	10	7	17	28	13	9	25	19	35	18
Pregnancies	7	9	13	6	14	32	21	11	29	10	26	12

7-4 Data were collected on 10 subjects who immigrated within the last 5 years, giving ability to deal with English (as measured by a test of English grammar); their incomes last year in $1000 units are as follows:

Subject	A	B	C	D	E	F	G	H	I	J
English	15	33	27	17	21	29	22	18	23	30
$1000	7	28	32	28	10	16	18	21	28	12

Correlate English skill with income.

7-5 A placement office followed up some of the people for whom they had found jobs. One of their counselors observed each person placed in retail sales jobs to rate them in proficiency (10 is low and 25 high). The foreman also rated each person on a 10-point scale on job skills (1 is low and 10 is high). Rank and correlate the two sets of ratings.

Person	A	B	C	D	E	F	G	H	I
Counselor	21	18	16	11	24	13	15	14	17
Foreman	9	10	8	4	9	5	6	5	3

7-6 A psychiatrist ranked 10 children on their adjustment to school (10 is low and 1 is high). The same children were also given a personality inventory (75 is high). Correlate these data.

Psychiatrist's rank	10	9	8	7	6	5	4	3	2	1
Inventory score	62	75	58	42	48	32	34	39	41	38

7-7 A psychometrist developing a new test wants test items (questions) to correlate with the total test score to show that the items reflect the same trait as the total test. Below are item scores for two of the items and total test scores for 24 students.

(*a*) Correlate item 1 with the total score.

(b) Correlate item 2 with the total score.

(c) Correlate item 1 with item 2 (R = right, W = wrong)

Student	Item 1	Item 2	Total	Student	Item 1	Item 2	Total
A	R	W	21	M	R	W	16
B	W	R	12	N	W	W	9
C	W	W	11	O	R	R	14
D	R	W	18	P	R	R	18
E	R	W	16	Q	W	R	11
F	W	W	10	R	R	R	15
G	R	R	12	S	W	W	8
H	W	W	7	T	W	R	6
I	R	R	19	U	R	R	19
J	W	W	9	V	W	W	10
K	R	R	13	W	R	R	17
L	W	R	10	X	R	W	16

7-8 A high school counselor looking at the records of last year's graduates sorted them out into went to college/did not go to college. Then he looked at their high school GPAs. Correlate GPAA with the decision to go on to college.

Student	College	GPA	Student	College	GPA
A	Yes	3.2	F	Yes	3.4
B	No	3.4	G	Yes	2.9
C	No	2.6	H	Yes	3.8
D	Yes	2.9	I	No	2.7
E	No	2.0	J	No	3.3

7-9 An animal science student is studying the relationship of body weight of sows and their likelihood of becoming pregnant. He observed 12 sows as they reached maturity, weighed each at the time of first insemination, and recorded whether they conceived. Correlate weight and pregnancy.

Sow	A	B	C	D	E	F	G	H	I	J	K	L
Weight	432	386	510	374	412	601	589	379	418	406	391	488
Pregnant	Yes	Yes	No	Yes	Yes	No	No	Yes	No	Yes	Yes	No

7-10 Two test items I want to correlate are both scored pass (+) or fail (0). Put the data below into a 2 × 2 contingency table like the sample shown below and calculate the correlation between the two items.

Person	Item 1	Item 2	Person	Item 1	Item 2	Person	Item 1	Item 2	Person	Item 1	Item 2
A	+	+	G	0	0	M	+	+	S	0	+
B	+	0	H	0	+	N	0	0	T	+	+
C	+	0	I	+	+	O	+	+	U	0	0
D	0	0	J	+	+	P	+	+	V	0	0
E	0	0	K	+	0	Q	0	0	W	0	0
F	0	+	L	0	+	R	+	+			

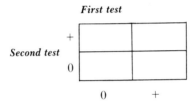

7-11 A psychologist theorized that people who are conventional will like "quiet" colors, rhyming poetry, musicals, and the *Chicago Tribune* but people who are not conventional will not like these items. He collected data on color preference (red, green) and on preference for stage productions (plays, musicals). If the theory is correct, color will correlate positively with stage production. The data are below. Correlate the two variables and test the theory.

		Production	
		Musicals	Plays
Color	Red	32	21
	Green	14	25

8

Regression: Predicting One Variable from Another

If two variables are positively correlated, higher values on one will be associated with generally higher values on the other. This is increasingly true as correlation coefficients approach 1.00. If two variables are negatively related, low values on one will be associated with high values on the other. These generalizations suggest that if I know your status on one variable, I may be able to predict your status on the other variable. If Y scores change in a systematic way with changes in X scores, we should be able to predict Y from X with better than chance accuracy. How these predictions are made and how their accuracy is determined are the topics of this chapter.

The Scatter Diagram

Let us begin with the scatter diagram. For each value of X (on the abscissa) we have a set of values of Y (on the ordinate). Figure 8-1 illustrates this arrangement. In terms of real variables, not everyone who weighs 98 pounds will be 48 inches tall: there will be some people less than 48 inches and some more, even though they all weigh 98 pounds. In Fig. 8-1 not all those who got a score of 3 on test X got the same score on test Y. Instead we have a small distribution of Y scores for each X score. This distribution is called an array.

If we calculate the mean for each array and plot a straight line through (or nearly through) these points, the slope of that line will tell us how much the Y scores (on the average) increase with each unit increase in X scores. For example, as we move from test X score 2 to 3,

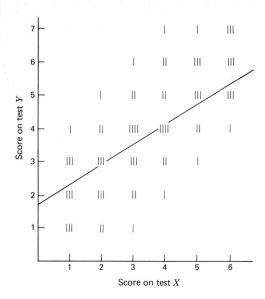

Figure 8-1

how much should we expect the test Y score to increase? The steeper the slope in the line, the more rapidly Y scores increase with unit increases in X scores. This is illustrated in Fig. 8-2, which is the same as Fig. 8-1 with the arrays omitted to emphasize the line. As we move from an X score of 2 to 3 (1 unit), the slope of the line shows us the increase in Y that is associated with that 1 unit in the X score.

The sloping line plotted above can be laid out more precisely by mathematical processes, but its function is the same, i.e., to indicate the increase in one variable associated with a given increase in another. We call it the *regression line*.

Whenever I have a set of scores and wish to guess the score for a given individual in that set, my best guess is the mean for the set. Across all individuals in the set I shall be least in error if I guess the mean as the score for each person in the set. This is essentially what we do when we predict one variable (Y) from another (X). If I know your X score, I predict you to be at the mean of the Y scores for all people who have the same X score as you; i.e., I predict everyone to be at the mean of their array.

If we designate a *predicted* score for the Y variable by $\hat{Y}$ (read "Y hat") and the actual score a given person obtains in that array by Y, the error in our prediction is $Y - \hat{Y}$. If we calculate the value of $(Y - \hat{Y})^2$ for each subject in a set of data and then average these squared values across all subjects, this average will be the smallest possible for the data we have

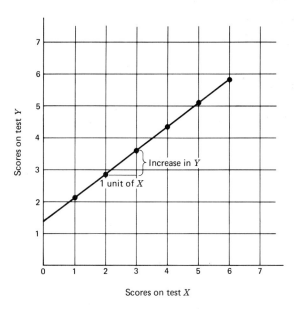

Figure 8-2

collected, i.e., if we predict *Y* scores at the mean of their array, no other *Y* score would give us a smaller amount of error across all scores in an array. For this reason the prediction procedure is often referred to as the *method of least squares*.

To summarize what we have just learned, we first locate a person's *X* score (IQ, aptitude, or whatever). We note that among people who received that *X* score there is a range of *Y* scores (the array). We predict everyone in the array to be at the mean, because this is the point which allows the least error in prediction across all the scores in this array. Of course everyone's *Y* score is not at the mean of the array, so we make some error in predicting to the mean of the array. Later in the chapter we shall see that by applying a kind of standard deviation to the array we can determine what portions of people will indeed be given distances above and below the predicted score, the array mean; but first the procedure for predicting *Y* from *X* must be developed.

The Regression Equation

In making a prediction of *Y* from *X* we need to know two things: (1) the slope of the regression line and (2) the value of *Y* when *X* is zero. Since this last bit of information is found at the point where the regression line cuts the *Y* ordinate (Fig. 8-2), this point is referred to as the *Y intercept.* We now calculate these two bits of information and put them into an equation for predicting Y from X. Our formula is

$$\hat{Y} = a + b_{YX}X \qquad (8\text{-}1)$$

where $\hat{Y}$ = predicted value of Y
   a = value of Y when $X = 0$
   b_{YX} = increase in Y when X increases 1 unit, i.e., slope of line
   X = value from which Y is predicted

The above formula is called a *regression equation* and it says that a predicted Y score, that is, $\hat{Y}$, is equal to a, the value of Y when X is zero, plus b_{YX}, the increase in Y which corresponds to a unit increase in X values, times the number of X units for a given individual.

Go back to Fig. 8-2. If your X score is 3, what is your expected Y score? We begin with the value of Y where X is zero, i.e., the value of a (here roughly 1.3). Next we calculate the increase in Y for each unit of X, that is, the b value (we might estimate it here to be about seven-tenths of a Y unit). Then begin with a and add to it one b_{YX} unit for each score along the X scale until we come to the score from which we are predicting, here 4. The steps will look like this with our approximations of a and b_{YX}:

$$\hat{Y}_4 = 1.3 + .7 + .7 + .7 + .7$$

or

$$\hat{Y}_4 = 1.3 + .7(4)$$
$$= a + b_{YX}(X)$$

In the example above we estimated a and b_{YX}. Our job now is to develop a mathematical procedure for calculating these constants accurately. The procedure for b_{YX} will be first. The resulting figure is called the *regression coefficient*. The procedure must be tied to values we already know, so we shall tie it to raw scores. If all X,Y coordinate points lie in a straight line, the value of b_{YX} can be found as

$$b_{YX} = \frac{Y_2 - Y_1}{X_2 - X_1} \qquad (8\text{-}2)$$

where X_1 and X_2 are two points on the X scale and Y_2 and Y_1 the two corresponding points on the Y scale. If X_2 and X_1 are adjacent scores, $X_2 - X_1 = 1$ and formula (8-2) gives the increase in Y, that is, $Y_2 - Y_1$, with a unit increase in X.

Although formula (8-2) illustrates what b_{YX} tells us, it is not a useful procedure for calculating b_{YX} because the actual X and Y coordinate points seldom lie exactly on the regression line. Instead formula (8-3)

$$b_{YX} = \frac{\Sigma XY - \Sigma X \Sigma Y/N}{\Sigma X^2 - (\Sigma X)^2/N} \qquad (8\text{-}3)$$

calculates b_{YX} by taking account of the fact that not all coordinate points lie on the line.

It should be evident at this point that values of b_{YX} are in units of the Y scale and tell us the amount of change in Y-scale units for a 1-unit change on the X scale. If we are predicting GPAs from IQ, b_{YX} will be in units of GPA and will indicate the amount of change expected in GPA with a 1-unit change in IQ, i.e., with one IQ point. If we are predicting job skill from an aptitude test, b_{YX} will be in units of job skill and will indicate the amount of change in job skill associated with a 1-unit change in aptitude scores.

The a value must be calculated next. It is found by formula (8-4)

$$a = \bar{Y} - b_{YX}\bar{X} \qquad (8\text{-}4)$$

The means of the Y and X distributions will be corresponding points on the scatter diagram. If one begins with the mean of Y and reduces the value by the amount of b_{YX} for each X score between the X mean and zero, i.e., $b_{YX}\bar{X}$, this reduction will locate the value of Y when X is zero. The value of a is often positive, but it can also be negative. If b_{YX} is negative, it simply says that Y is a point below the base line when X is zero. The a values, like b_{YX}, are in units of the Y scale.

We can now apply a and b_{YX} in solving a prediction problem. Assume that we have academic aptitude scores on 10 students and also have their GPAs (10 is too few but cuts down on arithmetic). We shall develop the prediction procedure for estimating Y values (here GPA) from X values (here aptitude scores). The data are in Table 8-1 (page 167). The b_{YX} value is .08. Since this is on the GPA scale, it means .08 GPA; i.e., as aptitude increases 1 unit, GPA increases .08 unit on the grade scale. Also, the a value, 1.68, is in GPA units. This value says that when the aptitude test score is zero the GPA is 1.68. And finally, for an aptitude score of 12, the predicted GPA is 2.64.

Exercises

8-1 The following data were obtained from Possumtrot High School. Physical strength is an average of several measures; awards refers to the number of track and field events in which the student reached first, second, or third place during the season. Develop a regression equation to predict the number of track and field awards. John has a strength index of 6, and Bill has a strength index of 14; predict the number of awards for each.

Student	Strength index X	Awards Y
A	12	9
B	8	3
C	11	6
D	5	2
E	4	4
F	10	6
G	9	7
H	15	10
I	7	5
J	16	8

8-2 An educational psychologist who believes that the number of days teachers are absent from work is correlated with their teaching skill has devised an observation schedule for assessing skill (low scores mean greater skill) and has recorded total days absent at the end of the school year. Develop a regression equation and predict the number of days absent for teacher M, who has a skill index of 12, and teacher N, who has a skill index of 8.

Teacher	Skill index	Days absent
A	6	2
B	9	0
C	7	3
D	17	10
E	13	12
F	5	5
G	10	9
H	14	8
I	8	4
J	15	6
K	11	3
L	12	7

An Equation when Descriptive Data Are Available

The above procedure is very useful when we have not calculated the correlation between X and Y or their standard deviations; however, if we have calculated these statistics, we can combine our prediction procedures into a single formula. Beginning with

$$\hat{Y} = a + b_{YX}X \tag{8-1}$$

Table 8-1 Predicting Y from X with Raw Scores

Formula	What it says to do
$\hat{Y} = a + b_{YX}X$ (8-1) $a = \bar{Y} - b_{YX}\bar{X}$ (8-4) $b_{YX} = \dfrac{\Sigma XY - \Sigma X \Sigma Y/N}{\Sigma X^2 - (\Sigma X)^2/N}$ (8-3)	1 Sum all values of X and Y to get ΣX and ΣY; square each X value and sum the squared scores to get ΣX^2, multiply each Y value by its corresponding X value and sum these cross products to get ΣXY; now put these values into formula (8-3)
	2 Calculate a from formula (8-4)
	3 Substitute these values into formula (8-1); our regression equation is now ready for the insertion of any X score
	4 Select the X value from which Y will be predicted, put it into formula (8-1), and solve for $\hat{Y}$

Example From the following aptitude test scores X and GPAs Y develop a regression equation and predict a GPA for an aptitude test score of 12.

Student	X	Y	X^2	XY
A	20	3.1	400	62.0
B	10	2.6	100	26.0
C	6	1.7	36	10.2
D	15	3.2	225	48.0
E	8	2.0	64	16.0
F	5	2.8	25	14.0
G	11	2.6	121	28.6
H	12	2.2	144	26.4
I	14	2.7	196	37.8
J	18	3.4	324	61.2
	119	26.3	1635	330.2

Step 1. $b_{yx} = \dfrac{330.2 - 119(26.3)/10}{1635 - 119^2/10} = .08$ GPA unit per unit of aptitude

Step 2. $a = 2.63 - .08(11.9) = 1.68$ in GPA units

Step 3. $\hat{Y} = 1.68 + .08X = 1.68 + .08(12) = 2.64 =$ predicted GPA for aptitude score of 12.

we can substitute from

$$a = \bar{Y} - b_{YX}\bar{X} \tag{8-4}$$

for a and end up with

$$\hat{Y} = \bar{Y} - b_{YX}\bar{X} + b_{YX}X$$

Rearranging the terms with X and $\bar{X}$ in them and factoring the b_{YX} gives

$$\hat{Y} = \bar{Y} + b_{YX}(X - \bar{X}) \tag{8-5}$$

It can also be shown that*

$$b_{YX} = r_{XY}\frac{s_Y}{s_X} \tag{8-6}$$

and so we can rewrite formula (8-5) as

$$\boxed{\hat{Y} = \bar{Y} + r_{XY}\frac{s_Y}{s_X}(X - \bar{X})} \tag{8-7}$$

With this formula we substitute the X value from which we are predicting Y and solve for $\hat{Y}$. Although some terms look complex (like $r_{XY}(s_X/s_Y)$) they can be solved for a single value and used for all successive values without recalculation.

For a moment let us return to the equation

$$b_{YX} = r_{XY}\frac{s_Y}{s_X}$$

for some intuitive verification. For the data in Table 8-1 the two variables correlate .72. Since the standard deviation is .55 for GPA (Y) and 4.93 for aptitude (X), we have

$$b_{YX} = .72 \frac{.54}{4.93} = .08$$

This is the same value we found for b_{YX} in Table 8-1.

We can carry this example a step further. The mean for GPA in Table 8-1 is 2.63 and for aptitude it is 11.90. With these data formula (8-7) becomes

$$\hat{Y} = 2.63 + .72\frac{.54}{4.93}(X - 11.90)$$

$$= 2.63 + .08(X - 11.90)$$

---

* The mathematical development of this entire procedure is given in Appendix B.

Table 8-2 Predicting Y from X when Description Data Are on Hand

Formula (8-7)	*What it says to do*
$$\hat{Y} = \bar{Y} + r_{XY}\frac{s_Y}{s_X}(X - \bar{X})$$	1 Calculate the basic descriptive data on each variable: $\bar{X}, \bar{Y}, s_X, s_Y, r_{XY}$
	2 Calculate the value $r_{XY}\dfrac{s_Y}{s_X}$
	3 Substitute all values into the formula (8-7)
	4 Insert the value of X from which Y will be predicted, solve for $\hat{Y}$

Example A social scientist wishes to predict repeated arrests for juveniles Y from scores on a delinquency proneness test X. The descriptive data are

$$r_{XY} = .60$$

$$\bar{X} = 37.3 = \text{mean test score} \qquad s_X = 4.0$$

$$\bar{Y} = \ \ 5.6 = \text{mean number of arrests} \qquad s_Y = 2.1$$

Predict the number of arrests for a juvenile whose delinquency-proneness score is 31:

$$\hat{Y} = 5.6 + .60\frac{2.1}{4.0}(31 - 37.3) = 3.6 \text{ arrests}$$

If we substitute an aptitude score of 12 in our formula, as we did in Table 8-1, we find

$$\hat{Y} = 2.63 + .08(12 - 11.90) = 2.64$$

the same value as in Table 8-1. A step-by-step procedure for applying formula (8-7) is given in Table 8-2.

Exercises

8-3 A journal article discussing the relationship between test anxiety and weekly hours of parental supervision children experienced reported these descriptive data:

	Mean	*s*	r_{XY}
Anxiety	14.5	3.0	.63
Hours	5.2	1.8	

Develop a regression equation and predict anxiety scores for each of the following students.

Student	A	B	C	D	E
Hours	8	0	7	2	5.2

8-4   A sociologist studied factors that appear to lead women to the selection of their husbands. One factor was the intelligence of the male compared with that of the female. The descriptive data are below.

	Mean	s	r_{XY}
Male	108.5	12.0	.71
Female	110.2	15.1	

Develop a regression equation and predict IQs for the husbands of the following women:

Woman	F	G	H	I	J
IQ	115	98	105	127	111

8-5   As you look over the data in the above problems, what do you conclude the predicted Y value will be when the X value is the mean of the X distribution?

Dealing with Errors in Prediction: The Standard Error of Estimate

Earlier we noted that not all people who got a given X score got the same Y score. Instead they distributed themselves above and below the regression line in a range of a few points in each direction. The array for $X = 2$ in Fig. 8-3 is an example. Our prediction is to the regression line that passes through the center of the array. The array itself looks very much like other score distributions on which we calculated means and standard deviations. With these statistics we figured out what proportions of the people were expected to be 1 standard deviation above (or below) the mean, 2 standard deviations above, etc.

If we take the predicted score $\hat{Y}$ as the mean of an array, we can find an indicator of array dispersion in calculations similar to those done in calculating the standard deviation. For standard deviation we began by subtracting and squaring

$$(X - \bar{X})^2$$

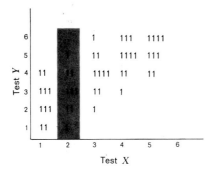

Figure 8-3

Then we averaged this value across all persons in the distribution and took the square root to get the standard deviation

$$s_X = \sqrt{\frac{\Sigma(X - \bar{X})^2}{N - 1}}$$

Although we can do something like this for scores in an array, it would be useful to have one indicator of dispersion that can be used with any array in our scatter diagram. After all, the assumption of homoscedasticity says that all arrays are equally distributed around the regression line. Such an indicator of dispersion that can be applied across all arrays starts with scores in the first array, where $(Y - \hat{Y})^2$ is calculated for each person. We then calculate the same value for each score in all the remaining arrays. All these array values are summed and averaged; then, as with standard deviation, the square root is taken

$$s_{Y.X} = \sqrt{\frac{\Sigma(Y - \hat{Y})^2}{N - 2}} \qquad (8\text{-}8)$$

where $s_{Y.X}$ (read "Y dot X") is used to show that the value is tied to the regression of Y on X and to distinguish it from the standard deviation of Chap. 5.

The $N - 2$ in formula (8-8) may appear as unusual. However, sample estimates of population values of the standard deviation were calculated with $N - 1$ degrees of freedom. Since we are dealing here with two restrictions on the data, the sample estimate is calculated with $N - 2$ degrees of freedom in the denominator.

The statistic $s_{Y.X}$, called the *standard error of estimate*, is used much like the standard deviation in determining portions of observations that lie given distances above and below the central point of a distribution, here the $\hat{Y}$. For example, Jolene is predicted to get a GPA ($\hat{Y}$) of 2.9. What portion of people predicted to get 2.9 actually achieve a 3.2 or higher?

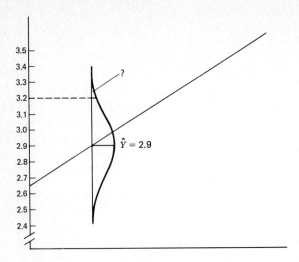

Figure 8-4

This problem looks like Fig. 8-4, where the predicted value 2.9 is laid out at the center of the array. The Y value of 3.2 is located above 2.9 within the distribution of scores in the array. It looks like problems solved in an earlier chapter with z; but to solve a problem with z scores, the standard deviation was necessary. Here the standard error of estimate will be used. The setup for the above problem will look like

$$z = \frac{Y - \hat{Y}}{s_{Y.X}} \qquad (8\text{-}9)$$

and when the data from the problem are fitted into the formula,

$$z = \frac{3.2 - 2.9}{s_{Y.X}}$$

At this point we must calculate $s_{Y.X}$. Look back for a moment at formula (8-8). If the $Y - \hat{Y}$ values are indeed squared and summed, it can be shown* that formula (8-8) is equal to

$$s_{Y.X} = s_Y \sqrt{1 - r_{XY}^2} \qquad (8\text{-}10)$$

Both r_{XY} and s_Y can be computed from raw data; indeed they will have been computed with the descriptive data most researchers will want to see. If we know that aptitude and GPA correlate at .52 and that the standard deviation of GPA is .75 GPA, the standard error of estimate

* See Appendix B for this development.

becomes

$$s_{Y.X} = .75\sqrt{1 - .52^2}$$

$$= .64 \text{ GPA}$$

This value, .64, is in GPA units, i.e., the standard error is .64 of a GPA unit. For our candidate above, the z value is

$$z = \frac{3.2 - 2.9}{.64} = .47$$

What portion of a normal distribution is beyond a z value of .47? Table C-2 provides that information. There it says that .3192, or 31.92 percent of the curve lies beyond a z value of .47. Therefore, the response to the above question is that about 32 percent of the students predicted to have a GPA of 2.9 will in fact achieve 3.2 or higher. Is one of them Jolene? No one can say. The best that can be said is that of the students who are predicted to get 2.9, about one-third (32 percent) will achieve 3.2 or more. We cannot identify which students will be in that 32 percent of the array.

Here is a similar problem. Ms. Kindhardt, a college counselor, knows that the correlation between an aptitude test and GPA is .50. The GPA mean for freshmen at the end of the first semester at her school is 2.83, and the standard deviation is .70 GPA. Ms. Kindhardt is now looking at the records of Charles, a promising athlete. He has an aptitude score for which formula (8-7) predicts a GPA of 1.90. Charles needs a GPA of 1.75 to stay eligible for the team. What portion of students with aptitude scores like Charles's will fall below 1.75 in GPA?

We begin with the predicted GPA as a mean of the distribution. The standard error becomes the standard deviation for the distribution. In this case, based on the data above, the standard error is

$$s_{Y.X} = s_Y \sqrt{1 - r_{XY}^2}$$

$$= .70 \sqrt{1 - .50^2} = .70(.87) = .61$$

We now apply our skill in using z scores and the table of areas under the normal curve. Charles has a predicted GPA of 1.90. The critical GPA is 1.75. We first find the z score for 1.75

$$z = \frac{1.75 - 1.90}{.61} = -.25$$

We now go to the table of areas under the normal curve to see how much of the curve lies below a z value of $-.25$. Table C-2 shows us that 40 percent of the cases predicted to have a GPA of 1.90 will in fact achieve 1.75 or less.

In solving problems with the standard error we begin with the predicted value $\hat{Y}$ as the mean of a distribution. The standard error of estimate becomes our standard deviation for the distribution. With these data we can then make speculations about how many cases will fall more than a given number of points above or below the predicted value. We do this by referring to the table of areas of the normal curve once we have found the appropriate z value.

Formula (8-10) not only is the array standard deviation, but also allows us to get a little more meaning out of r_{XY}. If we square both sides of the equation, we have

$$s_{Y.X}^2 = s_Y^2(1 - r_{XY}^2)$$

which can be rewritten

$$r_{XY}^2 = 1 - \frac{s_{Y.X}^2}{s_Y^2}$$

where r_{XY}^2 is the portion of variance in Y which is associated with X. The total variance of Y is 1.00, and if we take away $s_{Y.X}^2/s_Y^2$, the portion of Y variance which is not predicted by X, we have left r_{XY}^2, the portion of Y variance which is associated with X. Hence *the ratio of the error variance to the variance of the Y distribution tells us the portion of Y variance not associated with X variance.* The larger the r_{XY} the smaller the $s_{Y.X}^2/s_Y^2$ ratio.

The standard error also provides another way of interpreting a correlation coefficient, i.e., in terms of the accuracy with which Y can be predicted from X. The larger the value of r_{XY} the smaller the standard error and hence the greater the accuracy of prediction.

Exercises

8-6   I have found a correlation between IQ and GPA for college freshmen to be .60; GPA has a standard deviation of .50, and IQ has a standard deviation of 15.

(a) I predict a GPA of 3.0 for Mary Willslip. What portion of people with $\hat{Y}$ like Mary's will actually achieve a GPA of 3.5 or higher?

(b) I predict for John Plugger a GPA of 2.0. He must have a 1.8 to stay off the probation list. What portion of people predicted to be at 2.0 will actually fall below 1.8? What portion will actually make 3.0 or higher?

8-7   As an employer of machinists, I have conducted a study to find the correlation between success on the job and scores on a mechanical aptitude test. The r_{XY} is .70; test standard deviation is 10; and my job proficiency measure has a standard deviation of 5. I have decided that minimum proficiency is represented by a job success score of 30 but am willing to accept for employment all persons one standard error of estimate below that point. What is my minimum acceptable job success score?

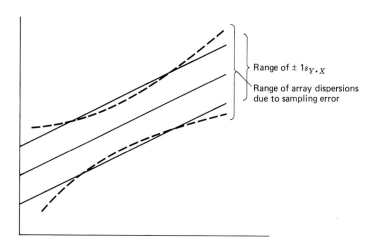

Figure 8-5

A word of caution is appropriate. We have interpreted the standard error of estimate as though each array were dispersed exactly equally with each other array. Although we assume this to be true in basic correlation analysis, in small samples from a population it will not be entirely true. Fewer cases fall into the extremely high and extremely low arrays, and the dispersion in these arrays will not typically look like the dispersion in other arrays. Our standard error is a useful guide to dealing with imprecision of predictions, but sampling errors will cause it to underestimate slightly the dispersion in extreme arrays. In Fig. 8-5 the dotted lines show deviance from $\pm 1 s_{Y.X}$ from the regression line and show how dispersion is affected by sampling error.

Although the standard error of estimate is a definite aid for appraising the accuracy of our predictions, when the data involved are based on samples, especially small samples (say up to 100 cases), the standard error misses its mark by a small amount when dealing with scores at the ends of the distribution.

Regression Effect

With the data given in a typical scatter diagram we can compute three standard deviations for Y scores: (1) one for the deviation of Y scores from $\bar{Y}$, (2) one for the deviation of Y scores from the array means $\hat{Y}$, and (3) one for the deviation of the array means $\hat{Y}$ from $\bar{Y}$. These three standard deviations and their distributions are schematically shown in Fig. 8-6. We speculate from the figure that

$$s_Y{}^2 = s_{Y.X}{}^2 + s_{\hat{Y}}{}^2$$

and this can be shown to be true. We can also see that the spread of

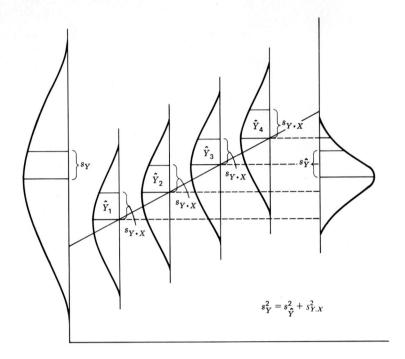

$$s_Y^2 = s_{\hat{Y}}^2 + s_{Y \cdot X}^2$$

Figure 8-6

predicted scores $\hat{Y}$ is less than that of actual scores Y. In fact we see $s_{\hat{Y}}$ equal to s_Y only when the correlation between X and Y is ± 1.0. In other words, we never predict scores as high or as low as actually will be attained. This tendency for predicted scores to fall toward the mean of the raw-score distribution is called the *regression effect*. Its result is to present a distribution of predicted scores $\hat{Y}$ which is more constricted in dispersion than actually obtained scores. The lower the value of r_{XY} the narrower the distribution of predicted scores compared with actual scores.

In other words, the lower the value of r_{XY} the greater the regression effect. In fact, if r_{XY} is zero, all Y scores will be predicted to be at the mean of the Y distribution. This can be illustrated with formula (8-7). If r_{XY} is zero and

$$\hat{Y} = \bar{Y} + r_{XY} \frac{s_Y}{s_X} (X - \bar{X})$$

everything to the right of the plus sign is zero, leaving

$$\hat{Y} = \bar{Y}$$

for any value of X we substitute into the formula.

Summary

If X and Y are related in some systematic way, it should be possible to predict Y from X with better than accidental accuracy. The mechanism for making these predictions, the regression equation, requires two pieces of information: (1) the value of Y when X is zero and (2) the amount of increase in Y scores per unit increase in X.

Since predictions contain a certain amount of error (unless $r_{XY} = \pm 1.00$), prediction procedures must contain a system for estimating the error. The statistic for this purpose, the standard error of estimate, is applied like the standard deviation to determine ranges above and below the predicted score and to relate deviations from the predicted score to portions of the normal curve above or below given points.

The standard error provides an additional way of interpreting correlation— in terms of the accuracy of predictions based on the relationship reflected in r_{XY}. The larger the value of r_{XY} the smaller the standard error of estimate and the more precise the prediction that can be made from X to Y.

The flowchart in Fig. 8-7 illustrates the application of regression procedures and serves as a guide to their use.

Key Terms

scatter diagram	regression line
array	regression coefficient
slope	regression equation
intercept	standard error of estimate
method of least squares	

Problems

8-1 Given scores on a creativity test and scores on an art-interest inventory, predict interest from creativity by

(*a*) Calculating the regression coefficient b_{YX}

(*b*) Calculating the Y intercept a

(*c*) Predicting an interest score for a creativity score of 17; of 8

Student	A	B	C	D	E	F	G	H	I	J
Creativity	18	10	13	6	9	14	12	15	11	16
Interest	11	12	10	7	8	13	10	16	14	15

8-2 In the situation in Prob. 8-1 the correlation between creativity and interest is .52; the mean for creativity is 12.40 and the standard deviation is 3.56; the mean for interest is 11.6 and the standard deviation is 2.95. Construct the regression equation using these descriptive data and predict the interest score for creativity scores of 17 and 8. Compare the results with those of Prob. 8-1.

error   .63

8-3   Twelve high school seniors are followed to see what jobs they hold 5 years after graduation. The hourly rate for these jobs is determined and the IQs taken from school records.

(a) Calculate the regression coefficient.

(b) Calculate the Y intercept.

(c) Predict hourly rates in dollars for IQs of 94 and 120.

Student	M	N	O	P	Q	R	S	T	U	V	W	X
IQ	94	101	98	118	128	104	113	109	90	97	103	121
Rate, $	6	5	4	17	10	9	6	8	5	7	6	12

8-4   With the data for Prob. 8-3 construct a regression equation with the descriptive data as follows: $r_{XY} = .70$; mean IQ is 106.33 and standard deviation 11.65; mean hourly wage is $7.92 and standard deviation $3.68.

(a) Predict hourly wages for students whose IQ is 115 and 106.

(b) How much does the rate (dollars per hour) increase with an increase of 1 point of IQ?

(c) What is the rate for an IQ of zero?

(d) Student Z is predicted to get $11 per hour. What portion of people like Z will actually get $13 or more?

8-5   A shop foreman has found a space-relations test that correlates .60 with product output for lathe operators. The mean test score for the lathe operators was 53.1 with standard deviation of 10.0; the mean output was 24 items per day with standard deviation of 4.3 items. Priscilla has a space-relations score that predicts an output of 18 items.

(a) What portion of people with scores like Priscilla's will have actual outputs of 20 or more items?

(b) What portion of these people will have outputs below 15 items?

(c) What range of outputs will include 95 percent of the people who are predicted to have an output of 18? *Hint*: Leave 2.5 percent in each tail of the distribution.

8-6   Henry has a college aptitude test score which predicts for him a GPA of 3.0 ($r_{XY} = .55$, $s_{GPA} = .70$, $\bar{X}_{GPA} = 2.75$). He must have a 3.2 to be eligible for a scholarship.

(a) What portion of people predicted to get 3.0 will indeed get 3.2 or higher?

(b) What portion of people predicted to get 3.0 will get 2.5 or lower?

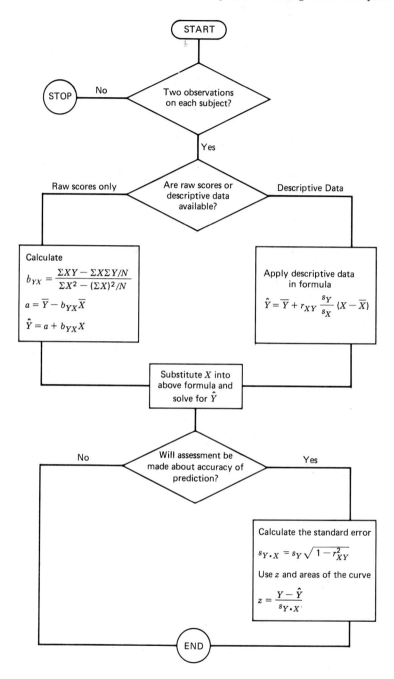

START

STOP — No ← Two observations on each subject?

Yes

Raw scores only ← Are raw scores or descriptive data available? → Descriptive Data

Calculate

$$b_{YX} = \frac{\Sigma XY - \Sigma X \Sigma Y/N}{\Sigma X^2 - (\Sigma X)^2/N}$$

$$a = \overline{Y} - b_{YX}\overline{X}$$

$$\hat{Y} = a + b_{YX}X$$

Apply descriptive data in formula

$$\hat{Y} = \overline{Y} + r_{XY}\frac{s_Y}{s_X}(X - \overline{X})$$

Substitute X into above formula and solve for $\hat{Y}$

No ← Will assessment be made about accuracy of prediction? → Yes

Calculate the standard error

$$s_{Y \cdot X} = s_Y \sqrt{1 - r_{XY}^2}$$

Use z and areas of the curve

$$z = \frac{Y - \hat{Y}}{s_{Y \cdot X}}$$

END

Figure 8-7

9

Inference and Probability: Basic Concepts

Previous chapters have dealt with ways to summarize data, i.e., basic descriptive procedures. The conclusions drawn from the statistics refer to the people on whom the data were collected. But research consists of much more than describing a set of people who are the focus of a collection of data. It also attempts to generalize from the people in a given study to all people of similar character. The objective is to develop theories of behavior that can be applied in making decisions across many groups of people in a common circumstance. It is not enough to observe a given instructional technique in one kindergarten; we want to arrive at principles of instruction that prove effective across the population of kindergartens. Sociologists do not wish to draw conclusions about the voting behavior of one sample of people but to develop principles of voting behavior that apply across populations. By observing samples of people behavioral scientists aim at developing principles of behavior that apply to broad populations from which the samples came.

Similarly, businesses collect opinions of a small set of consumers and generalize these opinions to the population to plan products which they hope will sell widely. Politicians send questionnaires to a small number of their constituents and from their responses generalize to the population of the home district.

We base many of our daily decisions on inference. Your car is a Hondota; it has had ignition problems. A friend also has a Hondota with ignition problems. You advise everyone to avoid Hondotas because they have ignition problems. From a sample you have made an inference to the population.

In all walks of life we make generalizations that are based on data from a sample from the relevant population. Inference is a standard method of predicting experiences. Scientists differ from other people only in that they have refined their procedures for making inferences in order to reduce uncertainty.

One procedure scientists use to increase the validity of their generalizations is *random* selection of individuals from the population. In *random samples* each individual must have the same chance of being chosen as any other individual. Therefore, if the sample to be studied is randomly chosen from the population, the character of the sample is expected to reflect the character of the population and conclusions across samples can be inferred to the population.

The "across samples" idea is important because now and then a completely random sample may not look at all like the population; so verifying data across other random samples is needed to project population characteristics.

In making inferences to populations, statistics are vital. The statistics that describe the sample are taken as estimators of the population—estimators because samples are expected to look like the population but cannot be guaranteed to be exactly like it. Therefore, estimates based on samples are expected to be approximations of the population value but may not be exactly *at* the population figure.

Good research will begin with an interest in reaching a fundamental conclusion about a population. Ideally, random samples will be taken from the target population, and selected observations will be made on the people who make up each sample. Basic descriptive data, the mean, standard deviation, correlation, etc., are computed. From the sample data we make inferences about the population from which the sample came. When we apply statistical procedures to sample data to reduce the uncertainty of the conclusions about the population, the techniques are called *inferential statistics*. This chapter and the remaining ones deal with these techniques.

Inferential techniques help us make probability statements about how likely it is to get a sample with a given characteristic from a defined population. For example, I have a sample of women, all age 25 (as nearly as can be established). Their mean height $\bar{X}$ is 5 feet 7 inches. But since the sample mean is only an estimate of the population value we ask: Is it likely that we could get a sample of twenty-five-year-old females like this ($\bar{X} = 5$ feet 7 inches) if the population average is 5 feet 5 inches? Inferential techniques help us answer such questions and in so doing help us zero in on population characteristics. It will help us greatly at this point if we can construct a *frequency distribution of sample means*. If we had a very large group of means, each from a random sample from the same population, we could make a frequency distribution of these means. Then, using a procedure much like that used with z scores, we could determine the likelihood of getting a sample whose mean is a given distance above or below the expected population mean. In this manner we could quickly decide which sample means were "frequent," i.e., highly likely to occur, and which sample means were not very likely. Much of this chapter deals with constructing such distributions and applying them to solving problems which ask: How likely is it to get a sample whose mean (or other characteristic) is X given that the population value is Y?

If sample statistics are going to estimate population values, they must be *unbiased*. A sample statistic is said to be unbiased if across a very large number of samples the average value of the statistic equals the population value. If we select all possible samples, calculate $\bar{X}$ for each sample, and average these $\bar{X}$ values across all samples, this average will equal the population mean μ. Remember that the Greek letter μ (mu) represents the population mean to distinguish it from the sample mean $\bar{X}$. Thus, $\bar{X}$ is an unbiased estimate of the population mean μ.

Similarly, s^2, the sample variance based on $\Sigma(X - \bar{X})^2/(N - 1)$ is an unbiased estimate of the population variance σ^2. Across many samples the average value of s^2 will equal the population variance. Note that unless $\Sigma(X - \bar{X})^2$ is divided by $N - 1$, instead of N, the average of the sample values will not equal the population variance.

Unbiased sample statistics prove to be useful estimates of the population parameters. Since whole populations can almost never be observed, we must rely on sample data to tell us about the population. To this end we find that larger samples are more accurate estimators than small samples for the statistics we most often use, $\bar{X}$ and s. A statistic that comes closer and closer to the parameter it estimates as samples get larger is referred to as a *consistent estimator*. Both the sample mean and standard deviation are unbiased and consistent estimators and are relied on heavily to tell us about the population.

We shall now describe some problems where unbiased and consistent estimators prove useful. First, we shall attempt to establish the population mean from sample data, and then we shall look at two samples to see whether they are from a common population.

The Single-Sample Model

We shall first lay out a research problem that uses a mean and standard deviation, unbiased and consistent statistics, to tell us about the population. To begin we shall make a hypothesis about the population mean, and then we shall see how close our sample mean is to the hypothesized value. If it is close, we say: Yes, this sample could have come from a population with a mean of that hypothesized value. This is referred to as the *single-sample model*.

Here is an example. Suppose our population is made up of all first-grade children who have a sight vocabulary of 20 words upon entering the first grade. We hypothesize (based on the literature or a given theory of cognitive development) that for this population the mean mental age will be 6.5 years. Then we take a sample from the population of 20 word recognizers and assess their mental age. We find the mean for our sample to be 6.3 years. Realizing that samples are only estimators, we know that our sample mean may not be absolutely accurate for the population and so we ask: Is this sample within the range of the means of random samples taken from a population whose mean is 6.5? In other words, if I had a very large number of samples from the population whose mean is 6.5, would samples with means of 6.3 be among the most frequent ones? Figure 9-1 illustrates this problem. The answer could be yes, that is, $\mu = 6.5$, or it could be no, i.e., $\mu \neq 6.5$. If the answer is yes, we know that the population mean could in fact be 6.5 and we have some information about the population we did not have before.

How do we decide that the sample mean and the hypothesized population mean are "close enough"? We use a variation of the z-score procedure described in Chap. 6. The magnitude of the difference between $\bar{X}$ (the sample value) and μ (the parameter) is assessed by a kind of standard deviation (which we shall soon develop), but use of this procedure depends on our having generated a frequency distribution of sample means that looks like a normal curve. We shall do something like this as soon as we have looked at some similar questions to which single-sample data apply.

A second approach to the single-sample model for discovering the population mean would be to begin as before with a sample. Since we know that sample means are expected to be at (or near) the population value, we observe the sample mean $\bar{X}$ and infer that this sample value $\bar{X}$ is very likely to be within a given number of points above or below the

Figure 9-1

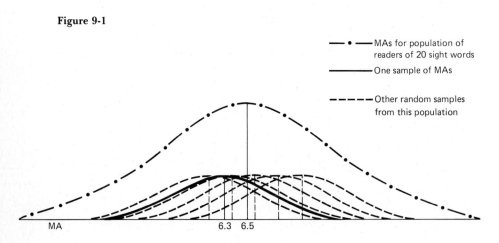

— • — MAs for population of readers of 20 sight words

——— One sample of MAs

- - - - Other random samples from this population

MA   6.3 6.5

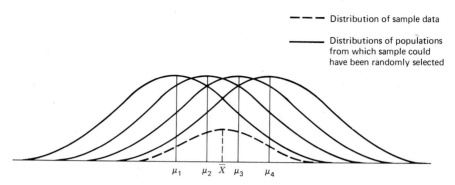

- - - Distribution of sample data

——— Distributions of populations from which sample could have been randomly selected

$\mu_1 \quad \mu_2 \quad \overline{X} \quad \mu_3 \quad \mu_4$

Figure 9-2

population value μ. This range of scores above and below the sample mean, the range that includes the population mean, is called the *confidence interval*. (The method of calculating it lies ahead.) With this method a range of possible values for μ is identified, and we conclude that μ could indeed be anywhere within that range.

Is this a useful procedure? It is, indeed! Before finding this range of possible values for μ we had no idea where the population value might be. With the range laid off, we have cut down the possible values of μ to a reasonable few, and hence have begun to identify the location of the parameter. This situation is shown in Fig. 9-2. Like the first procedure for estimating the population value, it depends on our ability to construct a frequency distribution of sample means so that we can identify the range of the most likely sample means in the population from which our sample came.

The Two-Sample Model

Another common kind of inference in research in social sciences is based on the *two-sample model*. Here two samples are randomly selected from a defined population. One of the samples is given a particular treatment. After the treatment a measure is taken on each sample and their respective means calculated. We now ask: Are these the means of two random samples from a common population? If the mean values are close to each other, we may wish to infer that the samples do still look like two random samples from the same population. If the means are not close together, the inference may be reasonable that the samples are not from a common population. This situation is illustrated in Fig. 9-3, where the two samples in Fig. 9-3a look like random samples from a common population; in Fig. 9-3b the two samples appear to be from different populations.

For example, two groups, each of 50 ninth-grade students, are

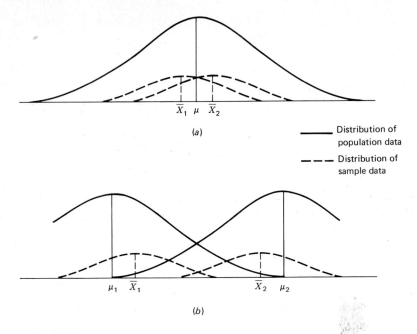

(a)

——— Distribution of population data

– – – Distribution of sample data

(b)

Figure 9-3

selected from all ninth-graders in a city school system. It seems reasonable to believe that all these students are from the same population, i.e., urban ninth-graders. One group is then given a special study method for learning mathematics while the other group is given nothing. Two weeks later students in both groups are tested over a common unit of mathematics that both have studied. The mean for each group is then calculated and the two means compared to see how different they are. If the means do not appear to be very different, we believe the special treatment made no difference in performance; but means that are quite different tell us that the two samples are from different populations of mathematics performers.

How do we assess the magnitude of a difference between two sample means? Once again we shall devise a kind of standard deviation and call upon the basic idea of z values. This procedure will be described just ahead. Again, a frequency distribution that looks normal must be available for interpreting the z value.

Testing Hypotheses

In both the single- and two-sample models we begin with a hypothesis, H. In the single-sample model we hypothesize that the population mean is a specified value. Our conclusions can be outlined as follows:

1. The population mean could very likely equal the specified value.

2. The mean is not likely to equal the specified value; hence
   a. The mean is likely to be larger than the specified value.
   b. The mean is probably smaller than the specified value.

The situation in 1 is called the *null hypothesis* H_0. The difference between the sample mean and the expected population value is simply a chance event of random sampling; i.e., no (null) difference exists between the sample mean and the hypothesized population value.

If H_0 is not true, i.e., the difference between the sample and expected population means is large, then H_1* must be true: the population mean is not likely to be the specified value. If this is so, then either

$$\mu > \text{the specified value}$$

or $$\mu < \text{the specified value}$$

When we use the two-sample model, the procedure is essentially the same. Two sample means are compared under the hypothesis H_0 that they look like random samples from a common population, i.e., that both sample means are estimates of the same population mean. If H_0 is not true, then H_1 must be true; i.e., the samples are estimates of means of different populations. If $\mu_1 \neq \mu_2$, then either

$$\mu_1 > \mu_2 \quad \text{or} \quad \mu_1 < \mu_2$$

These examples demonstrate the idea that sample data can be used to help us make conclusions about populations with some degree of certainty but of course not with absolute certainty. Since we cannot observe whole populations, we cannot calculate their parameters; but sample statistics can help us zero in on parameters, i.e., make inferences about the characteristics of populations.

Hypotheses about Dichotomous Data

With the above general description of inference we are ready to discuss how to devise frequency distributions that allow us to draw conclusions about populations once we have sample data. We begin with the simplest case, dichotomous situations like male/female, true/false, heads/tails, resident/nonresident, etc. Our procedure will be to make a frequency distribution of possible outcomes and then see how likely it is to find each of the outcomes. The single-sample model will be

---

* These designations (H_0 and H_1) arise from the notation of boolean algebra, in which variables can take one of only two values. These values are symbolized by 1 and 0, the digits of binary numbers.

applicable here. We shall take a sample, count the number of observations of a given kind, like heads/tails, and see where this number falls on a frequency distribution.

If the population probability for tossing coins is .50, that is, heads should come up half the time, and tails half the time, how likely is it to get a set of four coins which when tossed will produce four heads and no tails? Three heads and one tail? Exactly two heads and two tails? If I toss four coins and they come up four tails, how likely is it that they come from a population of coins in which the expected outcome is two heads and two tails? If we consistently get four tails, we may wish to conclude that the coins are not from the same population as "fair" coins.

Although decisions about fair coins are not particularly common in everyday life, they are convenient models for decisions in any situation where we are dealing with dichotomous variables. I have randomly selected four people from the telephone book. They are all male. If the population is half male and half female how likely is it that a random selection would produce four males? If it is rare to get four males, i.e., the probability is small, we may wish to infer that the population from which our sample came does not have half males and half females, i.e., a probability of .50. The coin example is a prototype of all dichotomies.

In our discussion of dichotomies the term *event* will be used to describe a single observation, e.g., the face of one coin or the sex of a single randomly selected person. *Outcome* will be used to describe the combination of events in a given sample. How many heads and how many tails in a single toss of five pennies? How many males and how many females in a random sample of 10 persons? These are examples of outcomes.

Before going on to make inferences about populations of dichotomous and continuous data we must learn how frequency distributions of various outcomes are constructed. We cannot speculate on the probability of getting a given outcome from a given population until we understand the procedure for developing the frequency distribution for outcomes.

In a sample of four events, some outcomes (like four heads or four tails) logically seem to be uncommon, while other outcomes (like two heads and two tails) appear to be quite common. For tosses of four coins the expected outcome would appear to be two heads and two tails. We realize that four tails can appear but is relatively unusual.

Suppose on several tosses of four coins we got four tails most of the time. We would first ask: If a head or a tail is equally likely for any coin, what are the chances of getting this outcome? Having pondered the fact that it is an unusual occurrence, we may conclude that a series of four tails is too unusual to enable us to believe that the four coins came from a population of fair coins.

Before we can make this decision we must calculate the probability of various combinations of heads and tails. Here *probability* is defined as a proportion determined by *the number of outcomes favoring an event compared with the total number of events*. Its formula is

$$P(A) = \frac{\text{no. of outcomes favoring event A}}{\text{total no. of outcomes}} \qquad (9\text{-}1)$$

where *P* is the probability and A identifies the event.

For example, I toss two pennies and count heads and tails. The two pennies can come up HH, HT, TH, TT; i.e., four outcomes can occur when two pennies are tossed. Out of these four outcomes only one is two heads. The probability of getting two heads on a toss of two coins is, by formula (9-1),

$$P(2H) = \frac{1}{4} = .25$$

Look around your classroom. (Suppose it to be such a large class that selecting one person will scarcely alter the ratio of females to males.) How many females are in the class? Suppose half of the class is female, $P(F) = .50$. You randomly select two members from the class. What is the probability of both being female?

The possible outcomes are FF, FM, MF, MM, i.e., four possible combinations, one of which has two females. The probability of selecting two females on a random draw is

$$P(2F, 0M) = \frac{1}{4} = .25$$

and
$$P(1F, 1M) = \frac{2}{4} = .50$$

$$P(0F, 2M) = \frac{1}{4} = .25$$

A plot of these probabilities is shown in Fig. 9-4*a*. Now we are beginning to develop the frequency distribution of various outcomes.

If my selection in the above group was of four people instead of two, what is the probability of getting four females? The possible outcomes are shown in Table 9-1 (page 191).

The probabilities for the various outcomes would be

$$P(4F, 0M) = \frac{1}{16} = .0625$$

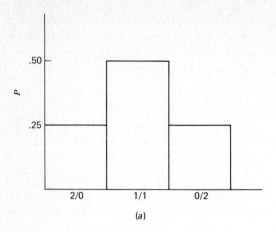

(a)

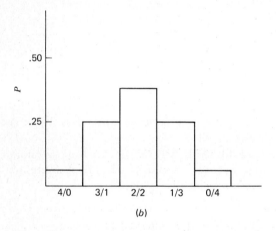

(b)

Figure 9-4 Number of females and males in (a) two and (b) four random samples.

$$P(3F, 1M) = \frac{4}{16} = .250$$

$$P(2F, 2M) = \frac{6}{16} = .375$$

$$P(1F, 3M) = \frac{4}{16} = .250$$

$$P(0F, 4M) = \frac{1}{16} = .0625$$

They are plotted in Fig. 9-4b. Here we are expanding the idea of the frequency distribution of various outcomes. The histogram begins to

**Table 9-1 Outcomes for a
Selection of Four**

Outcome	Number
FFFF	1
FFFM	
FFMF	
FMFF	4
MFFF	
FFMM	
FMFM	
FMMF	
MFFM	6
MFMF	
MMFF	
FMMM	
MFMM	
MMFM	4
MMMF	
MMMM	1
	16

look bell-shaped. As samples get large, say 30 cases per sample, the distribution may indeed be considered a normal curve, a fact we shall soon exploit.

The Binomial Model

When Fig. 9-4a is compared with b, we see that when the number of individuals in a sample (coins, persons, true-false items) increases, the number of outcomes increases in what appears to be a systematic manner. If this is so, we can develop a mathematical model (or formula) of the procedure that will tell us the probability of each outcome of interest without tediously laying out all the outcomes, as we did above.

The mathematical model used in this case is the *binomial expansion,* which takes the form

$$(p + q)^n \tag{9-2}$$

where p = proportion of the population for a given event
$\quad\quad q$ = proportion in the population for alternative to
$\quad\quad n$ = number of times that event can occur, i.e., sample size

When we make two random selections, as in the female/male problem above formula (9-2) would become

$$(p + q)^2$$

When we square this binomial, we find

$$1p^2 + 2pq + 1q^2$$

The coefficients 1, 2, 1 indicate the number of trials in which a given outcome occurs; the exponents correspond with the number of occurrences of a given event in an outcome.* If we applied this to the male/female problem cited earlier, where

$$p = \text{population proportion for females} = .50$$

$$q = \text{population proportion for males} = .50$$

the above calculation shows 1 outcome with two females p^2 and no males, 2 outcomes with one female p and one male q, and 1 outcome with two males q^2. At this point we are ready to substitute the value of .50 for p and .50 for q. Then we find

$$1p^2 \quad + \quad 2pq \quad + \quad 1q^2$$
$$1(.50^2) + 2(.50)(.50) + 1(.50^2)$$
$$.25 \quad + \quad .50 \quad + \quad .25$$

These are the same probabilities for the three outcomes we found earlier by the tedious process of laying out all possible outcomes and dividing the frequency of a given outcome by the total number of outcomes.

If we apply the binomial expansion to the situation where we randomly selected 4 persons at once, we find for $(p + q)^4$

$$p^4 \quad + \quad 4p^3q \quad + \quad 6p^2q^2 \quad + \quad 4pq^3 \quad + \quad q^4$$
$$.50^4 + 4(.50^3)(.50) + 6(.50^2)(.50^2) + 4(.50)(.50^3) + .50^4$$
$$.0625 + \quad .25 \quad + \quad .375 \quad + \quad .25 \quad + .0625$$

The probability of getting four females is .0625, of getting three females and one male is .25, etc. Again, these are the same probabilities we found above by the more tedious counting method.

Use of the binomial expansion model assumes two things: (1) only two kinds of events (p and q) are possible for each random selection,

---

* Here each female and each male is an *event*; a given combination of females and males is an *outcome*.

and (2) the outcome of one selection is *independent* of the outcome of any other selection. In the above example there were two kinds of events, female and male, and selection of a female on a given draw had no influence on the sex of the next person drawn; hence the events are independent.

We have applied the binomial expansion to selecting various combinations of males and females. It can also be applied to other dichotomies like true/false, voted/did not vote, attended/did not attend, etc. With males and females we took the population proportions to be .50; this would also be reasonable with some other dichotomies such as true/false. But the .50 expectation would not be reasonable with other variables. Although the binomial expansion will work quite well in providing a frequency distribution for various outcomes, the more the population values depart from .50 the more skewed the frequency distribution will become. For example, if the population proportions for males were .67 and for females .33 and we randomly selected four people from this population, the binomial expansion of $(p + q)^4$ would be

$$1(.67^4) + 4(.67^3)(.33) + 6(.67^2)(.33^2) + 4(.67)(.33^3) + 1(.33^4)$$
$$.201 + .397 + .293 + .096 + .012$$

Here the probabiliites on the left are somewhat greater than those on the right; i.e., the distribution is skewed. This will be the case as population proportions depart from $p = .50$.

Although the binomial expansion provides a model for calculating the probability of getting certain outcomes given population proportions, calculating the coefficients in the binomial expansion and the powers to which p and q are raised can be complex. Why hasn't someone figured these values for us and put them in a table? Fortunately someone has.

Blaise Pascal, a seventeenth-century French mathematician, devised a table of coefficients for each of the terms in an expanded binomial. They are given in Table 9-2, where the column under n tells us how many events there are in a single outcome. It also shows the power to which $p + q$ will be raised. If p stands for males and q for females and we are dealing with samples of three cases (three events in an outcome), the coefficients for the terms in the expanded binomial would be given by the line to the right of 3, that is, 1, 3, 3, 1.

Remember, our objective is to build a distribution of outcomes against which a given sample outcome can be evaluated to see how likely this outcome is. Above we selected samples (outcomes) of four people (events) at a time. On Pascal's triangle, n for this situation is 4, and the coefficients for the terms in the expanded binomials are (Table 9-2) 1, 4, 6, 4, 1.

Table 9-2   Pascal's Triangle up to $n = 10$

n													Sum
1					1		1						2
2				1		2		1					4
3			1		3		3		1				8
4			1	4		6		4		1			16
5		1		5	10		10		5		1		32
6		1	6	15		20		15	6		1		64
7	1		7	21	35		35	21		7	1		128
8	1	8	28	56		70		56	28	8		1	256
9	1	9	36	84	126		126	84	36	9	1		512
10	1	10	45	120	210	252	210	120	45	10	1		1024

With the coefficients from Pascal in place, our expanded binomial $(p + q)^4$ begins to take shape as follows:

$$1p^?q^? + 4p^?q^? + 6p^?q^? + 4p^?q^? + 1p^?q^?$$

We now need to insert the appropriate powers. In the first term on the left the power of p will be equal to the power to which we are raising $p + q$; here that power is 4. The power of q will be n minus the power of p, so the first term will be

$$1p^4q^{4-4} = p^4$$

In the second term we *reduce* the power of p by 1, and the power of q continues to be $n - r$, where r is the power of p. Then the second term will be

$$4p^{4-1}q^{4-3} = 4p^3q$$

We continue to reduce the power of p by 1 for each term to the right; the power of q continues to be $n - r$ through all the remaining terms. The general formula for the powers of p and q is

$$p^r q^{n-r} \tag{9-3}$$

where for the first term in an equation $r = n$ and r is reduced by 1 for each successive term to the right.

We are now ready to write the entire expansion of $(p + q)^4$ using the coefficients from Pascal's triangle and calculating the p and q powers as above. We find

$$(p + q)^4 = p^4 + 4p^3q + 6p^2q^2 + 4pq^3 + q^4$$

Since we assumed that the population was half p and half q, we insert .50 for p and .50 for q, raise each to its appropriate power, and multiply the resulting p and q values together. This gives

$$.50^4 + 4(.50^3)(.50) + 6(.50^2)(.50^2) + 4(.50)(.50^3) + .50^4$$

or   .0625 +   250   +   .375   +   .250   + .0625

These are the same values we calculated when we selected four cases at a time from a population where $p = .50$. We have a probability of .0625 of randomly selecting four females, .250 of selecting three females and one male, .375 of selecting two of each, .250 of selecting one female and three males, and .0625 of selecting four males. These values can be changed to percentages by multiplying by 100. Thus, we expect 6.25 percent of random samples from our population to have four females, 25 percent to have three females and one male, etc.

Not all populations contain events whose probability is .50. Once before we assumed that the proportion of males was .67 and females .33. If we randomly selected four people at a time from this population, we would have the binomial $(p + q)^4$, and from Pascal's triangle we find the coefficients 1, 4, 6, 4, 1. The terms, based on $p^r q^{(n-r)}$ as in formula (9-3), are p^4, $p^3 q$, $p^2 q^2$, pq^3, and q^4; inserting the coefficients we find that our expansion becomes, as before,

$$1p^4 + 4p^3 q + 6p^2 q^2 + 4pq^3 + 1q^4$$

We now substitute .67 for p, .33 for q and find

$$.67^4 + 4(.67^3)(.33) + 6(.67^2)(.33^3) + 4(.67)(.33^3) + .33^4$$

.201 +   397   +   .293   +   .096   + .012

In this case we expect 20.1 percent of the random samples to have four males and no females, 39.7 percent to have three males and one female, etc. (Note once more that as p departs from .50 the distribution becomes skewed.)

Let us apply this information we have just acquired. Suppose this fall in Keytown half of the registered voters actually voted. I go to the courthouse and randomly draw five voters from the registration list to survey their attitudes about the economy. I want to interview a "typical" sample of registered voters. I call each of these people and find that four voted and one did not. How likely is it to select randomly a sample in which four voted, and one did not?

This question can be answered by the binomial expansion. Using Pascal's triangle and formula (9-3), we generate the following figures:

$$(p + q)^5 = 1p^5 + 5p^4 q + 10p^3 q^2 + 10p^2 q^3 + 5pq^4 + 1q^5$$
$$= 1(.50^5) + 5(.50^4)(.50) + \cdots$$

The second term, $5(.50^4)(.50)$, is the one we are interested in, i.e., four voters and one nonvoter. It works out to be .156 out of a total of 1.00, or a little more than 15 percent of random samples will give us four voters and one nonvoter. Applying this result to the single-sample model, should we agree that our sample looks like a random sample from a population where $p = .50$? Since the p of .156 is quite different from .50, we may wish to conclude that getting a random sample of four voters and one nonvoter is too rare to believe that our sample represents a population where $p = .50$. In this case I may wish to conclude that my sample is not "typical" of registered voters.

A similar set of figures can be applied to true/false tests. Suppose we have a seven-item true/false test. Since we know nothing about the subject, we toss a penny: heads for true, tails for false. Across a large number of tosses half should be heads and half tails. Our test is scored; six items were correct and one incorrect. What is the probability of getting six items out of seven by chance alone? The binomial expansion is applied like this:

$$(p + q)^7 = 1p^7 + 7p^6q + \cdots$$

The second term, where we have p^6q, gives us the probability of six correct and one incorrect. It is equal to

$$7(.50^6)(.50) = .05$$

That is, .05 out of 1.00 or 5 percent of the time we expect a score of six correct out of seven items by tossing a coin to decide each mark. This is rare indeed. Would you believe a student who claimed to have guessed at every item but ended up with a score of 6 out of 7 total points? It is possible; it happens 5 percent of the time, but it is not a likely outcome if the population proportion is .50.

Exercises

9-1   Suppose I toss a group of three coins an infinite number of times. What proportion of the tosses will produce:

   (*a*) Three heads and no tails?

   (*b*) Two heads and one tail?

   (*c*) Three tails?

9-2   Assuming that there are as many boys as girls in a given school system, I randomly select a sample of eight cases. What is the probability of getting:

   (*a*) Eight boys and no girls?

   (*b*) Seven boys and one girl?

   (*c*) Four boys and four girls?

9-3   From the telephone book for Manhattan I randomly select seven people (ignoring businesses, etc.). Assuming that half such telephone listings are male, what is the probability of selecting:

(*a*) Six males and one female?

(*b*) Three males and four females?

(*c*) Seven females?

9-4 Among the very large rolls of the membership of the Federated Union for Supply and Storage I find that 40 percent have graduated from high school, 60 percent have not. I randomly draw a sample of four cases from the rolls.

(*a*) What is the probability that all four will be graduates?

(*b*) That two will be graduates and two will be nongraduates?

(*c*) Construct a histogram from the frequency of each outcome (like Fig. 9-4) and observe the nature of the distribution. If proportions in the population are not .50, what happens to the shape of the distribution?

Now let us look back at the forest of figures we have just passed through and ask what it was all about. We agreed earlier that it is not possible to observe whole populations, and so we work with samples. Usually we either want a sample to conform to population characteristics such as sex (if known) or we use it to speculate about what the population characteristics are. An example of the first can be found in a study of children learning to read. We have figures on how many males and females enter the first grade in the United States (the population is all children who enter the first grade). I select a sample of 50 children to study and find that 20 children, or 40 percent are males. How likely is it to get a sample of 50 cases where 20 are males and 30 females if the population proportion is .50?

In other circumstances where the population proportion is not known we often wish to know it. Random-sample statistics should provide a basis for making probabilistic statements about the value of population parameters. Suppose I do not know the proportions of Republicans and Democrats who voted in the last election. If I take a random sample of voters in these parties who actually voted, this sample's proportions should allow me to make a statement about the proportion of Republicans and Democrats in the population of voters.

What is the key element in both these applications? It is the availability of a frequency distribution of outcomes for samples in the defined population. My probabilistic statement of the likelihood of p is based on a frequency distribution of various outcomes. For example, with our first-grade children I cannot decide how likely it is to get a sample of 30 girls and 20 boys in a population where p is .50 unless I have expanded $(p + q)^{50}$. When I have done so, I have a probability for each possible outcome, including 30 girls and 20 boys. When I relate my p value to the distribution of all possible outcomes and see that my p value is a fairly common occurrence, i.e., near the modal point in the frequency distribution, I conclude that my sample could indeed have come from such a population. If my outcomes fall in the extremes of the frequency distribution, I may wish to conclude that my sample is so unlikely in the defined population that it may belong in another population.

The frequency distribution is vital to this decision. Without it we have no conclusions. The binomial expansion provides us a model with which to build frequency distributions. The values of each term in the expansion $(p + q)^n$ tell us what portion of the 1.00 events fall into each of the intervals in the frequency distribution. Noting these portions, we can make statements about how likely a given outcome is, as illustrated in Fig. 9-5, where we quickly see that if random samples of eight observations each are taken from a population where $p = .50$, across a large number of samples we should get about 27 out of 100 with 4 of the 8 events favoring p, that is, 4 females and 4 males; 4 heads and 4 tails; 4 true/false correct and 4 wrong. We can also read that 22 out of 100 samples will have 5 events favoring p, 3 favoring q, etc.

Figure 9-5 also appears to follow the line of a normal curve quite closely, as we have suspected all along. This is the first time n has been large enough to really demonstrate that fact. Indeed, in Chap. 10 we shall start using the normal curve for our probability distribution. We already know how to apply z values to identify areas under the curve,

Figure 9-5

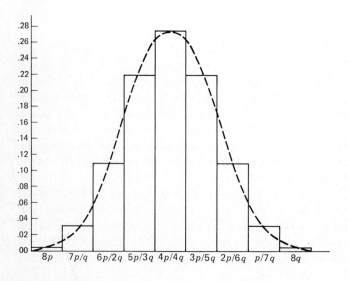

which can be interpreted as portions of 1.00 (the total area under the curve).

Classical Probability

At this point some classical rules for probability calculations are in order. Computations of probabilities follow two rules, the addition rule and the multiplication rule.

The Addition Rule

The *addition rule* starts with mutually exclusive events. Events are said to be *mutually exclusive* if the occurrence of one event excludes the occurrence of the other. If you flip one coin and it comes up heads, it is impossible to get tails on that same flip. *When two events are mutually exclusive, the probability that one or the other of the two separate events will occur is the sum of their separate probabilities.* For events A and B this rule becomes

$$P(A \text{ or } B) = P(A) + P(B) \tag{9-4}$$

which says that the probability of getting A or B is equal to the probability of A plus the probability of B.

Suppose in a group of 100 people you have 20 who are Black, 70 Caucasian, 8 Oriental, and 2 American Indian. What is the probability of selecting a person who is either Black or Oriental? The probability for Black is 20/100 = .20; for Oriental, it is .08. Then

$$P(B \text{ or } O) = P(B) + P(O)$$
$$= .20 + .08 = .28$$

If two events are *not* mutually exclusive, it is possible for both events to occur at the same time. For example, if we asked above: What is the probability of selecting a Black or a female?, we could get both events on a single random draw; i.e., we could select a Black female. In this case

$$P(A \text{ or } B) = P(A) + P(B) - P(A \text{ and } B) \tag{9-5}$$

Let us assume that half of each racial group is female and half male. In this case there will be 10 females among the 20 Blacks, with $P(B \text{ and } F)$ = .10. Our formula then becomes

$$P(B \text{ or } F) = P(B) + P(F) - P(B \text{ and } F)$$
$$= .20 + .50 - .10 = .60$$

For the above group the probability of selecting a Black or female is .60.

The Multiplication Rule

The addition rule deals with either/or situations. The multiplication rule deals with situations where we want to know the probability of *both* of two outcomes, often referred to as *joint probability*. When we have *two independent events*, i.e., one occurrence has no bearing on the other, *the probability that both will occur is the product of their separate probabilities:*

$$P(A \text{ and } B) = P(A)P(B)$$

I have two dice, one black and one white. What is the probability of getting a 3 on each die? Since the probability of a 3 on one die is 1/6 or .167,

$$P(3 \text{ and } 3) = .167(.167) = .028$$

Now let us turn to sampling items from a pool. Remember that the events must be independent. If I have, say, 10 red balls and 10 green balls in a sack, the probability of selecting a green ball is .50. But once I have removed that ball, the probability of selecting a second green ball is no longer .50, because there are now 9 greens and 10 reds left in the sack. For this reason we use *sampling with replacement.* We replace our selection in the pool to avoid altering the probability for successive selections. Thus the probability of a given ball being selected is the same for all events. When we select individuals from a class roster, a voter list, or similar pool, it is important to remember to use replacement.

The multiplication rule can also be applied when events are *dependent* or *conditional* on the occurrence of the other event. In this case replacement is not assumed. The procedure now becomes

$$P(A \text{ and } B) = P(A)P(B|A)$$

where $P(B|A)$ is read "probability of B given that A has occurred."

This rule can be illustrated for our sack of 10 red and 10 green balls. What is the probability that a sample of 2 balls will both be red (R)?

$$P(2 \text{ R}) = P(\text{R first draw}) \, P(\text{R second draw}|\text{R first draw})$$
$$= (10/20)(9/19) = .50(.44) = .24$$

The probability of getting red balls on each of two successive draws is .24, or 24 out of 100.

Here is another example. I am selecting a sample of first-grade children for a cognitive abilities study. There are 40 children in the first grade, 22 boys and 18 girls. I select a girl on the first draw and a girl (G) on the second draw. What is the probability of selecting 2 girls on 2 draws?

$$P(2G) = P(\text{G first draw})P(\text{G second draw} | \text{G first draw})$$
$$= (18/40)(17/39) = .45(.44) = .198$$

And if we ask: What is the probability of getting a girl on 3 draws?, we compute

$$P(3G) = P(\text{G lst draw})P(\text{G 2d draw} | \text{G lst draw})$$
$$P(\text{G 3d draw} | \text{G lst and 2d draws})$$
$$= (18/40)(17/39)(16/38) = .45(.44)(.42) = .08$$

These same procedures can also be used to develop frequency distributions for probabilities. We turn again to the pair of dice and ask: What is the probability of rolling a total of 2? This means that we must get a 1 on the white die and a 1 on the black die

$$P(2) = (1/6)(1/6) = \tfrac{1}{36} = .028$$

Next we calculate the probability for a sum of 3 dots on two dice. We can do this with a 2 on white and 1 on black or with a 1 on white and 2 on black

$$P(1W, 2B) = (1/6)(1/6) = \tfrac{1}{36}$$
$$P(2W, 1B) = (1/6)(1/6) = \tfrac{1}{36}$$

Since either of these events is acceptable, the addition rule applies

$$P(3) = P(1W, 2B) + P(2W, 1B) = 1/36 + 1/36 = \tfrac{2}{36} = .056$$

The next step is to calculate the probability of rolling a total of 4 dots on our two dice. We can get this by

$$P(1W, 3B) = (1/6)(1/6) = \tfrac{1}{36}$$
$$P(2W, 2B) = (1/6)(1/6) = \tfrac{1}{36}$$
$$P(3W, 1B) = (1/6)(1/6) = \tfrac{1}{36}$$

Then using the addition rule

$$P(4) = P(1W, 3B) + P(2W, 2B) + P(3W, 1B)$$
$$= 1/36 + 1/36 + 1/36 = \tfrac{3}{36} = .083$$

Similarly we can calculate the probabilities for rolling a 5, 6, 7, 8, 9, 10, 11, 12 and if plotted on a histogram these probabilities would look like Fig. 9-6. Again probability theory provides a model of the population against which we can test samples of data, and ask: Does my sample look sufficiently like a defined population for me to believe that it came from that population, or is it more likely to be from another population?

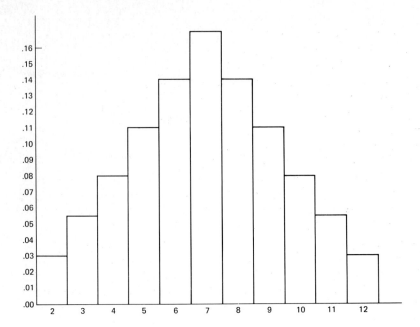

Figure 9-6

Exercises

9-5 From a sack of 10 red, 9 green, and 6 yellow balls what is the probability of randomly drawing:

(*a*) Either a red or a green ball?

(*b*) Either a red or a yellow ball?

(*c*) Either a green ball or a yellow ball?

9-6 From a class of 40 students I am selecting members to participate in an experiment. In the class there are 25 Caucasians, 10 Blacks, and 5 Orientals. What is the probability of randomly selecting:

(*a*) Either a Black or a Caucasian?

(*b*) Either a Black or an Oriental?

(*c*) Either a Caucasian or an Oriental?

9-7 From a standard deck of playing cards one card is drawn. What is the probability that it will be:

(*a*) A face card or a red card?

(*b*) A king or a black card?

9-8 From a study group of 52 students 12 are seniors and 40 are juniors; half of each class is male, half female. What is the probability of getting:

(*a*) A female or a senior?

(*b*) A male or a junior?

9-9 I have the list of 20 board members present for the recent teachers' union vote on contract provisions. Of the 20 members 12 voted yes and 8 voted no. What is the probability of selecting:

(*a*) A no voter and a yes voter (with replacement) on two random draws?

(*b*) A no voter on each of two random draws (without replacement)?

Summary

Many decisions are based on samples of evidence which are generalized to entire populations. We call this inference. The job of the statistician is to provide systematic models of how sample data relate to populations, so that we can indeed say something about a population once we know the characteristics of a sample. One such model is the probability distribution based on the binomial expansion. This model lays out the probability of getting given combinations of events in a sample of a predetermined size.

As sample sizes increase, the binomial distribution approaches the shape of the normal curve. In fact with sample sizes of, say, 30 cases or more, the normal curve can be used as a probability distribution.

Classical probability theory centers on the likelihood of getting given events in random draws from a defined population. There are two rules for calculating probability. The addition rule has to do with the probability of getting *one or the other* of two events. If the two events are mutually exclusive, the probability of one or the other's occurring is the sum of their separate probabilities. If the events are not mutually exclusive, the probability of getting one event or the other is the sum of their separate probabilities minus the chance that they will appear together.

The multiplication rule deals with the probability that both of two events may occur. If two events are independent, the probability of both occurring is the product of their separate probabilities. If the probability of one event is dependent, or conditional, upon the occurrence of the other event, the probability that both will occur is the product of the probability of the first event and the probability of the second event given that the first event has occurred.

Social scientists use probability to discover information about the world in several ways. In the single-sample model a population characteristic is set. Then a sample is taken and the characteristic observed in the sample. If the sample value is quite probable, as revealed by a probability distribution, we believe that indeed the population value could be the one set. An alternative is to begin with a sample, observe a characteristic, and project a range within which the population value is most probable. We call this range the confidence interval.

Probability is also used in making inferences in the two-sample model. Here two random samples are drawn, one is given a special treatment, and after the treatment a given characteristic of interest is observed. We then ask whether the two samples still look like random samples from a common population. If they do look like samples from a common population, the treatment had no effect. Again, a probability distribution is necessary to help decide whether the two samples appear to be from a common population.

In many studies the normal curve is used as a model for laying out probabilities in both the one- and two-sample models. In the next chapter we exploit the fact that the frequency distribution of probabilities approaches the normal curve as sample sizes get larger.

Key Terms

random sample	outcome
inferential statistics	independent
unbiased	binomial expansion
consistent estimator	Pascal's triangle
single-sample model	probability distribution
two-sample model	addition rule
confidence interval	mutually exclusive
null hypothesis	multiplication rule
probability	replacement
event	conditional

Problems

9-1   I am going to toss five pennies at a time 1000 times. Find the *expected* probability of getting:

(*a*) Five heads

(*b*) Four heads and one tail

(*c*) Three heads and two tails

(*d*) Two heads and three tails

(*e*) One head and four tails

(*f*) Five tails

(*g*) Add the probabilities across (*a*) to (*f*); do they sum to 1.0?

9-2   I believe the voter registrations show .60 females to .40 males. When I take a sample of 10 people from the voter rolls, I find that 4 of them are females and 6 males. How likely is 4 females and 6 males, i.e., what is its probability, if the proportions are in fact .60 and .40?

9-3   Of 7 persons selected at random from a university directory 5 live on campus and 2 do not. If my hypothesis is true that half of the students live on campus, what is the probability of getting a sample of 7 students, 5 of whom live on campus and 2 live off campus?

9-4 I am drawing a sample of 6 people from the telephone directory. Assuming that half of the people in the city are "liberal" and half "conservative":

(a) Construct a histogram to show the proportion of samples that will be at each of the possible outcomes, i.e., what proportion will have 6 conservatives, 5 conservatives and 1 liberal, etc.?

(b) I have taken a sample of 6 persons and found 5 to be liberals and 1 conservative. If the assumption of $p = .50$ for liberals is true, how likely is it to get a sample of 6 people with 5 liberals? 6 liberals? 5 or *more* liberals?

9-5 I draw a card from a standard deck of playing cards. What is the probability that it will be:

(a) A red card?

(b) A face card?

(c) An ace?

(d) A king or a queen?

(e) A face card or a red card?

9-6 In a single throw of two dice what is the probability:

(a) That each die will be a 6?

(b) That one die will show a 2 and the other a 3?

9-7 I have a jar with 10 red and 10 white poker chips in it. I draw two chips.

(a) What is the probability that both will be white, with replacement?

(b) What is the probability that both will be white, without replacement?

9-8 I have a group of students made up of 10 Blacks (6 women and 4 men), 20 Caucasians (half men and half women), and 10 Orientals (7 men and 3 women). If I randomly draw one person out of the group, what is the probability of getting a Caucasian or female?

9-9 In Prob. 9-8 what is the probability of drawing a single person who is a North American minority (Black or Oriental) or male?

10

Probability and Inference: The One-Sample Test

In Chap. 9 we saw how probability distributions can be devised and used to indicate how likely a sample outcome is, given a population value. We also saw that as n, the sample size, increases, the shape of the probability distribution of $(p + q)^n$ begins to look like the normal curve.

In this chapter we shall deal with larger samples (30 cases or more) and use the normal curve as our probability distribution. Although up to this point we have dealt with data in which individual observations could take one of only two alternatives (heads/tails, male/female), in problems in the social sciences individuals often take any of a range of values rather than just one of two. Nevertheless, the basic procedures illustrated with coins also apply to problems dealing with wider ranges of outcomes; i.e., we can set up a distribution of outcomes based on many samples randomly drawn from a given population and then determine the likelihood of getting a particular sample outcome for this population. The basic difference is that for coins and similar situations with only two alternative *proportions of samples* are our basic data, and we make a frequency distribution from these proportions; with situations that may take any one of a range of values we use sample *means* as our basic data, and our distribution will be made up of these sample means.

An example will illustrate the analogy. In Chap. 9 we took four coins and tossed them as a group repeatedly. We then could count the number of heads each time, make a proportion out of the number of heads by dividing it by 4, and plot these proportions on a frequency distribution. This distribution allowed us to establish the frequency of occurrence for various randomly determined outcomes in tossing four coins. We found that four heads or four tails is a rather infrequent event compared with the outcome of two heads and two tails.

Now suppose from the population of ten-year-old children I draw a random sample of four children, determine their IQs, and compute a mean IQ for the sample. I repeat this process 10,000 times just as I did with the coins. The means will be different from sample to sample, but we can make a frequency distribution of these sample means and on that basis decide the likelihood of getting a randomly drawn sample that has a given mean. In fact we shall see that our distribution of sample means will look something like the familiar bell-shaped curve, that we can compute a standard deviation for it, and that we can determine the number of samples that will appear between different points under the curve. What we have just observed follows from the *central-limit theorem*, which states that as sample sizes increase, the shape of the distribution of means becomes increasingly normal. Indeed, as sample sizes reach 30 cases, the curve becomes a good approximation of the normal curve whether the population is normal or not.

Whenever we deal with data which represent characteristics of samples rather than characteristics of individuals, we call the resulting frequency distribution a *sampling distribution*. In the above example of means from samples of ten-year-old children, if we plotted these means into a frequency distribution, we would have a sampling distribution of means.

Standard Error of the Mean

Since we treat these distributions much the same as any normal distribution, let us begin by computing the standard deviation for our sampling distribution of means. It is found by the formula

$$\sigma_{\bar{X}} = \frac{\sigma_X}{\sqrt{N}} \qquad (10\text{-}1)$$

where $\sigma_{\bar{X}}$ = standard deviation of sampling distribution of means
σ_X = standard deviation of population of X scores
N = number of cases in a sample

In order to distinguish this standard deviation from the one based on data from individual cases, we give it a different name. We call it the *standard error of the mean*, but it is simply the standard deviation of a distribution of a very large number of means of samples which were randomly drawn from a given population.

What is the mean of our sampling distribution of means? If a sampling distribution is indeed based on an infinite number of equal-sized samples from the population, the mean of the sampling distribution will be the mean of the population.

Solving Problems with Sampling Distributions

Let us apply what we have just observed. Suppose again we are dealing with the population of ten-year-old children. The population mean IQ for these people is 100, and the standard deviation is 15. Suppose I am going to study a group of 36 ten-year-olds randomly selected from the population. What would a sampling distribution based on samples of 36 cases look like? The mean would be 100, and the standard deviation would be

$$\sigma_{\bar{X}} = \frac{15}{\sqrt{36}} = 2.5$$

From our knowledge of the normal curve we immediately know that 68 percent of randomly drawn samples of 36 ten-year-olds will have mean IQs between 97.5 and 102.5. (This is 1 standard error above and 1 below the mean of the sampling distribution.) Figure 10-1 illustrates this point. Also we would expect only 2 percent of random samples of 36 cases each to have mean IQs beyond 105, or 2 standard errors above the mean of the sampling distribution. It should be clearly noted here that to this point we have not actually selected a single sample from our population. We need to know only the population mean, the population standard deviation, and the number of cases we expect to select in a sample we wish to study.

Figure 10-1

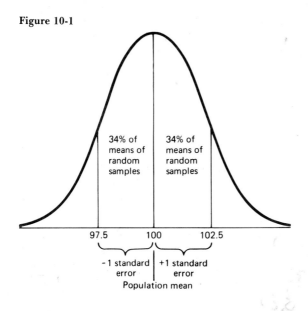

Knowing all this about our population of ten-year-old children, I decide to teach reading using a special method which I have not yet published. I want to use a sample of ten-year-olds as my subjects, and I want them to be "typical." I therefore randomly select 36 ten-year-olds and measure their intelligence. Their mean IQ is 107. Is it likely that I could randomly get such a sample from the population of children ten years of age? (Assume 100 to be the population mean IQ.)

We apply two bits of knowledge to the solution of this problem. We want to know (1) where in a sampling distribution a sample mean of 107 comes, and (2) what portion of sample means would fall beyond this point in the distribution. The first we find by applying the z-score procedure; the second comes from reference to Table C-2 of areas under the normal curve. The mean of 107 has a z score

$$z = \frac{\bar{X} - \mu}{\sigma_{\bar{X}}}$$

(10-2)

where $\bar{X}$ = mean of sample

μ = mean for total population

$\sigma_{\bar{X}}$ = standard error of mean (standard deviation for this sampling distribution of means)

We then find the necessary information

$$z = \frac{107 - 100}{2.5} = 2.8$$

Our sample mean is 2.8 standard errors away from the mean of our sampling distribution; from Table C-2 we find that much less than 1 percent of randomly drawn samples is expected to have means as large as 107. Can we say that we have a "typical" sample of ten-year old children? Our sample is in fact unusual among randomly selected samples from our population.

Sometimes we select a sample from a population, give the sample some kind of special treatment, and then ask whether it is still a sample from the original population. The assumption is that our treatment has altered the characteristics being observed in such a way that the sample now may no longer be representative of its original population. We therefore test the sample mean, after treatment, against the population mean to see whether our sample is now a rather rare one for a population like ours.

Suppose now that in your dark and musty cellar you have formulated a vitamin which appears to make white mice much more alert mentally. You believe it is ready to be tried out on a group of human subjects. A random sample of 100 cases is drawn from our population of ten-year-olds, and they are given the vitamin, followed 10 minutes later by an intelligence test. We find that the average IQ for the sample is 104. For a population of IQs we previously found the mean to be 100 and the standard deviation to be 15. Is our sample a likely one from this population?

The standard error for a sampling distribution of means of 100 case samples would be

$$\sigma_{\bar{X}} = \frac{15}{\sqrt{100}} = 1.5$$

and to find the number of standard errors our sample is from the mean of the distribution, we compute a z value

$$z = \frac{104 - 100}{1.5} = 2.66$$

From Table C-2 we find that less than 1 percent of the samples randomly drawn from our population has means as great as 104. Has our vitamin really changed the mental ability of our sample? Chances of randomly selecting this kind of a group are indeed remote, and so we certainly would want to take a closer look at our vitamin.

In the above examples we looked at sample values in terms of their similarity to the population values. How divergent, in terms of z scores, can a sample value be before we say it is too unlikely to be considered a sample from the basic population? Statisticians in the behavioral sciences have established two z-score points beyond either of which the sample is so unlikely that we do not believe it is from the population under consideration. These points are the z values 1.96 and 2.58. If we consult Table C-2, we see why these points have been selected. First, ±1.96 leaves 2.5 percent of the sample means in each tail of the curve (a total of 5 percent), while ±2.58 leaves 0.5 percent in each tail (a total of 1 percent). These values are known as the *95 percent level of confidence* and the *99 percent level of confidence*, respectively, because they lay off the range within which 95 and 99 percent of the sample means will fall, given the population mean. We call the 5 and 1 percent areas *alpha levels*. These criterion points are illustrated in Fig. 10-2.

In determining whether a sample is to be rejected as randomly drawn from the defined population, we are free to use either criterion point. We begin by hypothesizing that our sample mean is *not* different from the population mean at, say, the 5 percent level. As noted in Chap. 9, such a statement of no difference is called a null hypothesis, H_0. Then we test the difference to see whether the z value is indeed greater than (in this case) 1.96. If it is, we believe the sample mean to be sufficiently different from the population value to say that our sample

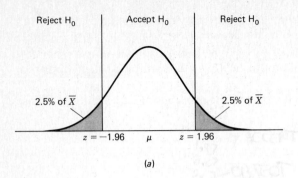

(a)

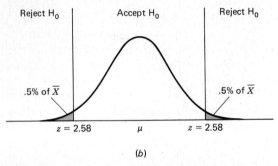

(b)

Figure 10-2 (*a*) **5 percent criterion;** (*b*) **1 percent criterion.**

is not a randomly selected group from the defined population; i.e., we
reject the null hypothesis.

In the example where we applied a vitamin to our sample of ten-
year-olds, our null hypothesis could be that there is no difference at the
1 percent level between our sample mean and the mean of the

Figure 10-3

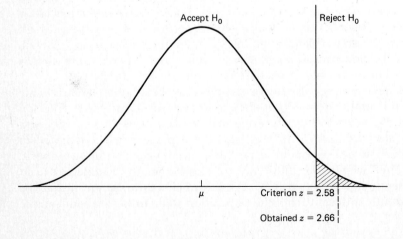

population. Since the difference produced a z value of 2.66, we reject the null hypothesis because a z value of 2.66 is *beyond* the table value of 2.58, the point that cuts off 1 percent of the area in the tails of the curve. In other words, we believe that it is unlikely such a sample would be randomly selected from our population.

Our decision conditions are laid out in Fig. 10.3. It appears likely that the sample we measured came from a population whose mean is greater than 100, the value against which we tested our sample. Here we believe that H_1, the alternative to H_0, is probably true.

Exercises

10-1 For all high school seniors in Forestville the average height is 68 inches and the standard deviation is 4 inches. I have a group of 49 students I wish to study in reference to the relation of nutrition and height. My sample has a mean of 66 inches. At the 95 percent confidence level can I consider that this is a random sample from the senior class in reference to height?

10-2 Suppose I have a group of 36 children with a mean height of 54 inches. Is it likely this is a sample from a population whose mean is 52.1 inches with a standard deviation of 3 inches? Test at the 95 percent confidence level.

10-3 The mean age for all voters in Mystic County is 42.4. The standard deviation 6.3 years. I have interviewed a sample of 49 voters whose mean age was 38.1. At the 99 percent level of confidence, is my sample a random sample from the county voters?

Type I and Type II Errors

In choosing the 1 percent or 5 percent confidence level we are faced with a dilemma. Since 1 percent of the samples randomly drawn from our population will produce z values beyond 2.58, if we reject *all* such samples we run the risk of making a mistaken conclusion in 1 out of every 100 replications of our study, since by sheer chance 1 out of 100 samples will indeed be this remote. We call this a *Type I error.* For example, suppose that I have a population of ten-year-olds and that the mean height of my population is 54 inches and the standard deviation 4 inches. For a sample of 64 children I find the standard error to be .50 inch. I know that by chance alone I could find a sample from my population with a mean greater than 55.29, that is, 54 inches + 2.58(.50), because 1 percent of random samples will be beyond $\pm 2.58 \sigma_{\bar{x}}$. Yet if I get a sample whose mean deviates this much from my population value, I have agreed to reject that sample based on the 99 percent level of confidence, as not being from my population. Using this cutoff point, I will reject 1 percent of the random samples even though they are legitimate samples from my population. These cases represent Type I errors.

A Type II error occurs when we accept the null hypothesis when it is false. When we conclude that a sample is from our specified population but it is in fact from another population, we commit a Type II error. For example look again at the height of our ten-year-olds. Our figures revealed that at the 99 percent level of confidence we would accept all samples with a mean height between 55.29 and 52.71, that is, $\pm 2.58\sigma_{\bar{X}}$. Now suppose I get my children mixed up, select a sample of 64 eleven-year-olds, and find their mean height to be 54.98 inches. At the 99 percent level of confidence, I may accept this sample as being from my ten-year-old population when it is not. This is a Type II error.

Here is another example of both error types. The average score for sixth-graders in my town on an arithmetic achievement test is 32 problems correct. For a sample of 50 sixth-graders I find the standard error to be 1.2 problems. I expect that by chance alone I will draw random samples (at the 95 percent confidence level) with means between 29.65 and 34.35, that is, $32 \pm 1.96(1.2)$. However, 5 percent of random samples from my population of sixth-graders will produce test means beyond these limits. Therefore, I could select from my population a random sample whose mean score was, say, 36.2. But using the 5 percent criterion I would reject that sample. When I falsely reject a sample from my population, I commit a Type I error.

On the other hand, suppose I had a sample of fifth-graders who got a mean arithmetic score of 29.85. Knowing nothing about this sample but their mean score, I might well accept them as being a random sample from my population when in fact they are not; they are from a population more skilled in arithmetic than my fifth-graders. This would be a Type II error.

Statisticians often speak of the *power* of a statistical test of the null hypothesis. The power of a test is the probability that the null hypothesis will be rejected when it is indeed false. Power can be written as

$$\text{Power} = 1 - \text{probability of Type II error}$$

This formula means that the power of a statistical test increases as the probability of the null hypothesis being correctly rejected decreases.

How can we increase the power of our test? Two ways are often cited: (1) increase the sample size N, and (2) use the .05 level as the criterion for rejecting H_0. Recall that the standard error is equal to $\sigma_X/\sqrt{N}$. Therefore, as N gets larger, the standard error gets smaller. As a consequence, as N gets larger, the range within which we shall accept the null hypothesis gets narrower, which decreases the likelihood of making a Type II error. Therefore, by increasing sample size we increase the value of $1 -$ the probability of making a Type II error and hence increase the power of our test. For example, suppose the mean spelling score on a given test for eight-year-olds is 30 and the standard deviation is 10. With samples of 64 cases the standard error $\sigma_x/\sqrt{N}$ will be 1.25 and the critical limits at the 95 percent confidence level will be 27.55 to 32.45, that is, $\pm 1.96(1.25)$. Any sample, even one not from my population, that has a mean score within this range will be accepted.

If I increase the sample size to 100, the standard error becomes 1.00, and at the 95 percent confidence level the critical limits will be 28.04 to 31.96, that is, $\pm 1.96(1.00)$. With the larger sample I have reduced the width of the critical limits, thus reducing the likelihood of accepting a sample from another population; i.e., I have reduced the likelihood of a Type II error and increased the power of my test.

The criterion for rejecting the null hypothesis is important here. If we set the 5 percent level as the point for rejecting sample values, we increase the likelihood of rejecting samples that are in truth random samples from our population; that is, 5 percent of all samples will fall by the wayside. On the other hand, the 5 percent level decreases the likelihood of accepting a sample as being from our population when in truth it is not. (Why?)

The 1 percent criterion reverses what we said in the last paragraph. Since only 1 out of 100 samples from our population will be erroneously rejected, this reduces the likelihood of rejecting samples which should in truth be accepted. However, this 1 percent point increases the likelihood of accepting samples which are from other populations and should be rejected; this decreases the power of the test. Figure 10-4 is a graphic description of this problem. Study this figure carefully.

The selection of the 5 percent or the 1 percent limit depends upon the kinds of errors we can best tolerate. Can we afford to reject 5 percent of the samples from our own population in an effort to reject more samples from other populations? If so, the 5 percent level is appropriate. But if we wish to reject as few samples as possible from our population and can afford to accept erroneously an occasional sample from other poulations, then the 1 percent level is to be chosen.

Type I errors occur when we reject a null hypothesis H_0 when in fact it is true. Type II errors occur when we accept H_0 when it is actually false.

Exercises

10-4   An investigator randomly selected a sample of rural residents, gave them a test on knowledge of nutrition, and found that they scored near the national average; hence he accepted H_0. He concluded that rural residents have at least average knowledge of nutrition. Later it was

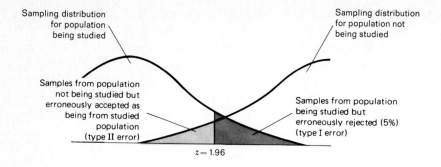

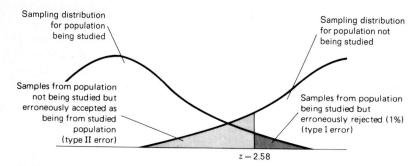

Figure 10-4

discovered that 60 percent of his sample had moved to the rural area from the city within the last 3 years and that the finding may well pertain to urbanites. Did the investigator make a Type I or Type II error?

10-5 A sportswriter found data to indicate that the average number of hits per baseball game for high school boys across 20 central states was 3.6. Since his home town's high school team averaged 2.7 hits, he attacked the local baseball program and claimed that the team did not belong in competition in the region. In defense the local athletic director found data on 200 regional teams, noted that 9 of them had hit records of 2.7 or less, and agreed that the local figures were low but within the range of the region. Did the sportswriter make a Type I or Type II error?

10-6 Reading that the average adult in American cities has 3.2 upper respiratory infections (URI) per year, a physician believed that this is too high and that the figure should be near 1.8. He sent an assistant out to get a sample of 100 adults and to find out how many URI they had in the last 12 months. The sample came from persons at "random" who visited a suburban shopping center. The average number of URI was 2.7. The physician concluded that the city-dweller figure indeed was probably near 3.2. Is Type I or Type II error involved here?

One-Sample t Test when σ Is Unknown

So far we have dealt with data for which we know the population standard deviation. Suppose we do not have this information. How do we compute the standard error? If we do not know the population standard deviation, we must estimate this parameter from sample data. The sample standard deviation s_X (calculated with $N - 1$ in the denominator) is our best estimate of the population value, and so $s_{\bar{X}}$ becomes

$$s_{\bar{X}} = \frac{s_X}{\sqrt{N}}$$

(10-3)

At this point we set up a formula very much like that for z, but to indicate that we are using sample estimates for σ we use the symbol t in place of z

$$t = \frac{\bar{X} - \mu}{s_{\bar{X}}}$$

(10-4)

and for large samples (30 or more cases) the resulting figure is evaluated as in Table C-2. In order to use this formula s must be computed as an estimate of the population standard deviation as described in Chap. 5.

Here is an example of what we have just done. Suppose I have a sample of 50 ten-year-old boys. I have a growth theory that says a combination of vitamins and exercise for 6 months will accelerate their growth in height by 1 year; i.e., if my theory is true, at the end of 6 months these boys should have a mean height greater than the mean for the population of boys who are 10.5 years old. I put the 50 boys through the treatment and find their mean height in 6 months to be 55 inches. The standard deviation is 4 inches. I also found data that said that the mean for 10.5-year-olds was 53 inches. Are my 10.5-year-olds, in regard to height, a random sample from the population of 10.5-year-olds? Here the standard error is

$$s_{\bar{X}} = \frac{4}{\sqrt{50}} = .57$$

and the t test goes

$$t = \frac{55 - 53}{.57} = 3.5$$

Table C-2 says that a value of 2.56 leaves 1 percent in the tails of the distribution, but since our value is larger than 2.56, our conclusion is that H_0, which says there is no difference between our sample mean and the population mean, is rejected at the 99 percent confidence level. We believe, with considerable certainty, that our sample is not a random sample from the population of 10.5-year-olds.

Confidence Intervals

So far we have been acting as if we knew the population mean and possibly the standard deviation. From these we decide whether a given sample is from a particular population. But often we do not know the population parameters and would like to know where the mean of the population is likely to lie. The following is the procedure for locating that mean.

We select a sample from the population and compute a standard error of the mean, using the sample estimate of the population standard deviation. We noted that only 5 percent of random sample means fell beyond $\pm 1.96\sigma_{\bar{x}}$. Therefore 95 percent of all means of random samples will fall *within* $\pm 1.96\sigma_{\bar{x}}$ of the population mean. Now I have taken a sample and have computed its mean and the estimated standard error of the mean $s_{\bar{x}}$. The mean of the population at the 95 percent confidence level should be within $1.96s_{\bar{x}}$.

But we do not know whether our sample mean is above or below the population mean. Therefore, we add to our sample mean $1.96s_{\bar{x}}$ and also subtract from it $1.96s_{\bar{x}}$. The resulting values (sample $\bar{X} \pm 1.96s_{\bar{x}}$) give the range within which (at the 95 percent confidence level) we expect the population mean to lie.

Let us see how this works with some real data. Suppose I have a sample of 36 six-year-old children. Their mean mental age (MA) is 74 months; the sample standard deviation is 4 months of mental age. Within what interval of scores (at the 95 percent confidence level) is the population mean likely to be?

First we compute the estimated standard error of the mean

$$s_{\bar{x}} = \frac{s_X}{\sqrt{N}}$$

$$= \frac{4}{\sqrt{36}}$$

$$= .67 \text{ month MA}$$

Now we multiply the $s_{\bar{x}}$ by the number of standard deviations (in this case standard errors) it takes to cut off 5 percent of the observations (in this case means of samples) in the tails of the curve

$$s_{\bar{X}}(1.96) = .67(1.96) = 1.31$$

Our population mean (at the 95 percent confidence level) should be within a range of ± 1.31 months of mental age from our sample mean, or 74 ± 1.31. That provides us a score interval of 72.69–75.31. This is the range within which we expect the population mean to lie, at the 95 percent level of confidence. We call this range the 95 percent confidence interval, and the procedure is called an *interval estimate*.

The 99 percent confidence interval is managed in the same way. Instead of using $\pm 1.96 s_{\bar{X}}$, we would use $\pm 2.58 s_{\bar{X}}$. This procedure provides us with the score range within which we could expect the population mean to fall at the 99 percent confidence level. Here is an example. Suppose I have collected a random sample of 49 regularly employed women who teach elementary school in Indiana. I measure their waistlines and find the mean to be 31.5 inches and the standard deviation to be 1.8 inches. The 99 percent confidence interval for the mean for waist measurements for the population is found as follows:

$$\bar{X} = 31.5$$

$$s_X = 1.8$$

$$s_{\bar{X}} = \frac{1.8}{\sqrt{49}} = .26$$

The 99 percent confidence interval is the mean plus or minus 2.58(.26) = .67, so that the 99 percent confidence interval is

$$31.5 \pm .67$$

$$= 30.83 \text{ to } 32.17$$

This is the range within which we expect the mean of the population to be, based on the 99 percent confidence criterion. The procedure is illustrated in Fig. 10-5, which shows that our sample, at the 1 percent criterion, could be from a population whose mean is as low as 30.83. We also could get this sample from a population whose mean is as high as 32.17. Of course, the sample could also come from a population whose mean is anywhere between these extremes.

The above procedure does not locate the mean exactly but gives us some idea where the population mean is likely to be. This information can be very useful and much less expensive and more manageable than measuring the entire population—a task which is often impossible.*

---

*The range within which this population mean is likely to fall is sometimes called the *fiducial limits*.

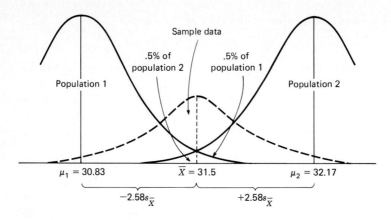

Figure 10-5

Exercises

10-7   From a random sample of 49 freshmen from a large university I wish to draw a conclusion about the mean score of the freshman class on a spelling test. I administer the test to the 49 students. Their mean is 27 words correct; the standard deviation is 5.

    (*a*) What is the 95 percent confidence interval for the class mean?

    (*b*) What is the 99 percent confidence interval?

10-8   I have a randomly selected group of 81 eleven-year-olds from an urban school system. I ask them to cross out as many vowels in a printed paragraph as possible in 2 minutes. The sample mean is 36 and the standard deviation is 6.

    (*a*) What is the 95 percent confidence interval for the population mean?

    (*b*) Make an interval estimate at the 99 percent level of confidence.

---

t Distributions

We have been talking about a sampling distribution of means based on an infinite number of randomly selected samples of at least 30 cases each. We have seen that the means of such samples when plotted produce a normal distribution. Suppose that our samples are composed of fewer than 30 cases, say 20 or even 15 cases each. How would this affect the shape of the sampling distribution?

When samples contain fewer than 30 cases, a noticeable change in the shape of the sampling distribution begins to appear. The nature of

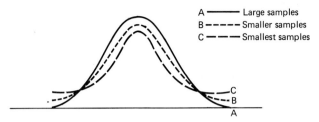

Figure 10-6

the change produces a curve which with smaller sample N's becomes increasingly leptokurtic, with larger areas in the tails of the curve. In other words, as sample sizes decrease, the sampling distribution of their means becomes more pointed in the middle and has relatively more area in its tails. Such a distribution is known as the *t distribution* or *Student's t distribution*.* Figure 10-6 compares the normal curve A with two t distributions B and C. Curve A is obtained by plotting means of large samples, curve B by plotting means of smaller samples, and curve C by plotting means of even smaller samples than B.

It should be emphasized that the t distribution is not a single curve but a *whole family* of curves, the shape of each being a function of sample size. The smaller the sample the larger the portions of the area under the curve that appear at the extremes of the distribution and the more peaked the middle of the curve; whereas the larger the sample size, up to about 30 case samples, the more nearly the t distribution approaches the shape of the normal curve (beyond 30 case samples the two distributions become very much alike).

Since we know the nature of Student's distribution, we can use it to solve problems just as we applied the normal curve; i.e., we locate a point on the base line in standard-error units and then find the percent of the area under the curve that lies beyond that point. Since the distribution is made up of means of many samples, the area beyond our point will represent a certain portion of sample means that are greater, or smaller, than the one we are considering.

Table C-3 is based on areas under the curve for t distributions given various sample sizes. But since we usually apply t to problems involving the null hypothesis and accept or reject the hypothesis that is at the established 95 or 99 percent confidence level, Table C-3 provides only the t value for those points.

Since the shape of the sampling distribution changes with different sample sizes, Table C-3 reports different values for different sample N's. If relatively more of the area under the curve is in the curve X tails

---

*Some years ago companies in England required employees to publish materials only under the company name, so company personnel often used pen names. W. S. Gosset used the name of "Student" when he published his first work with t in 1908.

than in the curve Y tails, then beginning at the mean we must move farther along the base line to locate the point in X beyond which 5 percent of the sample means will fall than we would in curve Y. Figure 10-7 is an illustration of this point. Therefore, if the samples represented in our sampling distribution are made up of 10 cases, it will take 2.262 standard errors to reach the point on the curve beyond which 5 percent of the sample means will fall, whereas if we have samples of 29 cases, it takes only 2.048 standard errors to reach the point beyond which 5 percent of the sample means will appear.

Students who have consulted Table C-3 have already found that the t values reported above actually correspond to the rows labeled 9 and 28 (not 10 and 29) case samples. The explanation lies in the *degrees of freedom*. The degrees of freedom associated with a given sample are determined by the number of observations free to vary. Suppose boys in a physical education class are shooting baskets with a basketball. Student A made 10 baskets. Can we tell from this how many baskets he has missed or how many he has attempted? No. But suppose we know that each student was allowed 15 tries. Now we can determine the number of misses, since the number of successes and the total number of attempts determine the number of misses. Only two of our three conditions are free to vary independently of the third condition. If we know the number of misses and successes, the total number of shots can be only one value; if we know the number of successes and the total number of attempts, the number of misses can be only one value. Thus, only two of the conditions are free to vary independently of the third. The third condition is always determined once we know the other two. We therefore have one restriction on our data and two degrees of freedom.

When we deal with a sampling distribution of means, we have a similar situation. For simplicity suppose that we have only two samples in our sampling distribution of means. The mean of a sampling distribution is the population mean, and the sum of deviations around the mean of any distribution is zero; i.e., for our sampling distribution,

$$(\bar{X}_1 - \mu) + (\bar{X}_2 - \mu) = 0 \quad \text{or} \quad \Sigma(\bar{X}_i - \mu) = 0$$

Figure 10-7

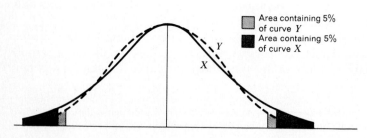

Area containing 5% of curve Y
Area containing 5% of curve X

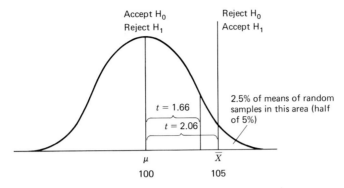

Figure 10-8

If we know $\bar{X}_1$ and μ, $\bar{X}_2$ can be only one value; or if we know $\bar{X}_2$ and μ, $\bar{X}_1$ can be only one value. Thus, since the distribution mean must be a given value, we have one restriction on the freedom of observations to vary and our degrees of freedom are $N - 1$.

Returning now to our problem with t, we look up t values in Table C-3 under degrees of freedom, which in the above problems is $N - 1$. (We shall see later that degrees of freedom may be other than $N - 1$.) Thus in our sample of 29 cases, we have $N - 1 = 28$ degrees of freedom; for 10 cases, 9 degrees of freedom, and these determine the row of Table C-3 which we consulted to get the 5 percent value of t.

The t formula (10-4) will be used to assess the position of $\bar{X}$ relative to μ. With large samples we assessed our outcome by relating it to areas of the normal curve as given in Table C-2, but for small samples the t values will be evaluated by use of Table C-3.

Suppose we have randomly drawn a sample of 25 six-year-olds, found their IQs, and found the mean to be 105 and the standard deviation to be 15. Is it likely that this is a random sample from the population whose mean is 100? We solve the problem with t as follows:

$$s_{\bar{X}} = \frac{s_X}{\sqrt{N}}$$

$$= \frac{15}{5} = 3$$

$$t = \frac{105 - 100}{3} = 1.66$$

Consulting Table C-3, with $N - 1 = 24$ degrees of freedom, we find the t value necessary for the 5 percent level to be 2.064. Since our computed value (1.66) is less than this amount, we accept the hypothesis that our sample could well be from the population with a mean of 100. The pertinent sampling distribution is shown in Fig. 10-8.

Estimating Population Mean from Sample with t

Up to this point we have been trying to determine whether a sample is from a population with a given mean. The procedure can also be reversed, in which case we have a sample mean and want to know the mean of the population from which the sample came. Unfortunately we cannot determine the exact mean for the population, but we can state the limits within which the population value is most likely to occur. In other words, we state an interval of values within which we believe the population mean will fall and the probability of finding the population value within that interval. We did this earlier with large samples where we called the procedure the interval estimate. The steps for making interval estimates with t are like the ones we used with z except that we consult a different table for significant areas.

Here is an example to illustrate the procedure. Suppose I have a sample of children randomly drawn from the population of lower income families. I tabulate the number of childhood diseases these children have had, but I wish to say something about the mean number of diseases for the population of lower income children. For my sample I find the mean to be 6 diseases and the standard deviation to be 2.1, and I have 25 cases in the sample. The first step is to find the standard error of the mean

$$s_{\bar{X}} = \frac{2.1}{\sqrt{25}}$$

$$= \frac{2.1}{5} = .42$$

Consulting Table C-3, with $N - 1 = 24$ degrees of freedom, I find the 95 percent level of confidence to be 2.064 standard errors above and below the mean. The 95 percent confidence interval is found by multiplying 2.064 by .42 and adding this to, and subtracting it from, the sample mean. Our 95 percent confidence interval then becomes 6 $\pm$ .867 = 5.133 to 6.867.

Exercises

10-9   I have samples of the following sizes and wish to do t tests against a given population value. For each sample list the number of standard errors above and below the mean that identify the range of the 95 percent confidence level.

(a) $N = 9$ (b) $N = 13$ (c) $N = 20$ (d) $N = 28$

10-10   From the results of Prob. 10-9 we can make an observation. As the degrees of freedom increase, what happens to the critical value of t?

10-11 I have a sample of 17 ten-year-olds who have a mean height of 52 inches with a standard deviation of 9 inches. Test at the 95 percent level the hypothesis that this is a random sample from a population whose mean height is 54.3 inches.

10-12 A social worker studying income and the amount of alcohol consumed by inner-city residents found that during the week of the study his 27 subjects consumed an average of 31.7 ounces of alcohol. The standard deviation was 5.8 ounces. Could this be a sample from a population in which the mean is 24.3 (at the 99 percent confidence level)?

10-13 I have a sample of 20 children whose mean height is 61 inches with a standard deviation of 4.5 inches. What is the range (at the 95 percent confidence level) within which we shall expect to find the mean height of the population from which my sample came?

10-14 A sample of 26 college freshmen have a mean number of errors on a finger-maze problem of 17.6 with a standard deviation of 3.2. What is the range within which we shall expect to find (at the 99 percent confidence level) the mean errors for the population from which my sample came?

Assumptions with t

When we apply the *t* test, several assumptions are made:

1 The population distribution of *X* is normal.

2 The *X*'s are a random sample from the population being studied.

3 The population standard deviation is unknown and must be estimated by the sample standard deviation.

4 The null hypothesis is true, an assumption the investigator may or may not wish to disprove, depending on the argument of the investigation.

Assumption 1, of normality, is of more importance in theory than in practice. If samples are large, the sampling distribution is only slightly affected by skewed populations, and even with small samples the area of rejection α is not greatly distorted.

The One-Tailed Test

When we test the null hypothesis H_0, we assume that our sample is a randomly selected group from a population whose mean is μ. The sample mean $\bar{X}$ is expected to be in the vicinity of the population value, but it may be either larger or smaller than μ. For this reason we use the two-tailed test, dividing the critical area (.05 or .01) into half in the left-hand tail and half in the right-hand tail. But consider H_1, the alternative hypothesis to H_0. Here we may wish to ask whether the sample comes from a population the mean of which is *larger* than the specified value. The hypothesis is then

$$H_1: \mu > \text{specified value}$$

We call this a *directional hypothesis*. Here all the critical area α is in the right-hand tail. This is shown in Fig. 10-9a. If we believed our sample came from a population whose mean is *less* than the specified value, the alternate hypothesis is then

$$H_1: \mu < \text{specified value}$$

as shown in Fig. 10-9b.

Our procedure for one-tailed tests follows formula (10-2) or (10-4) with one exception. If α is in one tail, it will take fewer standard errors to reach that point on the base line of the curve. The criterion values of z for large samples (or when σ is known) therefore become 1.64 for the .05 level and 2.33 for the .01 level. With small samples the criterion

Figure 10-9

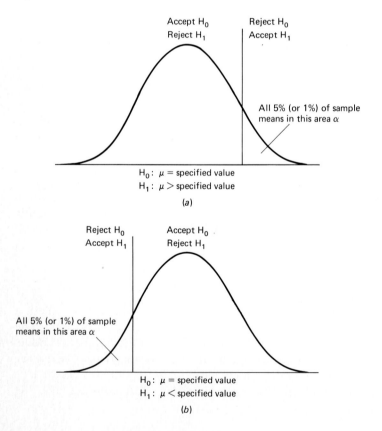

Accept H_0
Reject H_1

Reject H_0
Accept H_1

All 5% (or 1%) of sample means in this area α

$H_0: \mu = \text{specified value}$
$H_1: \mu > \text{specified value}$

(a)

Reject H_0
Accept H_1

Accept H_0
Reject H_1

All 5% (or 1%) of sample means in this area α

$H_0: \mu = \text{specified value}$
$H_1: \mu < \text{specified value}$

(b)

values will be found in Table C-3, but we must look under different column headings. The column that reports 5 percent in *one tail* is labeled .1, the 10 percent level. In the two-tailed test this column would have 5 percent in each tail. Since we are now dealing with only one tail and we want 5 percent in that tail, the column labeled .1 shows the t value that cuts off 5 percent in that one tail. Similarly the column in Table C-3 that identifies 1 percent in a single tail is labeled .02.

For example, a scientist studying learning has a procedure which she believes can only improve rapid acquisition of data by the learner. Since children can only improve under her system, we have a one-tailed test. She applies the method to teaching mathematics to 150 fourth-grade children, after which she administers a standardized arithmetic test. The national mean for fourth-graders is 36 items correct, the standard deviation is 4 items. For the experimental group the mean is 38.3 items. The null hypothesis is H_0: $\mu = 36$, and the alternative hypothesis is H_1: $\mu > 36$. The one-tailed test is

$$z = \frac{38.3 - 36.0}{4/\sqrt{150}} = 7.04$$

With large samples, the population values known, we use the normal-curve values of 1.64 at the .05 criterion and 2.33 at the .01 criterion. Since our calculated value exceeds both, we may wish to reject H_0 and accept the alternative hypothesis.

$$H_1: \mu > 36.0$$

On a pilot study the same scientist selected a sample of 25 fourth-grade children, applied her learning technique to them, and administered a standardized test of arithmetic, the "national" mean of which was 36 items correct. The 25 children had a mean of 37.4 and a standard deviation of 3.2. The one-tailed test here is a t test because we are estimating the population standard deviation and the sample is small. The test becomes

$$t = \frac{37.4 - 36.0}{3.2/\sqrt{25}} = 2.19$$

The criterion value must now be read from Table C-3. With $N - 1 = 24$ degrees of freedom we read in the column labeled .1 the value of 1.711 for the 5 percent criterion. If our scientist were using the 1 percent criterion, the criterion value would be in the column labeled .02, where we find that 2.492 corresponds to 24 degrees of freedom. In the pilot study

$$H_0: \mu = 36$$

is rejected but the alternative looks plausible at the .05 criterion; the reverse is true, however, at the .01 criterion.

Exercises

10-15   A physiologist has devised a gripping exercise which he believes will decrease writing fatigue. It will either have no effect or it will decrease fatigue. Twelve-year-old children complain of writing fatigue after 41.4 minutes of writing. A sample of 25 twelve-year-olds who have used the exercise for a week are asked to write until fatigue sets in. Their mean time to fatigue was 43.6 minutes; standard deviation 4.2 minutes. Do a one-tailed test to test the hypothesis that the children are from a population with a mean greater than 41.4 at the .05 criterion.

10-16   A sociologist has reason to believe that working in a factory will make managers more favorable toward labor unions. He takes a sample of 15 managers and puts them on the assembly line for 4 weeks. They then take a test of attitudes toward unions. Managers as a group have averaged 21.7 points on the test (high scores are favorable toward unions). The sample of 15 managers averaged 25.1, with a standard deviation of 2.4. At the .01 criterion, are these managers from a population more positively disposed to unions than managers in general?

Use of t to Test H_0: $\rho_{XY} = 0$

An additional use of the t test is in testing the hypothesis that a given correlation coefficient is based on a sample from a population in which $\rho_{XY} = 0$, where ρ_{XY} (rho) is the population correlation coefficient. In this case the observed r_{XY}, the one we calculate from the raw data collected from our sample of people, produces a value above (+) zero or below (−) zero. We hypothesize that the correlation in the population is indeed zero and that sampling error is the reason our calculated r_{XY} was not zero. Therefore the expected value of r_{XY} is zero.

To test the hypothesis H_0: $\rho_{XY} = 0$ we proceed as follows:

$$t = \frac{r_{XY} - 0}{\sqrt{(1 - r_{XY}^2)/(N - 2)}} \qquad (10\text{-}5)$$

where $r_{XY} - 0$ is the difference between the observed r_{XY} and expected value of ρ_{XY}. The standard error takes a little explaining.

So far we have been using as our standard error for t tests the square root of the estimate of the population variance divided by the degrees of freedom. Now look at the denominator in formula (10-5). The r_{XY}^2 value (as noted in Chap. 7)* is the proportion of variance in variable X associated with variable Y. Then $1 - r_{XY}^2$ is the proportion of variance

---

* A review of Chaps. 7 and 8 may be helpful. (Nothing in this book is intended to be forgotten.)

not associated with variable Y. It is based on the deviation of actual scores around the predicted score and reflects the distribution of scores within the various arrays of the scatter diagram. The portion of variance due to the distribution of these scores within the arrays is $1 - r_{XY}^2$. We divide this portion of the variance by degrees of freedom, $N - 2$, and take the square root to get the standard error for testing H_0: $\rho_{XY} = 0$. This is analogous to the procedure for finding the standard error in formula (10-3).

It can now be seen that the same reasoning that went into formula (10-4) also applies to formula (10-5). In both situations we have an observed sample value and an expected population value. We evaluate the difference by applying the standard error as a yardstick, i.e.; the difference is evaluated in terms of how many standard errors it is equal to. We then go to Table C-3 with our degrees of freedom to see whether we accept or reject H_0.

Here is an example of testing H_0: $\rho_{XY} = 0$. Suppose I have a sample of 25 students randomly drawn from applicants to the Graduate School of Fisheries at Grand Pacific University. I correlate their Graduate Record Examination (GRE) total scores with their first-semester GPAs. The $r_{XY} = .28$. Could this correlation be based on a sample from a population in which the GRE-GPA correlation is indeed zero? To answer this question we apply formula (10-5)

$$t = \frac{.28 - 0}{\sqrt{(1 - .28^2)/(25 - 2)}} = \frac{.28}{.20} = 1.40$$

With $N - 2$ degrees of freedom we go to Table C-3. At the .05 level the table value is 2.069. Since our computed value of 1.40 did not reach the table value, we accept H_0; that is, the correlation of .28 could be based on a sample from a population in which the GRE-GPA correlation is indeed zero. If this is so, we cannot place much confidence in predictions of GPA from these test scores.

Exercises

10-17 I read in the *Journal of Significant Research* that a sample of 15 cases shows the correlation between number of minutes spent in daydreaming per day and creativity to be .42. At the 5 percent criterion is this likely to have come from a population where the correlation is zero?

10-18 I have been studying the relationship between the number of minutes a child spends talking and reading ability. I have a sample of 28 ten-year-olds on whom I found a correlation between these variables to be .31. At the .01 criterion do these children come from a population in which the relationship is zero?

10-19   Calculate the t value for the difference of r_{XY} from zero for each situation.

	r_{xy}	N	t
(a)	.30	10	
(b)	.30	20	
(c)	.30	30	

What is the relationship between the size of sample N and t?

Summary

As the number of events in a sample gets large (about 30) the binomial probability distribution approaches the shape of the normal curve. Indeed at that point the normal curve can be used as a probability distribution. When a very large number of samples is taken from a population, their means plotted into a frequency distribution form a normal curve. We call this distribution a sampling distribution of means. Its mean is the population value μ, and its standard deviation is the standard error of the mean.

To decide whether a sample came from a given population whose mean and standard deviation are known we adapt the form of the z score (previously used with the normal curve) to tell us how far (in standard errors) a given sample mean is from population mean. Then we consult the table of areas of the normal curve to establish the likelihood of drawing a sample with such a mean from the population. We agree that a sample is not from a population if 1 percent or fewer (or 5 percent or fewer) of the samples will have a mean that far removed from the population mean. With the 5 percent criterion we risk making more Type I errors but do not make as many Type II errors; with the 1 percent criterion we make fewer Type I errors but make more Type II errors. The choice of criterion depends on the cost of one type error over the other.

When population values are not known, we estimate the population parameters from sample values. This procedure again uses an adaptation of the z score, but here we call it the t test. If samples are large (30 or more cases), we test t values against a normal curve, but if samples are small, we must use t distributions, each tied to the degrees of freedom associated with our sample size.

When population means are unknown, we can use the above procedures to estimate where the population value lies. In doing this we project a range of values above and below the sample mean and expect the population mean to be within that range. This range, called the confidence interval, provides us with a general location for the population mean but does not identify it at a specific point. In any case, knowing the range, we know more about the population than before the study began. A flowchart of decision points is given in Fig. 10-10.

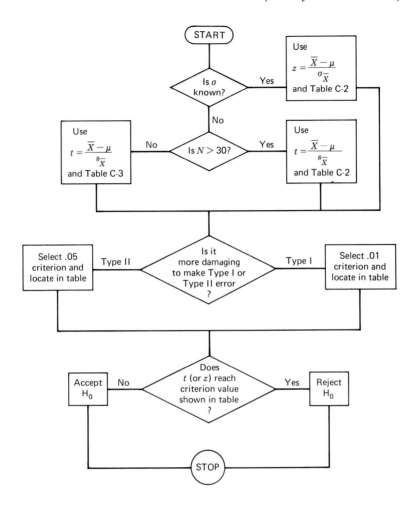

Figure 10-10

Key Terms

central-limit theorem power of a test
sampling distribution confidence interval
standard error of the mean t distribution
Type I error t test
Type II error degrees of freedom

Problems

10-1 In a given population of children the mean height is 43 inches and the standard deviation 6 inches. I have a sample of 36 children with a mean of 45.1 inches. Is it likely (at the 95 percent confidence level) that this is a random sample from the described population?

10-2 A government study concluded that the mean weight of inner-city children at age six was 42.1 pounds. The standard deviation was 1.4 pounds. I have taken a random sample of 100 inner-city children (age five) and put them on a special vitamin supplement for 12-months. At age six their mean weight was 43.8 pounds. Are these children a random sample from a population whose mean is 42.1 pounds (.05 criterion)?

10-3 At a large state university the mean GPA for entering freshmen was 2.70, standard deviation .80 GPA. The population data are normally distributed. I take a random sample of 24 applicants and give them 6 weeks of reading skill instruction. At the close of the first semester these 24 have a mean GPA of 2.72. Are they from a population with a mean GPA of 2.70? Test H_0 at the .05 criterion.

10-4 Professor Aye has a theory about recall and has developed a memory procedure out of the theory. Across many users at the college level a set of 20 standardized nonsense syllables has been shown to have a mean 24-hour retention of 12.3 and standard deviation of 2.6. The population data look normal. Professor Aye asks a sample of 17 college students to memorize the list of syllables following his procedure. Their mean 24-hour retention is 11.2 syllables. Test H_0 at the 95 percent confidence level.

10-5 I have a sample of 46 children. Their mean IQ is 112; the standard deviation is 11. Could this be a random sample (99 percent confidence level) from a population with a mean IQ of 115? What is the range within which the population mean will fall (95 percent confidence level)?

10-6 A sample of 10 college women had a mean of seven dates per month, with a standard deviation of two dates. Is it likely (at 95 percent confidence level) that the population of women from which these 10 came has a mean of nine dates per month? What is the interval within which the mean falls for the population from which these women are a sample (99 percent confidence level)?

10-7 I have a sample of 10 college freshmen who have been tested on a test-anxiety scale. Their mean score was 26 and standard deviation 6 points. What is the score range (fiducial limits) within which the population mean is likely to fall at the 95 percent confidence level?

10-8 A sample of 26 elementary school boys produced a mean mental age of 8 years 2 months. The standard deviation was 7 months. At the 99 percent confidence level could this sample be from a population whose mean is 8 years 6 months? Test the hypothesis at the 95 percent level of confidence.

10-9 The Educational Science Laboratory has a theory for teaching arithmetic to illiterate adults. The hypothesis is that these adults will be from a population different from illiterates once they have experienced the process based on this theory. After 3 months of training a sample of 50 illiterates took a mathematics test. The mean was 14.2 correct out of 30 items, with a standard deviation of 4. Across many samples of illiterate adults the mean was 12.7. Is the sample from a population whose mean is greater than 12.7? Test the hypothesis at 95 percent confidence level.

10-10 Professor Zed, who has a counseling procedure for reducing test anxiety, has taken a sample of 26 students, put them through the procedure, and given them a test-anxiety questionnaire. The mean for a national sample is reported in the questionnaire's manual as 36.7. Professor Zed believes that students who go through the procedure will be from a population whose text anxiety will be lower than students in general. His sample produced a mean of 33.1 with a standard deviation of 10.3. Test the hypothesis at the 95 percent level of confidence that the experimental sample's mean is less than that for students in general.

11

Testing Differences between Means

In the study of human behavior we often develop hypotheses about ways of altering behavior. We believe that a given method of teaching will develop in our students a designated skill in reasoning; a new set of materials will facilitate the acquisition of specified competency in mathematics; a particular treatment will improve emotional adjustment. When the available information allows us to hypothesize reasonably that a specified treatment will alter a given behavior, the next step is to test this hypothesis. One widely used procedure for testing hypotheses is to select two samples randomly from a given population. One sample is given the special treatment (the experimental group) and the other is not (the control group). After the treatment we apply a test or other measure to determine the status of each group on the condition we sought to change. Then we compare mean scores for the two samples to see whether our treatment group is different on the measure from the nontreatment group. The procedure for making this comparison is the subject of this chapter.

Basic Concepts

Here is an example of research in which sample means are to be compared. Professor Dubblyue had been studying young children's verbal skills and decided that we are not correct in our methods of introducing children to reading. Therefore, he devised a set of activities in a prescribed sequence for every child to go through before being introduced to formal reading instruction. Do these activities really prepare children to be better readers? We cannot say until they have been tested in an experimental design.

In a given school district Professor Dubblyue randomly chose 60 entering kindergarten children to be given his set of activities; 60 others were randomly assigned to ordinary kindergarten work; 6 months later the activity list was completed, and children in both groups were working on sight vocabulary. The professor then gave a test of 20 simple sight vocabulary words to each group, calculated the mean number correct for each group, and compared the means.

The experimental group had a mean of 13.2 words correct out of 20, and the control group had 11.7 words correct. It appears that the experiment showed the professor's activities to be effective. But were they? Whenever we take two random samples from a population, they may look very much alike on any variable we wish to observe but probably will not be identical. Our question is: How different can two samples be and still be accepted as random samples from a common population?

We can profit from our previous experience with z scores and with testing the hypothesis that a sample is from a given population. In these situations we began with an observed value and compared it with an expected value by applying the yardstick of a standard deviation or standard error

$$z = \frac{\text{observed} - \text{expected}}{\text{sampling variation of observations}} \qquad (11\text{-}1)$$

With a single distribution we located scores with

$$z = \frac{X - \bar{X}}{s_X}$$

where X is the observed score and the mean was the expected value of any randomly selected score. The standard deviation was the yardstick.

When we were dealing with samples and asking whether they came from a given population, we used

$$z = \frac{\bar{X} - \mu}{s_{\bar{X}}}$$

where $\bar{X}$ = sample mean = observed value
μ = population mean = expected value for the mean of any randomly selected sample
$s_{\bar{X}}$ = standard error of mean, used as yardstick

Now we shall be dealing with a difference between means of two samples and asking whether this difference is greater or smaller than that expected from two random samples from a common population.

The formula, only a slight variation of the z-score procedure, is

$$z = \frac{\text{observed difference} - \text{expected difference}}{\text{sampling variation of differences}} \quad (11\text{-}2)$$

Let us deal with each of these terms separately. The observed difference is simply $\bar{X}_1 - \bar{X}_2$. We subtract one sample mean from the other. (Since we shall relate this difference to a bell-shaped curve, where one side of the curve is a mirror image of the other, it does not matter which sample is labeled 1 and which 2.) The expected difference between pairs of random samples is the mean difference of a large number of pairs of random samples. Sometimes the first sample in a pair will be the larger, sometimes the second. The same is true for the second sample's means. As we sum the $\bar{X}_1 - \bar{X}_2$ differences across a large number of pairs of random samples from a common population, we expect $\Sigma(\bar{X}_1 - \bar{X}_2)$ to be zero. Since the expected difference is usually zero, it is normally dropped out of the formula. Now we turn to the denominator term, i.e., the sampling variation of differences.

The Standard Error of the Difference: Unpooled-Variance Procedure

In Chap. 9 we saw that if we successively draw random samples from a given population, compute their means, and put these means into a frequency distribution, the result will be a normal curve. We called this a sampling distribution of means, and with it we could state the chances of getting a sample whose mean was above or below a given value.

Now, instead of selecting only one sample at a time, let us select *pairs* of samples, compute the mean for each sample in the pair, subtract the first sample mean from the second, and record this *difference*. We repeat this process many times, recording only the differences between the two sample means. This procedure produces a list of numbers (differences) which can then be arranged into a frequency distribution. We shall see that our distribution looks like a normal one, and we can calculate the dispersion of scores around its central point. We then have a *sampling distribution of differences* between means, and just as we can state how many scores are expected to fall beyond given points in a raw-score distribution, we can state how many differences between means will fall beyond given points.

Now we shall actually do what we have just described. From a population of college freshmen we take 15 pairs of random samples of 5 cases each. (This is neither enough pairs nor enough cases in a sample to build an adequate sampling distribution of mean differences, but it is sufficient for our example.) We collect IQs on the samples we select, compute the mean IQ for each sample in a pair, find the difference between means for the paired samples, and record this difference in a distribution (Table 11-1).

Table 11-1   Differences in IQ for 15 Sets of 5 Pairs

	Pair 1		Pair 2		Pair 3		Pair 4		Pair 5	
	113	116	126	120	121	118	113	116	114	115
	115	113	103	117	118	124	117	120	115	113
	112	115	117	126	117	119	109	112	120	120
	117	126	121	108	119	108	114	117	113	108
	113	110	118	119	120	116	112	115	108	114
$\bar{X}_i$	114	116	117	118	119	117	113	116	114	114
$\bar{X}_1 - \bar{X}_2$	−2		−1		2		−3		0	

	Pair 6		Pair 7		Pair 8		Pair 9		Pair 10	
	125	109	117	120	119	125	113	115	116	117
	120	119	119	115	125	104	114	112	114	113
	118	121	116	119	109	118	116	114	118	114
	114	121	111	116	117	120	115	116	115	112
	113	114	117	110	120	118	112	118	117	109
$\bar{X}_i$	118	117	116	116	118	117	114	115	116	113
$\bar{X}_1 - \bar{X}_2$	1		0		1		−1		3	

	Pair 11		Pair 12		Pair 13		Pair 14		Pair 15	
	109	117	117	121	112	117	114	119	114	115
	117	120	122	115	115	122	117	115	117	116
	119	115	119	126	120	113	118	115	116	113
	113	110	104	109	112	110	119	116	113	114
	122	118	122	119	121	118	117	116	115	117
$\bar{X}_i$	116	116	117	118	116	116	117	116	115	115
$\bar{X}_1 - \bar{X}_2$	0		−1		0		1		0	

If we put the differences between means $\bar{X}_1 - \bar{X}_2$ into a frequency distribution, we get Table 11-2. Although only 15 pairs of randomly selected samples probably would not produce a sampling distribution of differences between means quite as symmetrical as Table 11-2, for illustration purposes our sampling was especially fortunate. Our distribution of differences between means does indeed appear as if it might turn out to be a normal curve when more differences are added to the data. Indeed, if an infinite number of pairs of sufficiently large samples are selected and their mean differences recorded, the distribution will be normal; and if the samples are truly randomly drawn from a common population, the mean of this sampling distribution will be

Table 11-2 Distribution   Differences between Means

$\bar{X}_1 - \bar{X}_2$	f
+3	1
+2	1
+1	3
0	5
−1	3
−2	1
−3	1

zero. Remember that the data in Table 11-2 are differences between means of pairs of random samples.

Suppose I select two samples of college students, five cases each, and find that the difference between their mean IQs is 3. How likely is it these samples are a random selection from our population of freshman students? The distribution in Table 11-2 suggests that such a pair of samples would occur once out of 15 times; i.e., the probability of getting two random samples whose $\bar{X}_1 - \bar{X}_2$ difference is 3 is .067, not very likely. Because it is not very likely that two random samples would give us a difference between means of 3, we may wish to conclude that they are not samples from the same population.

From the distribution of differences in Table 11-2 we could actually compute a mean and a standard deviation and decide just what percent of pairs of samples would have a difference between ±1 standard errors, between ±2 standard errors, etc., just as we do with any normal distribution. To do this with assurance that our data are reliable, however, we would first have to collect many pairs of samples in order to construct our frequency distribution against which to test our pair of samples. This is a prohibitive job. Nevertheless, we shall want to take a pair of samples, calculate means for each, and assess the difference between those means by placing that difference on a probability distribution somewhat like that in Table 11-2. A z-score approach is useful here

$$z = \frac{\bar{X}_1 - \bar{X}_2}{\sigma_{\text{diff}}}$$

where $\bar{X}_1$ and $\bar{X}_2$ are the sample means and σ_{diff} is the standard deviation of the distribution of differences (like Table 11-2) for an infinite number of pairs of random samples. Because we cannot calculate σ_{diff} directly from the distribution of differences, the alternative

$$\sigma_{\text{diff}} = \sqrt{\frac{\sigma_1{}^2}{n_1} + \frac{\sigma_2{}^2}{n_2}} \tag{11-3}$$

has been developed, where $\sigma_1{}^2$ and $\sigma_2{}^2$ are the variances of the populations from which samples 1 and 2 came and n_1 and n_2 are the number of cases in each sample.

How do we get the population variances? Since one seldom measures an entire population, the calculation of $\sigma_1{}^2$ and $\sigma_2{}^2$ is usually impossible, but we do have an option. In Chap. 5 we saw that if we use $N - 1$ instead of N in the denominator when calculating the standard deviation we have an unbiased estimate of the population standard deviation. With our sample data we can calculate the standard deviations as estimates of the population parameters and apply them in formula (11-3). We then have

$$s_{\text{diff}} = \sqrt{\frac{s_1{}^2}{n_1} + \frac{s_2{}^2}{n_2}} \tag{11-4}$$

Everything we need to apply formula (11-4) can be developed from samples. The calculation of this standard error is given in Table 11-3.

Test of Significance: Unpooled Variance

We are now ready to test the difference between samples to see whether we should accept or reject the null hypothesis. First, however, one further point is necessary. From Chap. 10 we recall that in the one-sample problem when we used sample estimates of the population standard deviation the distribution took the form of t rather than z, the normal curve. The same is true in two-sample problems. Our test of the null hypothesis then takes the form

$$t = \frac{\bar{X}_1 - \bar{X}_2}{s_{\text{diff}}} \tag{11-5}$$

As long as samples are large, the t distribution approximates z. Therefore as long as we have samples of 30 or larger, we can establish

Table 11-3   Computation of the Standard Error of the Difference

Formula (11-4)	What it says to do
$$s_{\text{diff}} = \sqrt{\dfrac{s_1{}^2}{n_1} + \dfrac{s_2{}^2}{n_2}}$$	1 Compute the variance for each sample (Chap. 5)
where $s_1{}^2$ = variance of sample 1 $s_2{}^2$ = variance of sample 2 n_1 = number of cases in sample 1 n_2 = number of cases in sample 2	2 Divide the variance of sample 1 by n_1 and the variance of sample 2 by n_2 3 Add the results of step 2 4 Take the square root of the figure obtained in step 3

Example   We have two samples of 40 girls each randomly selected from all those who have just enrolled as college freshmen. The mean age for sample 1 is 220 months, and the standard deviation is 6 months; for sample 2 the mean age is 223 months, and the standard deviation is 7 months. What is the standard error of the difference between mean ages?

$$s_{\text{diff}} = \sqrt{\frac{6^2}{40} + \frac{7^2}{40}} = \sqrt{2.12} = 1.46$$

That is, remembering that the distribution of differences between means approximates a normal curve when samples are large, we find that 68 percent of the pairs of random samples ($N = 40$) drawn from our population would be expected to have *differences* between their means not to exceed ± 1.46 months.

---

the 95 or 99 percent confidence cutoff points by using the normal-curve values of 1.96 and 2.58. (We shall deal with small samples shortly.) We already know how to compute the denominator [formula (11-4)]; it is given in Table 11-3, and the numerator is simply the difference between the means of the two samples.

Now we are ready to work through a problem. I have two samples of fifth-grade children randomly drawn from Elm City schools. Each sample is made up of 65 children. I also have a speed-reading machine, the use of which I believe will increase reading speed significantly. I select one sample (the experimental group) to get the practice (30 minutes per day for 10 days) and the other group to have free reading during the same periods. After 10 days both groups take a standard reading-speed test. We get the following data for the reading rates of the two groups (where wpm stands for words per minute):

Experimental	Control
$\bar{X}_1 = 174$ wpm	$\bar{X}_2 = 162$ wpm
$s_1 = 36$	$s_2 = 40$
$n_1 = 65$	$n_2 = 65$

We now look at the difference between the means in terms of our yardstick, the standard error of the difference. To do this we pose the null hypothesis that there will be no difference and then select a critical difference beyond which we shall reject the null hypothesis. Let us use the 95 percent level of confidence here. We noted earlier in Table C-2 that a difference of 1.96 standard errors left 2.5 percent in each of the tails of the normal curve, or a total of 5 percent. Therefore, if our two means differ by more than 1.96 standard errors, we shall reject the null hypothesis H_0 for the problem above.

We first compute the standard error of the difference

$$s_{\text{diff}} = \sqrt{\frac{36^2}{65} + \frac{40^2}{65}}$$

$$= \sqrt{\frac{1296}{65} + \frac{1600}{65}}$$

$$= \sqrt{19.94 + 24.62}$$

$$= \sqrt{44.56}$$

$$= 6.67$$

Now we have the standard error of the difference, the yardstick against which to test the obtained difference between the two means, How many units of 6.67 separate $\bar{X}_1$ and $\bar{X}_2$? If our answer is 1.96 or more, we shall reject the null hypothesis H_0, but if the difference is less than 1.96 of these units we shall accept H_0. Here are the remaining computations:

$$t = \frac{\bar{X}_1 - \bar{X}_2}{s_{\text{diff}}} = \frac{174 - 162}{6.67} = \frac{12}{6.67} = 1.80$$

Since our t value did not reach the critical value of 1.96, we accept the null hypothesis.

What does this mean? It means that in terms of reading speed that the two groups still look like random samples from the same popula-

tion of fifth-graders; at least at the 95 percent level of confidence we
cannot deny that they are two samples from the same population.
Now suppose we test reading comprehension and find

	Control	*Experimental*
	$\bar{X}_1 = 30$ questions correct	$\bar{X}_2 = 43$ questions correct
	$s_1 = 7$	$s_2 = 8$
	$n_1 = 65$	$n_2 = 65$

Are the two groups still from the same population of reading comprehenders? We test the null hypothesis again with t as follows:

$$s_{\text{diff}} = \sqrt{\frac{7^2}{65} + \frac{8^2}{65}} = \sqrt{1.74} = 1.32$$

$$t = \frac{43 - 30}{1.32} = \frac{13}{1.32} = 9.85$$

Now if we again choose the 95 percent confidence level for our critical
point against which to test our t value, the t value must be at least 1.96
or larger in order to reject the null hypothesis. Our t value was 9.85,
considerably larger than 1.96 and so we do indeed reject the null
hypothesis. Our two samples do not appear to be from the same
population of comprehenders.

In this problem we might have reason to choose the 99 percent level
of confidence. If so, our value of t must be equal to or greater than
±2.58. This is the point on the base line of the curve beyond which 1
percent of the cases fall. Our t value (9.85) is still much larger than the
critical value at the 1 percent limit, and so we reject the null hypotheses
and again conclude that the two samples appear to be from different
populations of reading comprehenders.

Exercises

11-1 Given the following means from pairs of random samples, compute
their differences d and arrange them into a distribution.

(a) Does it appear likely that from this population we would get a pair of
samples with a difference between their means equal to at least two
points?

(b) Does it appear likely that we would randomly select from this population a pair of samples whose means differ as much as 6 points?

$\bar{X}_1$ $\bar{X}_2$ d	$\bar{X}_1$ $\bar{X}_2$ d	$\bar{X}_1$ $\bar{X}_2$ d	$\bar{X}_1$ $\bar{X}_2$ d
27 26	23 28	29 25	24 27
29 24	26 27	27 24	23 24
24 27	27 24	26 25	25 25
26 26	26 27	25 23	24 28
26 24	25 27	26 26	27 27

11-2 In a learning experiment a psychologist selected two samples of 30 sixth-grade boys each. She first wanted to know if they were samples from a common population of learners, and so she compared them on IQs. The first sample had a mean IQ of 102 and a standard deviation of 10 IQ points, and the second sample had a mean of 100 with a standard deviation of 11 points.

(*a*) Compute the standard error of the difference.

(*b*) In your own words, what does this figure tell us?

11-3 I wish to compare the results of studying biology by use of a programmed textbook with those of the regular text-lecture method. Of the students who enroll in Biology 101 at Eks University this fall I randomly assign 50 students to programmed textbook work and 50 others the regular text-lecture work. At the close of the semester all students take a common final examination.

(*a*) Which group is the experimental group? Which is control?

(*b*) The final test data were:

	Experimental	*Control*
$\bar{X}$	39.5 items correct	36.1 items correct
s	9	8

At the 95 percent confidence level, test the null hypothesis.

(*c*) Do the above samples at the end of the semester still look like random samples from the same population of biology students?

11-4 Since I have reason to believe that the noise of traffic may affect the auditory acuity of traffic police, I select a random sample of traffic officers and compare their auditory acuity with that of a random sample of officers assigned to other duties. I find the following results for tones of 4000 hertz:

	Traffic officers	Nontraffic officers
$\bar{X}$	46.5	51.3
s	5	8.1
n	35	40

(a) Test the null hypothesis at the 99 percent level.

(b) In your own words what does the result of this test say?

11-5 I have two samples of first-grade children. One was drawn from all children who attended Montessori preschools, and the other from public school kindergartens. I wish to know whether these children differ in sight recognition of printed words. I test them all on a set of 30 words selected from first-grade texts and find the following test results in terms of number of words recognized:

	Montessori	Public
$\bar{X}$	21	18
s	4	5
n	36	30

(a) Test the null hypothesis at the 95 percent level.

(b) What does this say about our two samples?

Test of Significance: The Pooled-Variance Procedure

We have been testing differences between means for large samples—30 or more subjects per sample, but we often find that our samples are not that large. What can we do then?

Earlier we saw that as samples get smaller the sampling distribution of the mean begins to be more leptokurtic and takes on the characteristics of the t distribution rather than those of the normal curve. Now in this chapter we are dealing with the sampling distribution of differences between means. Again, as samples get smaller in size, the sampling distribution looks like t rather than the normal curve.

When we work with small samples, we shall continue to compute the standard error of the difference and interpret the $\bar{X}_1 - \bar{X}_2$ difference in terms of units of this standard error, but the method of computing the standard error is slightly different.

If our two samples are either from a *common* population or from two populations with equal variances (an assumption we *must* make if our sample sizes are small, i.e., less than 30 cases), we estimate the population variance best by *pooling* the data from the two samples. We do this by first finding the sum of the squared deviations of scores around the mean of sample 1, that is, $\Sigma(X_i - \bar{X}_1)^2$, where i represents the 1 to n scores in the first sample. Next we calculate the sum of squared deviations around the mean for the second sample, $\Sigma(X_j - \bar{X}_2)^2$, where j represents the 1 to n scores in the second sample. We then add the two values

$$\Sigma(X_i - \bar{X}_1)^2 + \Sigma(X_j - \bar{X}_2)^2$$

which gives us a pooled sum of squared deviations across the two samples. If we divide this sum by the total degrees of freedom across the two samples ($n_1 - 1$ for the first sample and $n_2 - 1$ for the second, or $n_1 + n_2 - 2$), we have an estimate of the population variance based on the two samples.

This procedure is necessary because with small samples our estimate of the population variance is less precise than with large samples. By pooling the squared deviations across two samples we do a better job of estimating the population variance than if we did not pool.

Next we substitute the pooled variance into the formula for the standard error of the difference as shown in the following development.

$$s^2 = \frac{\Sigma(X_i - \bar{X}_1)^2 + \Sigma(X_j - \bar{X}_2)^2}{n_1 + n_2 - 2} \tag{11-6}$$

Our estimate of the population variance, s^2, is substituted for s_1^2 and s_2^2 in formula (11-4)

$$s_{\text{diff}} = \sqrt{\frac{s_1^2}{n_1} + \frac{s_2^2}{n_2}}$$

to get

$$s_{\text{diff}} = \sqrt{\frac{\dfrac{\Sigma(X_i - \bar{X}_1)^2 + \Sigma(X_j - \bar{X}_2)^2}{(n_1 + n_2 - 2)}}{n_1} + \frac{\dfrac{\Sigma(X_i - \bar{X}_1)^2 + \Sigma(X_j - \bar{X}_2)^2}{(n_1 + n_2 - 2)}}{n_2}}$$

Factoring out the common variance estimate, we have

$$s_{\text{diff}} = \sqrt{\frac{\Sigma(X_i - \bar{X}_1)^2 + \Sigma(X_j - \bar{X}_2)^2}{n_1 + n_2 - 2} \left(\frac{1}{n_1} + \frac{1}{n_2} \right)} \tag{11-7}$$

Since in Chap. 5 we saw that $\Sigma(X - \bar{X})^2 = \Sigma X^2 - (\Sigma X)^2/N$, formula (11-7) is equal to

$$s_{\text{diff}} = \sqrt{\frac{\Sigma X_i^2 - (\Sigma X_i)^2/n_1 + \Sigma X_j^2 - (\Sigma X_j)^2/n_2}{n_1 + n_2 - 2}\left(\frac{1}{n_1} + \frac{1}{n_2}\right)} \qquad (11\text{-}8)$$

This is the formula for a pooled-variance approach to computing the standard error of the difference. Calculation of this value is illustrated in Table 11-4.

If sample variances s^2 are already computed, we can pool variances by

$$s_{\text{diff}} = \sqrt{\frac{(n_1 - 1)s_1^2 + (n_2 - 1)s_2^2}{n_1 + n_2 - 2}\left(\frac{1}{n_1} + \frac{1}{n_2}\right)} \qquad (11\text{-}8a)$$

which can be shown to be equivalent to formula (11-7).

We now have all the procedures necessary to deal with the question: Is the performance of one group statistically different from another? The general approach to answering this question is to begin by observing the difference between $\bar{X}_1$ and $\bar{X}_2$ and evaluating it by using the standard error of the difference as a yardstick. The question then becomes: Which standard error, unpooled or pooled-variance procedure? As noted earlier, if we have large samples, the unpooled-variance approach is typically satisfactory and we look to the normal curve for our critical values, but if the samples are small, the pooled-variance procedure is required. Also, when we have small samples, the sampling distribution of differences follows Student's t distribution. Like any t distribution, it varies with sample size. Our job is first to sort out the degrees of freedom for a given small-sample problem and look up the critical values in Table C-3. The number of degrees of freedom for the pooled-variance procedure is

$$(n_1 - 1) + (n_2 - 1) = n_1 + n_2 - 2$$

The entire procedure for working through a t test for pooled variances is given in Table 11-5. When we have calculated the difference between $\bar{X}_1$ and $\bar{X}_2$, found the standard error of the difference, and looked at $\bar{X}_1 - \bar{X}_2$ in terms of that standard error, we find a t value of 4.33, which says that the difference between $\bar{X}_1$ and $\bar{X}_2$ is 4.33 standard errors.

Is this a large enough difference to believe that the two samples are from different populations, or is it no more than we would expect from sampling error? The answer comes from consulting Table C-3. Here, with $10 + 9 - 2 = 17$ degrees of freedom, we find that the critical value

Table 11-4 Pooling Variances to Compute a Standard Error of the Difference

Formula (11-8)

$$s_{\text{diff}} = \sqrt{\frac{\Sigma X_i^2 - (\Sigma X_i)^2/n_1 + \Sigma X_j^2 - (\Sigma X_j)^2/n_2}{n_1 + n_2 - 2}\left(\frac{1}{n_1} + \frac{1}{n_2}\right)}$$

where s_{diff} = standard error of the difference
X_i = scores in first sample
X_j = scores in second sample
n_1 = number of cases in first sample
n_2 = number of cases in second sample

What it says to do

1 Compute the sum of the squared deviations around the mean for the first sample: $\Sigma X_i^2 - (\Sigma X_i)^2/n_1$

2 Repeat step 1 for the second sample: $\Sigma X_j^2 - (\Sigma X_j)^2/n_2$

3 Add the results of step 2 to step 1

4 Divide the result of step 3 by the degrees of freedom $n_1 + n_2 - 2$

5 Add $1/n_1$ and $1/n_2$ and multiply this result by the result of step 4

6 Take the square root of the result of step 5

Example Compute s_{diff} for the following data:

	Sample 1	Sample 2
	$\Sigma X_i = 148$	$\Sigma X_j = 172$
	$\Sigma X_i^2 = 1932$	$\Sigma X_j^2 = 2448$
	$n_1 = 24$	$n_2 = 20$

$$s_{\text{diff}} = \sqrt{\frac{1932 - 148^2/24 + 2448 - 172^2/20}{24 + 20 - 2}\left(\frac{1}{24} + \frac{1}{20}\right)}$$

$$= \sqrt{\frac{1019.33 + 968.80}{42}(.04 + .05)}$$

$$= \sqrt{47.34\,(.09)} = \sqrt{4.26} = 2.06$$

Table 11-5 Computation of the t Test

Formula (11-5)	*What it says to do*
$$t = \frac{\bar{X}_1 - \bar{X}_2}{s_{\text{diff}}}$$ where $\bar{X}_1$ = mean of first sample	1 Compute the mean for sample 1 and the mean for sample 2
$\bar{X}_2$ = mean of second sample	2 Determine their difference
s_{diff} = standard error of difference	3 Compute the standard error by the appropriate method [formula (11-4) or (11-8)]
	4 Divide the difference between the means by the standard error

Example Two samples of students have practiced a letter-cancellation task. One sample practiced for three 10-minute sessions and the other for six 5-minute sessions. Their final scores are given below. Compute the t value for the difference between the means of the two samples.

Sample 1	*Sample 2*
10 14 13 11 9	20 12 17 16 18
11 15 12 13 12	14 16 19 15
$\Sigma X_i = 120$	$\Sigma X_j = 147$
$\Sigma X_i^2 = 1470$	$\Sigma X_j^2 = 2451$

$$s_{\text{diff}} = \sqrt{\frac{1470 - 120^2/10 + 2451 - 147^2/9}{10 + 9 - 2}\left(\frac{1}{10} + \frac{1}{9}\right)} = .994$$

$$t = \frac{16.3 - 12.00}{.994}$$

$$= 4.33$$

With $10 + 9 - 2$ df, significant at the 99 percent confidence level, the null hypothesis is rejected

(at the .01 criterion) is 2.898. That is, a difference between means which is 2.898 standard errors from the expected difference, i.e., zero, leaves .5 percent of sample mean differences in one tail of the curve and .5 percent in the other (Fig. 11-1). Recall that the 1 percent criterion with large samples was 2.58 standard errors.

Since the t value computed in Table 11-5 is 4.33, it is beyond the

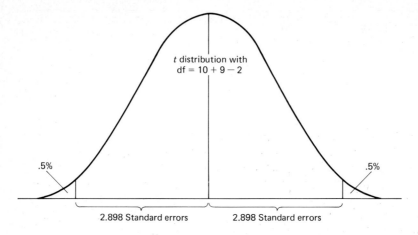

Figure 11-1

critical value of 2.898. Therefore, we conclude that it is quite unlikely that the two samples are from a common population. The difference between the means is so great (over 4 standard errors) that only in very rare circumstances could it occur between two randomly drawn samples from the same population.

When using the *t* test with small samples, as we did in the example in Table 11-5, we must remember to use the pooled-variance approach in calculating the standard error of the difference. We shall get our critical values for the .05 or .01 levels by consulting Table C-3, with n_1 + n_2 − 2 degrees of freedom. We do this because for every value of the degrees of freedom the *t* distribution has a slightly different shape. Therefore, the number of standard errors it will take to get to the point where .025 (or .005) percent is in each tail will vary with each curve and hence with each different degree of freedom.

It is also useful to remember that this *t* distribution is a sampling distribution of differences between pairs of means based on pairs of random samples from the population. The expected value of this difference, $(\bar{X}_1 - \bar{X}_2)$, is zero. Although the mean of the sampling distribution is zero, many pairs of random samples from a given population do not produce a difference between their means equal to zero. How far can the difference between means of two samples depart from zero and still be regarded as coming from the population whose mean difference is zero? That is the secret the *t* test reveals.

Testing Variances for Differences before Pooling

Wherever the pooled variance procedure is used we assume that both samples come from populations with a common variance. To be sure that we meet this assumption we should first test our sample variances for the significance of their difference before pooling the variances. This is called a *test of homogeneity of variance*. We make this test with the formula

$$F = \frac{s_i^2}{s_j^2} \tag{11-9}$$

where s_i^2 is the larger of the two variances and s_j^2 is the smaller, and determine its significance by entering Table C-4 with $n_i - 1$ degrees of freedom for the numerator and $n_j - 1$ degrees of freedom for the denominator, where i is the sample with the larger variance and j is the sample with the smaller variance. F values that are larger than the table values show a significant difference in the sizes of the two variances in the ratio. We should note, however, that the F table involves only *one* tail of the curve, and so values listed in the table as the .05 level are, for our present test, actually the .10 level, and those listed as .01 values are the .02 level.

We test the hypothesis that both samples in Table 11-5 came from populations with the common variance thus

$$F = \frac{6.25}{3.33} = 1.88$$

We then enter Table C-4 with $n_i - 1 = 8$ degrees of freedom for the numerator and $n_j - 1 = 9$ degrees of freedom for the denominator. Moving down the v_1 (df = 8) column in the table to the v_2 (df = 9) row, we find that an F value of 3.23 is needed to be significant. But since our value of 1.88 is less than the table value, we accept the hypothesis that our samples come from populations with common variances and rest assured that we were correct in pooling our variances. However, F tables are one-tailed and so we double the value given at the heading of Table C-4. Our test was at the 90 percent level of confidence. If we wish to make a test at the 95 percent level, we would use Table C-5; values at the 2.5 percent alpha when doubled will be at the 5 percent alpha.

Exercises

11-6 Given the following data for the number of simple additions made by 10 subjects in each of two samples in a 4-minute period:

Sample 1		Sample 2	
10	7	7	9
9	6	8	11
9	5	6	5
7	4	7	4
7	3	8	3

(*a*) Compute the standard error of the difference using the pooled-variance method and figuring the sum of the squared deviations for each sample by the formula $\Sigma(X_1 - \bar{X}_1)^2 + \Sigma(X_2 - \bar{X}_2)^2$.

(b) Compute the standard error of the difference using the formula $\Sigma X_1^2 - (\Sigma X_1)^2 n_1 + \Sigma X_2^2 - (\Sigma X_2)^2/n_2$.

(c) Which procedure was faster?

11-7 Two randomly selected groups of 20 subjects each were chosen from beginning classes in educational psychology. One group was given training in estimating the area of rectangles; the other group was a control group with no training. At the end of the training period both groups were given a test in estimating areas of circles. The number of correct estimations by each subject in each group is given below:

Training group					Control group				
18	15	14	13	17	17	16	13	12	18
9	15	14	12	16	11	14	13	12	16
8	13	16	21	14	16	12	9	15	8
9	10	15	13	14	11	14	15	10	14

(a) Test the variances for homogeneity.

(b) Test the hypothesis that there is no difference between the means of the populations from which the sample came. (Note the sample size before selecting the method for computing the standard error of the difference.) Use 5 percent alpha level.

11-8 In a university health center conducting a weight-loss clinic 12 applicants were assigned to a diet plan and 13 to diet plus counseling. After 6 weeks the weight in pounds lost by each person was recorded. At the .05 criterion test the differences between the means.

Diet				Diet and counseling				
3	7	5	10	10	5	11	12	7
4	9	7	6	9	13	9	8	10
7	5	8	8	17	6	10		

An Approximation Test when Variances Are Unequal and Samples Small

Suppose that when we complete our F test we find that the variances are indeed significantly different and our sample sizes are too small to apply formula (11-3) as it now stands. Must we abandon a test of the significance of the difference between means?

Luckily we can test the hypothesis that the samples came from populations with equal means irrespective of hypotheses concerning the population variances. In this case we take s_1^2, the first sample variance, as the best estimate of σ_1^2, and we take s_2^2, the second sample variance, as the best estimate of σ_2^2 and compute a t value

$$t = \frac{\bar{X}_1 - \bar{X}_2}{\sqrt{s_1^2/n_1 + s_2^2/n_2}} \tag{11-10}$$

However, the value found by formula (11-10) *cannot* be evaluated by consulting Table C-3. Instead we compute an approximation of the t value for the 5 percent cutoff point

$$t_{.05} = \frac{t_1 s_1^2/n_1 + t_2 s_2^2/n_2}{s_1^2/n_1 + s_2^2/n_2} \tag{11-11}$$

where t_1 is the table value of t at the 95 percent confidence level with $n_1 - 1$ degrees of freedom and t_2 is the corresponding value for $n_2 - 1$ degrees of freedom. This computed t value is then used in place of the table value. Our null hypothesis in this case is that the samples have been chosen from two populations with equal means. If the value obtained from formula (11-10) is *larger* than that obtained from formula (11-11), the difference between the means is significant at the 95 percent confidence level and the null hypothesis is rejected; i.e., the samples are not believed to have come from populations with equal means.

For example, I have randomly assigned students to two groups: 16 cases in one group, 15 in the other. One group ($n = 16$) took a test at temperature 70°F, humidity 20 percent; the other ($n = 15$) took the test at the same temperature, but at a humidity of 95 percent. Out of the 30 items on the test, the low-humidity group had a mean of 24.1, standard deviation of 6.0; the high-humidity group had a mean of 20.3, standard deviation of 2.2.

The F test gives

$$F = \frac{6.0^2}{2.2^2} = 7.44$$

Consulting Table C-4 with 15 and 14 degrees of freedom, we find that the table value at the 5 percent point is 2.4630. Since our table value is smaller than the calculated F, we reject the hypothesis that the samples are from populations with common variances. We now do an approximation test as follows. We find the critical t value (.05) for $16 - 1$ degrees of freedom, and again for $15 - 1$ degrees of freedom. We enter them in formula (11-11) and calculate the critical value of t to be

$$t_{.05} = \frac{2.131 \, (6.0^2/16) + 2.145 \, (2.2^2/15)}{6.0^2/16 + 2.2^2/15}$$

$$= \frac{4.79 + .69}{2.25 + .32} = 2.13$$

The critical value for our t is 2.13 standard errors. We now do the t test between means using formula (11-10):

$$t = \frac{24.1 - 20.3}{\sqrt{6^2/16 + 2.2^2/15}} = \frac{3.8}{1.60} = 2.38$$

Since our t value of 2.38 is larger than our computed critical value of 2.13, we reject H_0 at the 95 percent level of confidence. It should be remembered, however, that the above is an approximation procedure and warrants some caution in use.

Testing the Difference between Means when Observations Are Correlated between Samples

Suppose I want to conduct a study in which teaching method A is compared with method B. I quickly see that if one group of subjects is more intelligent than the other, my results will be distorted. I therefore match on intelligence each subject in the A group with a subject in the B group. This control eliminates the possibility of getting a mean difference on achievement which is merely a result of intelligence; however, since the selection of a child for one group influences the selection of a child in the other group, the observations in group A will be correlated with group B to the extent that intelligence is a factor in getting scores on the achievement test given at the close of the study.

Other possibilities for correlating observations also exist. For example, we may wish to compare husbands with wives on an attitude scale or fathers with sons on an intelligence test. Or we may wish to give a test of some type to a group of subjects, administer an experimental treatment, then readminister the test to the same people. The pretest and posttest will no doubt be correlated because the same individuals are responding to the items both before and after the treatment and certainly not all personal traits reflected by the test are presumed to be altered by the treatment.

When the data in one sample are correlated with the data in the other, we must alter our procedure for the t test to acknowledge the fact that changes in one set of data will be reflected in the other. Typically in studies of this type a subject in one sample is matched in some way with a subject in the other sample. The extent to which the samples are different then depends on how the subjects in one sample differ from their matched partner in the other sample; i.e., the difference between sample means will be equal to the mean of the matched-pair differences. The data in Table 11-6 illustrate this fact.

Table 11-6 Illustration of $X_1 - X_2 = d$

	IQs for subjects		$IQ_A - IQ_B$
	Sample **A**	Sample **B**	
	115	110	5
	112	113	−1
	110	110	0
	108	105	3
	100	97	3
$\Sigma X/n$	$545/5 = 109$	$535/5 = 107$	$10/5 = 2$

$$\bar{X}_A - \bar{X}_B = 109 - 107 = 2 \qquad \bar{d} = 2$$

We have five matched pairs of subjects. The mean of their score differences is equal to the difference of the group means, $\bar{X}_1 - \bar{X}_2 = \bar{d}$. Therefore, in paired data, we can substitute $\bar{d}$, the mean of differences between pairs of subjects, for $\bar{X}_1 - \bar{X}_2$ in doing a t test.

It now appears that the differences between the scores of paired subjects in matched groups can be put into a distribution themselves. If so, we can compute a standard deviation of those differences just as we compute any standard deviation, but we must substitute difference scores for the raw scores we used in describing data in Chap. 5. The formula is

$$s_d = \sqrt{\frac{\Sigma d^2 - (\Sigma d)^2/N}{N - 1}} \tag{11-12}$$

and we saw in Chap. 10 that the standard error is found by dividing a standard deviation by the square root of N. Therefore, if we divide s_d by $\sqrt{N}$, the number of *differences* (or the number of matched *pairs* in our samples), we have a standard error—of what? Since the basic data are differences, it must be a standard error of the difference, but this procedure takes into account the correlation in observations of the two samples. Previous standard errors of the difference have assumed that the correlation between groups is zero.

The procedure for doing a t test with correlated data like those in Table 11-6 is condensed into the formula

$$t = \frac{\bar{d}}{s_{\text{diff}}} \tag{11-12a}$$

Its application is laid out in Table 11-7.

Table 11-7   *t* Test for Paired Data

Formula (11-12a)	*What it says to do*
$$t = \frac{\bar{d}}{s_{\text{diff}}}$$ where $\bar{d}$ = mean of differences in scores between pairs of matched subjects $$s_{\text{diff}} = \frac{s_d}{\sqrt{N}}$$ s_d = standard deviation of difference scores N = number of differences (or matched pairs)	The procedure applies to situations where each subject in sample A is paired with a subject in sample B

The procedure applies to situations where each subject in sample A is paired with a subject in sample B

1   Compute the score differences for all pairs of subjects, $X_{1_A} - X_{1_B}$, $X_{2_A} - X_{2_B}$, etc.

2   For these differences d compute a mean $\Sigma d/N$

3   Compute a standard deviation of differences as follows:

   (*a*) Square all d values and find their sum $d_1^2 + d_2^2 + \cdots + d_N^2$

   (*b*) Sum all d values $d_1 + d_2 + \cdots + d_N$

   (*c*) Apply (*a*) and (*b*) in formula (11-12) to get the standard deviation s_d of d values

4   Divide s_d by $\sqrt{N}$, where N is the number of pairs, to get s_{diff}

5   Divide $\bar{d}$ by s_{diff} to get the t value and enter Table C-3 with $N - 1$ df

Exercises

11-9   An undergraduate student in a psychology laboratory wished to test the hypothesis that adolescent boys are brighter than their fathers. He collected IQs on 10 fifteen-year-old boys and also on their fathers. The data are below. Test the null hypothesis for these data.

Sons	115	114	114	110	108	107	105	100	97	95
Fathers	120	117	112	118	102	95	107	106	93	99

11-10   An opinion scale was developed to assess attitudes about the desirability of attendance at a given private college for women. The scale was scored so that a low value reflects a negative attitude toward attendance, a high value a positive attitude. It was then administered to 15 mothers

Table 11-7 Continued

Example Two samples of 10 high school seniors each have been *matched* on IQ before beginning an experiment in learning words in a fictitious language. During a 20-minute study period Group A studied in pairs, one student reading the words to the other while the students in group B studied alone. Afterwards a test of words learned was given. Did the different procedures result in different achievement? Our null hypothesis is that there is no difference between the groups in number of words learned.

Words learned			
A	**B**	*d*	*d²*
10	11	−1	1
9	7	2	4
9	8	1	1
8	9	−1	1
8	6	2	4
7	6	1	1
7	8	−1	1
5	4	1	1
4	3	1	1
4	4	0	0
		$\Sigma d = 5$	$\Sigma d^2 = 15$

$$\bar{d} = \frac{5}{10} = .5$$

$$s_d = \sqrt{\frac{15 - 5^2/10}{9}} = 1.18$$

$$s_{\text{diff}} = \frac{1.18}{\sqrt{10}} = .37$$

$$t = \frac{.5}{.37} = 1.35$$

With 9 df the *t* value is insignificant and the null hypothesis is accepted

and their 15 teenage daughters. The results are below. Test the null hypothesis for these data.

Mothers	53	59	32	43	37	46	39	48	42	45	49	47	46	55	50
Daughters	48	57	46	59	42	40	45	36	56	42	39	55	39	58	45

One- or Two-Tailed Test

In estimating the value of the population mean it is sometimes reasonable to believe that the population mean can only be on one side of the sample mean. In such cases we are dealing with only one tail of the distribution. In our table of *t* the column listed as the 5 percent level provides values which leave 2.5 percent in each of the two tails of the curve. If we are doing a one-tailed test at the 5 percent level we use the column in Table C-3 listed as the .10 level. The data in this column represent points beyond which 5 percent of the area under the curve lies in each tail of the distribution.

Similarly when we test the difference between two means, it is sometimes logical to believe that differences can go only in one direction, i.e., that a given mean can change only in one direction. For example, suppose I have a learning method which a series of pilot studies shows never was a detriment to learning but which sometimes produced a real advantage. I believe that the only direction in which my experimental group can go is up, i.e., increase their learning scores. I randomly assign students to two groups of 10 each. The experimental group gets my special learning method; the control group gets advice on study habits. After three 1-hour sessions with the groups I give them a special learning task and check the number of items correctly solved out of a total of 75. Here are the results:

	Experimental		*Control*
$\bar{X}$	49.6		46.2
n	10		10
s_{diff}		1.88	
t		1.81	

Typically, we cannot say what the impact of a treatment will be, and so when we use the 5 percent level of confidence, 2.5 percent of the differences between pairs of random samples is in one tail of the curve and 2.5 percent is in the other tail. But in the preceding experiment we had reason to believe that differences would occur only in one direction, and so the 5 percent area of the curve is concentrated all in one tail.

This requires us to reason with Table C-3. The data given in each column are the points on the base line that cut off half of the desired area *in each tail* of the curve. For example, with $10 + 10 - 2$ degrees of freedom, at the 5 percent level (.05) the critical table value is 2.101. This says that beyond $+2.101$ standard errors of the difference we shall have 2.5 percent of the cases and below -2.101 standard errors we shall have the other 2.5 percent. Figure 11-2 illustrates this. For 18 degrees of freedom $+2.101$ standard errors is at point A and -2.101 standard errors is at point A'. The area above A is 2.5 percent of the curve; the area below point A' is 2.5 percent. But since in our study above we believe that the difference can move only in one direction, we want the 5 percent to be all in one tail of the curve. How do we find the point that cuts off 5 percent all in one tail?

The Table C-3 values at the .10 criterion cut off 5 percent in each tail

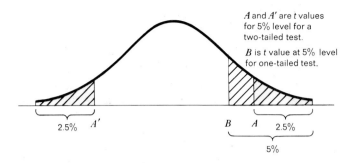

Figure 11-2

of the curve; therefore we can use the .10 criterion to identify the critical value for the 95 percent confidence level. The area above point B in Fig. 11-2 includes the one-tailed test at the 5 percent confidence level.

Turning again to our experiment above, we look up the one-tailed 5 percent cutoff point in Table C-3 with $10 + 10 - 2 = 18$ degrees of freedom by reading in the *.10 column*. The value is 1.734. In the right-hand side of the curve this is the number of standard errors above a mean difference of zero that cuts off 5 percent of the cases in the upper tail, and so 1.734 is the critical t value for our study.

Our calculated t value is 1.81, larger than the table value of 1.734, and we therefore reject the null hypothesis. If we had conducted a two-tailed test, we would have looked under the column labeled .05, with 18 degrees of freedom, and found the value 2.101. In this case we would have had to accept the null hypothesis.

However, for one-tailed tests at the 95 percent confidence level we look under the .10 column in Table C-3; at the 99 percent confidence level we look under the .02 column.

Statistical techniques are devices made and selected to conform to our conceptions of the world around us. Sometimes we cannot speculate about the impact of an experimental treatment since the subjects may improve from it or may get worse; in these circumstances the two-tailed test is relevant. Occasionally we have good reason to believe that our treatment will produce an effect in one direction only; then our conception of the world indicates that a one-tailed test is appropriate, so we select our statistical procedure to fit that conception.

Additional Applications: Proportions and Correlation Coefficients

Two additional applications of the basic procedure for testing differences between samples will be demonstrated: testing differences between proportions and testing differences between correlation coefficients.

Testing Differences between Proportions

Sometimes our data are not continuous like height or age but are simply categorical, like male/female or yes/no. In such cases we often want to know whether there are differences between the proportions of cases in the two categories. For example, suppose we go to the nearest shopping mall and stop people at random and ask whether they approve of their congressman's vote on a recent issue. The data turn out like this:

	Yes	No	
Men	40	65	105
Women	58	42	100
	98	107	205

A larger proportion of women approved than men, but is this within sampling error? To answer this question we test the null hypothesis H_0: $\pi_1 = \pi_2$, where π is the population parameter, with the following procedure:

$$z = \frac{p_1 - p_2}{\sqrt{pq/n_1 + pq/n_2}} \tag{11-13}$$

where p_1 = proportion of women who said yes
p_2 = proportion of men who said yes
p = best estimate of proportion of population who said yes
$q = 1 - p$

Since the product pq is the variance for proportions, dividing these variances by n makes the denominator of formula (11-13) analogous to formula (11-3).

How do we estimate the population proportion? Since our null hypothesis is that there is no difference between the sexes, we combine the two to get as large a sample as we can and calculate the proportion of yes sayers across both sexes. This is our best estimate of the population proportion p. In our case it is $98/205 = .478$. Then q becomes $1 - .478 = .522$. Applying formula (11-13), we calculate the proportion of yes-saying women ($58/100 = .58$) and yes-saying men ($40/105 = .38$) and multiply p by q, $.478(.522) = .245$, to estimate the population variance for our proportion. Then

$$z = \frac{.58 - .38}{\sqrt{.245/100 + .245/105}} = \frac{.20}{.07} = 2.86$$

This value of z is tested against the normal-curve values at the .05 or .01 criterion, whichever was chosen as appropriate before collecting the data. We recall that a z value of 1.96 leaves 5 percent in the tails of the curve; 2.58 leaves 1 percent. Since our calculated value of 2.86 exceeds the value necessary for the 1 percent criterion, we can reject the null hypothesis. Our data say that it is quite unlikely that men and women come from a common population with regard to approval of their congressman's recent vote.

This test for proportions is suited only to large samples. The sampling distribution of the $p_1 - p_2$ differences approaches normal only when sample sizes are relatively large.

Testing Differences between Correlation Coefficients

If we have two correlation coefficients that have been computed on separate samples and hence are independent, we may wish to ask: Are these two coefficients based on samples from a common population? The null hypothesis, $H_0: \rho_1 = \rho_2$, where ρ is the population correlation, can be tested to answer this question. The procedure is basically the same as that used throughout this chapter. We calculate the difference between the two observed values in which we are interested and evaluate that difference in terms of the standard error of the difference.

One adjustment must be made in correlation coefficients. As we move away from population coefficients of zero, the sampling distribution of r_{12} becomes increasingly skewed. Therefore, before making a test of our null hypothesis, we transform the correlation coefficients into z values, which are normally distributed. Then we can proceed with our test. The procedure looks like this:

$$z = \frac{z_{r_1} - z_{r_2}}{\sqrt{1/(n_1 - 3) + 1/(n_2 - 3)}} \qquad (11\text{-}14)$$

Suppose we take a sample of 58 inner-city teenagers and measure their delinquency proneness and their sociometric status (number of selections by peers). We do this also for a group of 43 suburban teens, and for each group we correlate the two measures. For the inner-city group the correlation is .47; for the suburban group it is .21. Do these coefficients come from samples from a common population?

To pursue this question we apply formula 11-14, but first we must change the coefficients to z values. From Table C-10 our values of r (.47 and .21) become z values of .510 and .214. Our test then is

$$z = \frac{.510 - .214}{\sqrt{1/(58 - 3) + 1/(43 - 3)}} = \frac{.296}{.208} = 1.42$$

This z of 1.42 is evaluated by reference to the normal curve, where a z of 1.96 leaves 5 percent in the tails and 2.58 leaves 1 percent. Since our computed value does not reach 1.96, we accept the null hypothesis.

Exercises

11-11 I have surveyed 110 women and 95 men to see whether they agree that school funds for athletics should be equal for girls and for boys. Their responses are below. Test H_0 at the .01 level.

	Agree	Disagree	
Women	84	26	110
Men	26	69	95
	110	95	205

11-12 A public opinion poll was taken to see whether local citizens support a building bond issue. Subjects were divided into older and younger age groups. Does the support of younger differ from the support of older residents? Test H_0 at the .05 criterion.

	Against	For	
Older	43	27	70
Younger	21	33	54
	64	60	124

11-13 In a sample of 48 college freshmen from Tuff Institute the correlation between GPA and the College Aptitude Test (CAT) was .67. In Laidbach Tech it was .31 for a sample of 39 students. Are these correlations based on samples from a common population? Test H_0: $\rho_1 = \rho_2$ at the .05 criterion.

11-14 In a sociological study between income and days of vacation available each year the investigator found a correlation between these variables of $-.43$ based on 85 cases. A repeat of the study with 52 cases showed a correlation of .30. Are these correlations from samples from a common population? Test H_0: $\rho_1 = \rho_2$ at the .01 level.

Summary

Just as the means of an infinite number of randomly drawn samples from a population produce a normal curve when plotted in a frequency distribution, so do the differences between means of randomly drawn samples. We call this distribution of differences between means a sampling distribution of differences.

Like the data from any normally distributed characteristic, the sampling distribution of differences has a standard deviation. This standard deviation can be used as a yardstick for determining the number of differences between means of randomly drawn samples that are expected to fall beyond given points. We can then establish the likelihood of getting by chance two means with a given difference.

If we are observing two samples of individuals, we calculate the ratio of this difference between their means and the standard error of the difference. We then compare this ratio with points in the distribution beyond which either 5 or 1 percent of the sample differences are expected to fall. If our two sample means differ more than the number of standard errors reported as the 5 or 1 percent criterion points, we reject H_0 and conclude that the samples are likely not to be from the same population.

The selection of the correct procedure for computing the standard error of the difference is especially important when one performs a t test.

1 When sample n's are large, we apply the z procedure and typically compute the standard error of the difference with the unpooled variances.

2 When sample n's are small, we must assume that the samples have come from populations with common variances. If this assumption is reasonable, the pooled-variance technique is the appropriate one for computing the standard error of the difference.

3 If samples are small and cannot be presumed to come from populations with a common variance, unpooled procedures can be applied, but table values of t must be revised before we can test the significance of the difference between the means being observed. Revision of the table values can be made by means of formula (11-11).

4 When the data in the two samples are correlated, revisions in the t procedure must be made to accommodate the fact that values in one sample partially determine values in the other. Special techniques must be employed for these situations.

The basic idea of this chapter can also be applied to testing differences between types of data other than test scores and similar measurements. Two examples are (1) testing differences between proportions and (2) testing differences between correlation coefficients.

A flowchart of decisions in testing differences between means is given in Fig. 11-3.

Key Terms

standard error of the difference	pooled variance
large-sample test	correlated observations
small-sample test	z transformations of r
homogeneity of variance	proportion

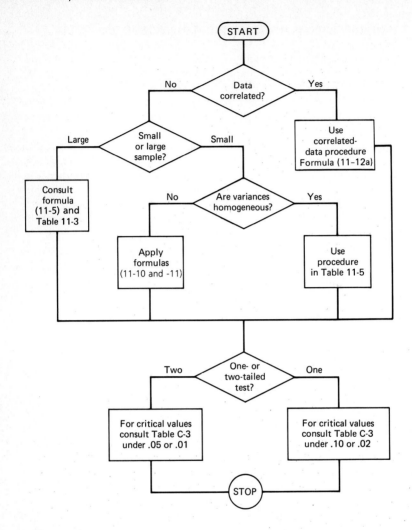

Figure 11-3

Problems

11-1   A curriculum supervisor is studying two methods of teaching reading in which 50 first-graders are assigned randomly to method A and 50 to method B. After 6 weeks of instruction the following statistics were computed from an achievement test in reading:

	$\bar{X}$	s
A	47	9
B	51	10

Test with the 1 percent criterion the hypothesis that there is no difference between the means of the populations from which the samples came.

11-2 In Prob. 11-1 there were 36 girls in method A and 34 girls in method B. The basic data were as follows:

	$\bar{X}$	s
A	51	9
B	54	11

Test the hypothesis that the two samples are random samples from a common population. Use the 95 percent level of confidence.

11-3 A psychologist studying the effect of drugs on accuracy of hand-eye coordination uses a star-tracing problem as her measure of accuracy. She has 30 subjects in the drug X group and 34 subjects in the placebo group. The number of errors made were as follows:

	$\bar{X}$	s
Drug X	34.3	9
Placebo	37.6	7

Test the null hypothesis for these two groups of subjects.

11-4 A school psychologist who believes that experience with intelligence tests will increase the scores of inner-city children randomly assigns children to two groups. One gets weekly experience with tests for 4 weeks and then takes the criterion test; the other reads current events for 4 weeks during the testing period and then takes the criterion test. The descriptive data for the groups are below. Test the null hypothesis with the .05 criterion.

	Test	No test
$\bar{X}$	111.3	109.6
s	14.1	13.7
n	110	84

11-5 On a repeat of the study in Prob. 11-3 the psychologist had a group of 10 subjects who took drug X and a group of 12 who took the placebo. The number of tracing errors made by the subjects was as follows:

Drug X

Subject	A	B	C	D	E	F	G	H	I	J
Errors	17	19	21	27	25	31	30	28	24	36

Placebo

Subject	M	N	O	P	Q	R	S	T	U	V	X	X
Errors	29	21	34	36	29	37	38	26	28	39	35	28

(*a*) Test these samples for homogeneity of variance.

(*b*) Test the null hypothesis for differences between means at the 1 percent level. *Hint:* If you are computing without a calculator, be sure to use Table C-1 as an aid for squaring numbers.

11-6 The psychologist in Probs. 11-3 and 11-4 repeated her study on a third group of 12 subjects. Half the group got drug X, half a placebo, and they were tested. Two days later the half who had gotten the placebo took drug X, and those who had first taken the drug got the placebo. Again the test was given; thus, two scores were available on each subject, as follows:

Subject	A	B	C	D	E	F	G	H	I	J	K	L
Drug X	28	24	29	36	34	37	31	39	29	28	35	33
Placebo	26	31	31	35	35	39	32	41	31	29	36	33

Test the hypothesis that there is no difference in performance under drug and placebo.

11-7 An educational psychologist developed a new theory about learning verbal material. Out of this theory evolved a method for developing reading readiness. The psychologist tested the method by training an experimental group, while the control was in kindergarten as usual. After these children spent 1 month in grade 1, the psychologist, using flash cards, tested the number of words in each child's sight vocabulary. The scores were as follows. Test the null hypothesis at the 1 percent level.

Experimental

Subject	A	B	C	D	E	F	G	H	I
Words	7	8	4	12	9	10	14	6	9

Control

Subject	M	N	O	P	Q	R	S	T	U	V
Words	4	0	6	10	5	14	6	5	6	7

11-8 A sociologist believes that a strong alcohol-education program for teenagers will reduce the incidence of drinking. He arranges with the local school to present 10 lectures on the effects of alcohol to the experimental group; no lecture is given to the control group. An independent observer 6 weeks later surveyed the groups to see how many drinks each student had in the last 2 weeks. Test the null hypothesis (.05). Test for homogeneity of variances (.05) before use of t.

Experimental		Control	
1	1	2	5
3	3	4	4
0	2	1	12
5	3	0	8
4	2	7	5

11-9 The experiment in Prob. 11-8 was repeated on twins, one twin getting the special treatment, one not. The results (twins' data paired together) were as follows. Test the null hypothesis at the 5 percent level. (Will these data be correlated?)

Experimental	7	9	4	3	5	11	5	6	7
Control	6	10	4	2	3	9	4	6	5

11-10 A psychologist, wishing to control for the effect of intelligence matched pairs of five-year-olds on their mental age (making them correlated pairs). Then one member of each pair was randomly assigned to seriation exercises while the other had a play period. After 1 week both groups were tested on seriation. Their scores are below. Since training was believed only to improve performance, test H_0 at the .05 level with a one-tailed test.

Twin set	A	B	C	D	E	F	G	H	I	J
No Training	10	7	4	8	11	5	9	6	8	7
Training	11	7	5	10	13	4	11	8	7	9

12

Introduction to Analysis of Variance

The t test, as we have just seen, is a procedure for determining whether the means of samples A and B are sufficiently different to say that such a difference is unlikely as a result of chance selection of random samples. But the t test is limited primarily to situations in which we are dealing with only two samples at a time. Suppose we have three samples, all being observed simultaneously. For example, we are studying the effect of alcohol on hand-eye coordination. Group A takes 2 ounces of alcohol 5 minutes before tracking a stylus along a wire, group B has 1 ounce, and group C has none at all. Are these three groups from the same population of stylus trackers?

At first we might think that a reasonable solution would be to apply the t test to each of the possible pairs of group means, that is, A and B, A and C, B and C. In the present case, where we have only three tests to make, the labor is not extensive. But suppose we had five groups. Ten tests would have to be made. And suppose we had 8 groups, 12 groups, 15. Clearly the t test is an inefficient procedure here.

Furthermore, the t test is tied to a distribution of differences between *pairs* of sample means. At the 95 percent level of confidence we expect that 5 pairs of random samples out of every 100 will fall in the tails of the distribution beyond the points where we reject the null hypothesis. If we make a large number of t tests on samples from a common population, we shall find by chance alone some differences which fall beyond the critical value. Hence if we make many t tests on many samples from a common population, the likelihood of getting at least one significant difference by chance increases markedly over what it would be if we took only two samples.

Another and probably more serious problem arises from multiple *t* tests done on the same body of data. The various differences between means are not independent of each other. For example, suppose we have three samples of children and have means for nonsense-syllable learning on these samples. The difference between means A and B is 3 and between A and C is 4. Now what is the difference between B and C? Did you say 1? How did you decide that? The difference between B and C is clearly tied to the differences between A and B and A and C. In other words, one difference in the three is not independent of the other differences, and with this nonindependence of differences the *t* test, which is tied to a distribution of differences between pairs of means of independent, random samples, does not appear to be entirely satisfactory.

What we need is a procedure for simultaneously testing the differences between several groups. This procedure is found in the *analysis of variance*.

As its name implies, the analysis of variance deals with variances. Variance is simply the arithmetic average of the squared deviations of scores from their mean; i.e., a variance is the square of the standard deviation. Therefore in the procedures ahead we shall be repeatedly dealing with $\Sigma(X - \bar{X})^2$, which in Chap. 5 we saw is equal to $\Sigma X^2 - (\Sigma X)^2/N$. These values are the sum of squared deviations of scores from their mean often called simply the *sum of squares* (SS).

Up to this point we have dealt with standard deviations (and variances) based on two kinds of data: (1) the standard deviation of raw scores from the mean of the group and (2) the standard deviation of group means from the population mean (which we called a standard error of the mean). In the analysis of variance we simply borrow these two ideas to make up a single procedure.

Partitioning Variance

Suppose you are standing among a group of people clustered around the 30-yard line on a football field. How far are you from the 50-yard line? We can find out in either of two ways. We can measure in yards how far you are from the 50-yard line, or we can measure the distance you are from the 30-yard line and add this algebraically to 20 (the distance from the 30-yard line to the 50-yard line). In the latter case there are two components of the desired distance to the 50-yard line. One is the distance you are from the 30-yard line, and the second is the distance between the 30-yard line and the 50-yard line.

Suppose we have collected IQs on five randomly selected samples of ten-year-old children, 20 in each sample. Our best estimate of the population mean (we shall call it the *grand mean* or *total mean* to distinguish it from sample means) is the sum of the entire 100 cases divided by N. Now let us look only at Susan, who is a member of sample 1. How far does her IQ deviate from the grand mean? We can find out in one of two ways. We can subtract the grand mean $\bar{X}_t$ from Susan's IQ score $X_s - \bar{X}_t$. Or we can find how far the mean of sample 1 (Susan's sample) deviates from the grand mean $\bar{X}_1 - \bar{X}_t$ and then algebraically add to this the amount by which Susan's score deviates from her group mean $X_s - \bar{X}_1$. The sum of these two deviations tells us how far Susan's score deviates from the grand mean

$$X_s - \bar{X}_t = (\bar{X}_1 - \bar{X}_t) + (X_s - \bar{X}_1)$$

In other words, the deviation of a given score from the grand mean can be partitioned into two clearly distinct segments: the deviation of the score from the mean of its group and the deviation of that group mean from the grand mean. This is shown in Fig. 12-1, where the individual's score deviation from the grand mean $\bar{X}_t$ is shown as being made up of two parts: the deviation of the individual's score X from the mean for her group $X_1 - \bar{X}_1$ and the deviation of that group's mean from the grand mean $\bar{X}_1 - \bar{X}_t$.

Whenever we have data of any kind varying around a central point, we can compute an indicator of dispersion. One useful indicator of dispersion is the variance $\Sigma(X_i - \bar{X}_i)^2/\mathrm{df}$, where df stands for degrees of freedom. When we have several samples from a common population, we have the opportunity to compute two such variances, one from that portion of scores which makes up the deviations around sample means $\Sigma(X_i - \bar{X}_i)^2$ and another based on deviations of sample means around the grand mean $\Sigma(\bar{X}_i - \bar{X}_t)^2$. By partitioning a score deviation from the grand mean into two components, $X_i - \bar{X}_i$ and $\bar{X}_i - \bar{X}_t$, we provide ourselves with the data necessary to compute two variances. What shall we do with them?

Before we decide we must digress for a moment. Suppose I

Figure 12-1

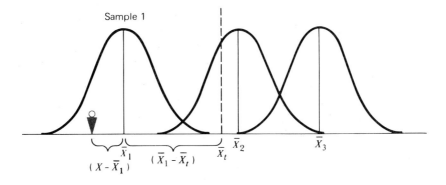

randomly select four teams of 15 boys each (total 60) from a freshman physical education class. I ask each boy to stand at the free-throw line on a basketball court and throw 20 free throws. I then compute a mean number of successful free throws for each of the four teams and the mean for all 60 boys combined.

How far does John's score deviate from the grand mean for 60 boys? This deviation has two parts. We first subtract John's score from the mean of his team $X_i - \bar{X}_i$ and add this algebraically to the deviation of his team's mean from the grand mean $\bar{X}_i - \bar{X}_t$.

Now let us look at these two components of variance. Suppose John made three more baskets than the average for his team. Does this tell me anything about how far his team ranked above, or below, the grand mean for all 60 boys? It does not. John's position above or below the mean for his team has no relation to the position of his team's mean above or below the grand mean. John could make three more baskets than average for his team and still be in the lowest scoring team, the highest scoring team, or either of the other teams.

The point is that the two components of variance cited above, $X_i - \bar{X}_i$ and $\bar{X}_i - \bar{X}_t$, are independent of each other. The variances we compute out of them are also independent.

We are now ready to decide what to do with the two variances we shall compute. We shall put them into a ratio called the F ratio. If the two variances are estimates of a common population variance, their ratio will be near 1.00. If the F ratio deviates very far from this value, we must assume that the two variances are not estimates of a common population value. We shall elaborate on this below, but for now let us go back to the two variances we can compute from partitioning deviations around the grand mean.

Recall that we compute a variance by first finding the deviation of each score from the mean, squaring each of these deviations, and then adding these squared deviations together. This provides the sum of squared deviations, or simply the sum of squares. To get the variance we divide the sum of squares by the degrees of freedom.

Since we now have two components of a score's deviation from the grand mean, we shall go through the above steps for each component. We call the variance based on deviation of sample means from the grand mean the *between-groups variance*, and the one based on deviations of scores from sample means the *within-groups variance*. The rationale of analysis of variance lies in the fact that *both the within-groups and the between-groups variance are estimates of the population variance*.

Having found these two variances, we ask: Do the sample means vary around the grand mean *more* than the individual scores vary around the sample means? If the sample means *do* vary around the grand mean *more than* the individual scores vary around their sample means, the samples are comparatively widely dispersed from each other; but if the sample means vary around the grand mean about as much as, or less than, individual scores vary around their sample means, the samples are very much like each other in score values.

Figure 12-2*A* and *B* show three samples of individuals. The individuals in the samples in Fig. 12-2*A* are dispersed around the *sample means* to exactly the same extent as individuals in the samples in Fig. 12-2*B*; i.e., the within-group variance in Fig. 12-2*A* is equal to the within-group variance in Fig. 12-2*B*. On the other hand, the *sample means* in Fig. 12-2*A* are much more widely dispersed around the *grand mean* than those in Fig. 12-2*B*. There is almost no overlap between samples in 12-2*A*. In other words, the between-group variance in Fig. 12-2*A* is much greater than the between-group variance in Fig. 12-2*B*. Considering these two items of information, we may wish to conclude that the samples in Fig. 12-2*A* are not random samples from the same population, but those in 12.2*B* probably are from the same population. The difference is the relatively large between-groups variance in 12-2*A*.

The conclusion whether the samples come from the same or different populations is based on the size of the between-groups variance compared with the within-groups variance. We can state this comparison as the *F* ratio

$$F = \frac{s_{bg}^2}{s_{wg}^2}$$ (12-1)

Figure 12-2

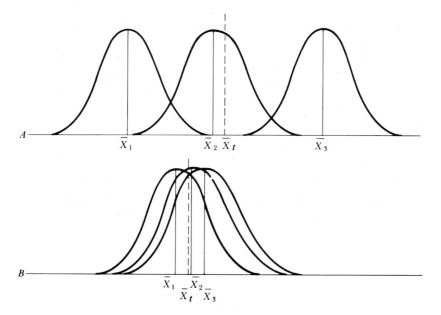

where s_{bg}^2 is the between-group variance and s_{wg}^2 is the within-group variance. As noted above, if s_{bg}^2 and s_{wg}^2 are really estimates of the variance of the same population, the F ratio will approach 1.0; in a given problem if they are estimates of variances of different populations, the F ratio will clearly deviate from 1.0.

Within-Groups Variance

How do we compute these variances? If you remember how pooled variances were found for the t test, you have the basic technique for computing s_{wg}^2. Let us deal with it first. For a three-sample problem the procedure is

$$s_{wg}^2 = \frac{\Sigma(X_1 - \bar{X}_1)^2 + \Sigma(X_2 - \bar{X}_2)^2 + \Sigma(X_3 - \bar{X}_3)^2}{(n_1 - 1) + (n_2 - 1) + (n_3 - 1)} \qquad (12\text{-}2)$$

If we square each value in the numerator, as we did in computing the variance in Chap. 5,* we come up with

$$s_{wg}^2 = \frac{\Sigma X_1^2 + \Sigma X_2^2 + \Sigma X_3^2 - [(\Sigma X_1)^2/n_1 + (\Sigma X_2)^2/n_2 + (\Sigma X_3)^2/n_3]}{N - 3}$$

$$(12\text{-}3)$$

where X_1, X_2, X_3 = scores from samples 1, 2, 3
n_1, n_2, n_3 = number of cases in samples 1, 2, 3
N = total number of cases = $n_1 + n_2 + n_3$

  In reading research where analysis of variance has been used, one often sees the term *mean square*. If the values $(X - \bar{X}_i)^2$ are averaged, we have a mean of the squared deviations around the mean, or in short, a *mean square*. To average anything we add all the values (Σ) and divide by their number (here by degrees of freedom) and get a mean. In this case for a single sample i it is

$$\frac{\Sigma(X - \bar{X}_i)^2}{n_i - 1}$$

When we combine these terms across all samples, we have formula (12-2). That formula is for the within-group variance, which is also referred to as the *mean square within groups* MS_{wg}. The terms are synonymous.

  The generalized procedure for dealing with k samples is given in Table 12-1.

* If this procedure is not familiar, students should review pages 89–91.

Table 12-1   Computational Procedure for Within-Groups Variance

---

Formula (12-3)

$$s_{wg}^2 = \frac{\Sigma X_1^2 + \Sigma X_2^2 + \cdots + \Sigma X_k^2 - [(\Sigma X_1)^2/n_1 + (\Sigma X_2)^2/n_2 + \cdots + (\Sigma X_k)^2/n_k]}{N - k}$$

where $\quad s_{wg}^2$ = within-group variance = mean square within groups

$X_1, X_2, \ldots, X_k$ = individual scores in samples $1, 2, \ldots, k$

$n_1, n_2, \ldots, n_k$ = number of cases in samples $1, 2, \ldots, k$

N = total number of cases = $n_1 + n_2 + \cdots + n_k$

k = number of samples

What it says to do

---

1   For each sample, square each score and sum these squared scores to get $\Sigma X_1^2, \Sigma X_2^2, \ldots, \Sigma X_k^2$

2   For each sample sum all scores and then square these sums to get $(\Sigma X_1)^2$, $(\Sigma X_2)^2, \ldots, (\Sigma X_k)^2$

3   For each sample divide the $(\Sigma X)^2$ value by the n value for that sample

4   (a) Sum all ΣX^2 values

    (b) Sum all $(\Sigma X)^2/n$ values

    (c) Subtract the value of step (b) from the value of step (a)

5   Divide the result of step 4 by $N - k$

Between-Groups Variance

Now let us turn to the numerator of the F ratio, the between-groups variance s_{bg}^2. Since this figure indicates the extent to which the sample means vary around the grand mean, a first impression may suggest that it is found by applying the formula $\Sigma(\bar{X}_i - \bar{X}_t)^2/(k - 1)$. But remember that we found for every $(X - \bar{X}_t)$ deviation there are the two components $(X - \bar{X}_i)$ and $(\bar{X}_i - \bar{X}_t)$. Therefore we have an $(\bar{X}_i - \bar{X}_t)$ component for the deviation of each X score from the grand mean; so in computing the sum of squares for between-groups variance we must repeat the $(\bar{X}_i - \bar{X}_t)$ for each case within a sample $n_i (\bar{X}_i - \bar{X}_t)^2$ and then for all samples $\Sigma n_i (\bar{X}_i - \bar{X}_t)^2$. The procedure is

$$s_{bg}^2 = \frac{\Sigma n_i(\bar{X}_i - \bar{X}_t)^2}{k - 1} \tag{12-4}$$

For our three-sample example this becomes

$$s_{bg}^2 = \frac{n_1(\bar{X}_1 - \bar{X}_t)^2 + n_2(\bar{X}_2 - \bar{X}_t)^2 + n_3(\bar{X}_3 - \bar{X}_t)^2}{3 - 1}$$

When each of the enclosed values is squared, this becomes

$$s_{bg}^2 = \frac{(\Sigma X_1)^2/n_1 + (\Sigma X_2)^2/n_2 + (\Sigma X_3)^2/n_3 - (\Sigma X_t)^2/N}{k - 1} \tag{12-5}$$

The generalized procedure for between-groups variance is given in Table 12-2.

We began this chapter by pointing out that the deviation of a given

---

Table 12-2   Computation of Between-Group Variance

Formula (12-5)

$$s_{bg}^2 = \frac{(\Sigma X_1)^2/n_1 + (\Sigma X_2)^2/n_2 + \cdots + (\Sigma X_k)^2/n_k - (\Sigma X_t)^2/N}{k - 1}$$

where s_{bg}^2 = between-groups variance = mean square

$\Sigma X_1, \Sigma X_2, \ldots \Sigma X_k$ = sums of individual scores on samples $1, 2, \ldots, k$

ΣX_t = sum of individual scores in all samples combined
$= \Sigma X_1 + \Sigma X_2 + \cdots + \Sigma X_k$

$N = n_1 + n_2 + \cdots + n_k$ = number of cases in samples 1, 2, $\ldots, k$

k = number of samples observed

What the formula says to do

1   For each sample, sum all scores to get $\Sigma X_1, \Sigma X_2, \ldots, \Sigma X_k$; square these values and divide each by the number of cases that went into that sum; sum all resulting $(\Sigma X_i)^2/n_i$ values

2   Sum the sample sums $\Sigma X_1 + \Sigma X_2 + \cdots + \Sigma X_k$ to get the sum of all cases ΣX_t; square this value and divide it by the total number of cases in all samples combined

3   Subtract the result of step 2 from the result of step 1

4   Divide the result of step 3 by 1 less than the number of samples in the analysis $k - 1$

case from the grand mean can be divided into two parts: (1) the deviation of the individual from his sample mean and (2) the deviation of this sample mean from the grand mean. The deviations around the grand mean for a group of scores for combined samples can then be partitioned into the within-group deviations and the between-groups deviations. In other words, the total sum of squared deviations of scores around the grand mean for all groups combined should be equal to the sum of the two partitioned sums of squares; i.e., the sum-of-squares total equals the sum of squares between groups plus the sum of squares within groups

$$SS_t = SS_{bg} + SS_{wg}$$

We can compute the sum of squares by the procedure

$$SS_t = \Sigma(X - \bar{X}_t)^2$$

or the equivalent

$$SS_t = \Sigma X_t^2 - \frac{(\Sigma X_t)^2}{N}$$

We can then use this figure to simplify other computations. For example,

$$SS_{wg} = SS_t - SS_{bg}$$

We can also check the computation since the sum of SS_{bg} and SS_{wg} should equal the independent calculation of SS_t.

The F Test

Now that we have the components in hand, we are ready to complete the F test. As noted earlier, the F test is based on the ratio of the between-group variance and the within-group variance. The former tells us how widely the group means are distributed around the grand mean; the latter provides us with a yardstick evaluating that dispersion. If both variances are about equal, i.e., are estimates of the same population variance, the samples are not dispersed beyond what we expect from sampling variations. In this case the data may resemble Fig. 12-2B. On the other hand, if the between-group variance is somewhat larger than the within-group variance, we may wish to conclude that the samples are spread out sufficiently to make us believe that they are not all from a common population. Here the data would look more like Fig. 12-2A.

To decide whether our samples are all from a common population we test the null hypothesis

$$H_0: \mu_1 = \mu_2 = \cdots = \mu_k$$

If we accept the null hypothesis, we believe the groups look like random samples from a common population. If we reject the hypothesis, we believe that not all the samples came from the same population. At least one, and possibly more, came from a population different from the population from which the other samples came.

An inspection of the components of the variances in the F test may help explain this concept more fully. The within-group variance is a direct estimate of the population variance σ^2, but the between-group variance is made up of two components. Since one is an estimate of the population variance σ^2 and one is tied to the treatment effect, i.e., "real" differences between the groups, the F ratio actually looks like

$$F = \frac{\sigma^2 + \dfrac{\Sigma n(\mu_i - \mu)^2}{k - 1}}{\sigma^2}$$

If the differences between the groups, as shown by the second term in the numerator, are zero, the F test becomes a comparison of two estimates of the population variance and will be near 1.0. However, if the differences between groups are important, i.e., some systematic difference does exist between the groups, the second term in the numerator will be greater than zero and will add variance. Then the numerator will be larger than the denominator.

How much larger does it have to be? We decide this by relating our F ratio to a distribution of F ratios made up of many samples. The distribution of F is typically not symmetrical like z or t distributions; in fact, it can be quite skewed (Fig. 12-3).

Like the t distribution, the F distribution is not a single curve but a large family of curves, varying with the degrees of freedom. Therefore

Figure 12-3

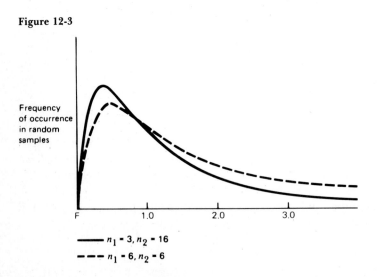

Frequency of occurrence in random samples

F   1.0   2.0   3.0

——— $n_1 = 3, n_2 = 16$
- - - $n_1 = 6, n_2 = 6$

the proportion of F ratios expected by chance to fall beyond given values varies with the degrees of freedom. Knowing this, from such a frequency distribution we can determine the likelihood of getting F values of a given magnitude for various combinations of degrees of freedom, which brings up our next point.

In a t test we have only one value for degrees of freedom, but in an F test we have degrees of freedom for s_{bg}^2 and s_{wg}^2. In Table C-4 we find the critical value that leaves 5 percent in the tail of the curve by first locating in the left marginal column the degrees of freedom associated with s_{wg}^2, the denominator of the F test. The degrees of freedom here are $N - k$. Then moving across that row we find the column at the top which is the number of degrees of freedom associated with s_{bg}^2, the numerator, i.e., $k - 1$. This column-row cell will contain an F value at the 95 percent level of confidence. Table C-6 contains F values for the 99 percent level of confidence. If our F ratio exceeds one or both of these criterion values, we reject the null hypothesis at the level of the point surpassed. If it is less than the 95 percent value, the null hypothesis is accepted. In making these comparisons we have completed the F test. An example is laid out in Table 12-3.

Table 12-3 Completion of an F Test on Three Randomly Selected Samples of College Freshmen*

	Sample 1	Sample 2	Sample 3	
	5	5	6	
	3	3	4	
	1	3	4	
	4	1	2	
	2	4	3	
ΣX_i	15	16	19	$\Sigma X_t = 50$
ΣX_i^2	55	60	81	$\Sigma X_t^2 = 196$

$$s_{wg}^2 = \frac{55 + 60 + 81 - (15^2/5 + 16^2/5 + 19^2/5)}{15 - 3} = 2.30$$

$$s_{bg}^2 = \frac{15^2/5 + 16^2/5 + 19^2/5 - 50^2/15}{3 - 1} = .86$$

$$F = \frac{.86}{2.30} = .37$$

With $N - k = 12$ df in the denominator and $k - 1 = 2$ df in the numerator, we consult Table C-4 and find that we must accept the null hypothesis.

* The data represent the number of algebra problems solved in 3 minutes by each student.

Exercises

12-1   Three samples of nine-year-old boys are selected from a grade school for the purpose of testing physical training methods. Sample 1 was a control group and received no special training; sample 2 had 1 hour of work a day in such group sports as volleyball, touch football, and basketball; sample 3 had calisthenics 30 minutes daily. At the end of 2 weeks a physical fitness test was administered to all groups. Their scores are below. Test the null hypothesis (.05) for these samples.

Sample 1	Sample 2	Sample 3
10	10	7
11	9	9
9	5	6
6	6	5
8	8	3
7	7	2

12-2   Compute the mean for each sample of Prob. 12-1. Except for rounding errors does

$$\frac{n_1 \bar{X}_1^2 + n_2 \bar{X}_2^2 + n_2 \bar{X}_3^2 - N \bar{X}_t^2}{k - 1}$$

equal the value you found for the between-group variance? If means of samples are already available, is this method faster than formula (12-5)?

12-3   A psychologist has developed a training regimen which be believes can improve one's ability to memorize factual material. To test his procedure he randomly selects three groups of subjects from the freshman class in introductory psychology. One class gets the training regimen, one class reads a book on how to improve your mind, while the third group does nothing special. Afterwards all subjects take a memory test. The scores (number correct) are below. Test the null hypothesis at the 99 percent level of confidence.

Person	Training	Person	Book	Person	Control
A	8	G	7	N	4
B	14	H	6	O	9
C	19	I	9	P	7
D	11	J	4	Q	14
E	9	K	15	R	5
F	13	L	11	S	10
		M	8		

Assumptions in Analysis of Variance

Although the distribution of F values is known to take on certain characteristics with different combinations of degrees of freedom, these characteristics are known only under certain conditions which we assume to be present:

1 All populations from which samples have been drawn are normally distributed.

2 The variances for the populations from which samples have been drawn are equal.

3 The individuals being observed have been randomly selected from the populations represented by the samples.

The values in Tables C-4 to C-7 are precise only when we have met the above assumptions. However, in actual practice it has been observed that one or more of these assumptions can be "bent" without appreciable loss in the adequacy of the F test. Researchers strive to meet the assumptions of the F test but usually find that if their data are reasonably close to meeting the assumptions, their conclusions based on the F test are not markedly affected. This is especially true if sample sizes are equal or very nearly so.

Summary

The t test is an adequate procedure for testing the null hypothesis when we must deal with means of only two samples. But since we often have more than two samples to consider at one time, an alternative procedure is needed for testing the hypothesis that all samples could likely be from the same population. The analysis of variance is appropriate for such a test.

The procedure is based on the fact that the deviation of a given score from the population mean can be divided into two parts: (1) the deviation of the score from its sample mean and (2) the deviation of the sample mean from the population mean. Therefore, out of the total deviation of individuals around the population mean we can compute two variances: (1) one for deviation of scores around their sample means and (2) one for deviation of sample means around the grand mean.

We then put these two variances into an F ratio. If the variance of sample means around the grand mean is conspicuously greater than the variance of scores around the sample means, the samples must be, relatively speaking, widely dispersed around the grand mean, very likely not representing random samples from the same population. However, if the dispersion of sample means around the grand mean is similar to the dispersion of sample scores around the sample means, the samples are likely all to be random samples from a common population.

Key Terms

partitioning variance	*F* ratio
within-group variance	*F* distribution
between-group variance	sum of squares
grand mean	mean square

Problems

12-1   A curriculum supervisor has undertaken a study of the value of grade reports to parents in promoting achievement. Three groups of fourth-grade students have been randomly chosen from all fourth-graders in the L. B. Johnson Elementary School. Group 1 gets a conventional report card every 8 weeks; a parent-teacher conference is held every 8 weeks for group 2; and group 3 is given no report at all. At the end of the year tests are given in all subject areas. Raw scores for arithmetic problems are given below. Test, at the 95 percent confidence level, the hypothesis that the three samples are from a common population.

Sample 1		*Sample 2*		*Sample 3*	
12	10	14	11	8	5
10	7	8	13	11	6
11	9	19	12	13	8
11	10	15	9	9	7
8	6	10	12	7	10

12-2   In the study described in Prob. 12-1 the spelling-test scores were as below. Test the null hypothesis for spelling (.05).

Sample 1		*Sample 2*		*Sample 3*	
8	5	5	4	10	4
2	6	10	7	4	7
10	5	2	12	5	8
4	6	6	5	2	6
7	4	8	1	6	7

12-3   A medical team studying the relationship between cigarette smoking and various diseases selected five groups of subjects; four groups consisted of patients in a city hospital, and one was made up of a random selection of visitors to the hospital. The number of cigarettes smoked per day by each subject is given below. Test the null hypothesis (at the 99 percent level of confidence) for these data.

Lung diseases	Heart disease	Digestive diseases	Kidney disorders	Visitors
32	19	17	12	12
17	26	26	15	15
28	30	30	10	36
24	17	35	20	17
21	34	20	18	20
38	15	15	30	25

12-4 In Table 12-3 we computed an F ratio. Before we divided SS_{wg} by its degrees of freedom, $15 - 3$, we had the value 27.6. Before dividing SS_{bg} by its degrees of freedom, $3 - 1$, we had the value 1.73. Now compute the SS_t by the formula $\Sigma X_t^2 - \Sigma X_t/N$.

(a) Does $SS_{wg} + SS_{bg} = SS_t$?

(b) Can we compute SS_{bg} if we already have SS_{wg} and SS_t? Is it less effort to compute SS_{bg} this way?

12-5 In the secretarial pool of a large company I administered a vocational-interest inventory and on the basis of its results identified three distinct groups of people: (1) those who had high clerical interests; (2) those who had moderate to low clerical interest but high interest in other areas traditionally identified as "female," e.g. nursing; and (3) those who had moderate to low clerical interest but scored high in an area traditionally thought of as "male," e.g., carpentry. I then administered a job-satisfaction inventory. The scores are below. Test the null hypothesis at the 95 percent level.

High clerical	High female	High male
18	20	7
29	17	15
28	21	9
26	19	13
24	29	5
31	26	11
30	25	
27		

12-6   Compute the means for the separate groups in Prob. 12-5. (You already have the sums for each group.) Speculate between which groups the important differences appeared. (This is an example of post hoc data snooping.)

12-7   After dividing the members of a Little League baseball program into three groups by body weight, I recorded for each child the number of hits made at the next 20 times at bat. At the .05 criterion is there a difference between weight groups in the number of hits?

Low weight	Medium weight	High weight
10	14	4
4	9	6
1	11	3
8	7	2
6		9
		5

12-8   An art instructor who wanted to observe the impact of two methods of teaching appreciation for line in paintings randomly assigned students to two experimental groups and one control group. The next 2 weeks were spent by one group in observing and discussing famous paintings and by the other group in drawing and discussing these drawings. Controls looked at magazines during the same period. Then a set of three paintings known to have definite line elements were rated by the students. Their ratings (based on a maximum of 10 points per painting, or 30 total) are below. At the .01 criterion, is there a difference between the groups in ratings?

Control	Famous	Own
20	18	24
12	12	27
14	16	14
10	10	21
17	15	18
9	8	20

12-9   A psychologist randomly assigned children in an "emotionally disturbed" group to two treatments and a control. Treatment A received a behavioral modification program; B received daily group therapy; the control received no special program. After a month, independent observers rated the children on emotional control in their classrooms. The ratings are below. Test H_0: $\mu_A = \mu_B = \mu_C$ at the .05 level.

A	B	*Control*
8	6	3
6	4	5
9	5	4
7	8	2
10	7	3

12-10 A journalist studied the accuracy of four publishers A to D, each of whom had handled a book in mathematics, physics, chemistry, and biology at the same level of complexity. He counted the number of printer's errors in the first 100 lines of proof for each of the four books for each publisher, listed below. Test H_0: $\mu_A = \mu_B = \mu_C = \mu_D$ at the 95 percent confidence level. (Publishers used different typesetter for each book.)

	A	B	C	D
Mathematics	18	15	21	10
Physics	4	11	13	2
Chemistry	12	21	26	8
Biology	8	26	15	6

13

Post Hoc Tests Following Analysis of Variance

The null hypothesis in analysis of variance is H_0: $\mu_1 = \mu_2 = \cdots = \mu_k$. If we reject the null hypothesis, we believe that not all our groups are from a common population. This means that (1) all groups can be from different populations (Fig. 13-1a), (2) several could be from one or more populations while the rest are from a common population (Fig. 13-1b), or (3) only one sample is from a different population (Fig.13-1c). The F test itself does not sort out samples by populations. It merely says they are or are not all from a common population. Therefore, once we have rejected H_0 with analysis of variance we typically want to perform additional tests to see which samples cluster into various populations. These tests are called *post hoc* or *a posteriori tests* and are the subject of this chapter.

We shall explore two types of post hoc tests, the Tukey and the Scheffé procedures.

The Tukey Procedure

Tukey proposed a method for testing all pairs of means to see which pairs are from common populations and called it the *honestly significant difference test* (HSD). It resembles the t test in that we first calculate a difference between two means and assess the magnitude of that difference by a standard error of the difference. However, the procedure departs from the t test in two ways: (1) the method of calculating the standard error uses the within-group mean square from our analysis of variance, and (2) the probability distribution to which we relate our t varies with the number of sample means that enter into the

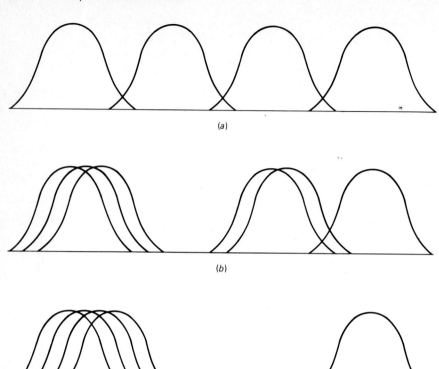

(a)

(b)

(c)

Figure 13-1

total problem. For example, if there are three samples being observed, the distribution of t values would be different from that for five samples.

Let us begin with some data taken from a real problem. Three random samples of five college freshmen each have been used in a learning study. After treatment each group was given a test of 20 nonsense syllables to learn, and trials to perfect performance were recorded for each student. The data and the analysis of variance on them are shown in Table 13-1. Since Table C-4 says that 3.8853 is the .05 criterion, H_0: $\mu_1 = \mu_2 = \mu_3$ is rejected.

Having rejected H_0, we now turn to the individual group means to sort them out. We believe that not all samples are from a common population, but are they all from different populations or are two of them from the same population while the third is from another

Table 13-1   Data for Learning Study and Analysis of Variance

	Sample 1	Sample 2	Sample 3
	21	10	15
	17	7	18
	16	9	10
	15	5	19
	13	6	17
ΣX	82	37	79
ΣX^2	1380	291	1590

Source	SS	df	MS	F
Between groups	253.2	2	126.5	14.71
Within groups	103.2	12	8.6	
Total	356.4	14		

population? Tukey's test will be applied here to help answer this question. The group means in the study are:

Sample 1	Sample 2	Sample 3
16.4	7.4	15.8

We are now ready to begin testing differences between pairs of means to see which came from different populations. To do this we need a yardstick to assess the magnitude of each $\bar{X}_i - \bar{X}_j$ difference; as with the t test, this will be a standard error of the difference. But here that standard error will use the MS_{wg} from the analysis of variance because it is our best estimate of the population variance. The formula is

$$s_q = \sqrt{\frac{MS_{wg}}{n}} \tag{13-1}$$

where s_q = standard error of difference for the post hoc comparison of means

MS_{wg} = mean square within groups from Table 13-1

n = number of cases in a single sample

290 Elementary Statistical Procedures

For the Tukey test, *samples must be of equal size*. For our data

$$s_q = \sqrt{\frac{8.6}{5}} = 1.31$$

We are now ready to do the Tukey test, which we shall call Q. The complete procedure is

$$Q_{i-j} = \frac{\bar{X}_i - \bar{X}_j}{\sqrt{MS_{wg}/n}} \qquad (13\text{-}2)$$

Since we have already calculated the denominator, we shall compute Q for our data as follows:

$$Q_{1-2} = \frac{16.4 - 7.4}{1.31} = 6.87$$

$$Q_{1-3} = \frac{16.4 - 15.8}{1.31} = .46$$

$$Q_{2-3} = \frac{7.4 - 15.8}{1.31} = 6.41$$

With which of these Q values can we reject the null hypothesis, H_0: $\mu_i = \mu_j$? We turn to Table C-9, Percentage Points of the Studentized Range. This range is based on the expected values of

$$Q = \frac{\bar{X}_{\max} - \bar{X}_{\min}}{\sqrt{MS_{wg}/n}} \qquad (13\text{-}3)$$

With more random samples from a population the expected difference between $\bar{X}_{\max}$ and $\bar{X}_{\min}$ will increase. Since the sampling distribution of the $\bar{X}_{\max} - \bar{X}_{\min}$ difference is known for various numbers of samples and for various degrees of freedom, we can test our computed values of Q against the table values at the .05 or .01 criterion and decide whether our means are indeed more widely dispersed than expected from random sampling.

Table C-9 is now entered with k groups, and $N - k$ degrees of freedom. We had three groups and $15 - 3$ degrees of freedom (associated with the mean square within groups). Table C-9 provides a value of 3.77 at the .05 criterion. Our computed values of Q_{i-j} must reach this value for the .05 critical value. Two of our differences reached this value, $\bar{X}_1 - \bar{X}_2$ and $\bar{X}_2 - \bar{X}_3$. The sign of the computed

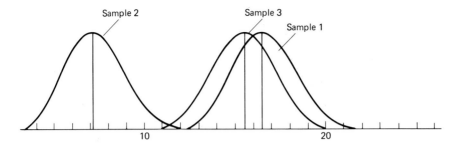

Figure 13-2

value is not of great importance since we are dealing with a symmetrical curve.

The distributions of scores look like Fig. 13-2. Here we see that sample 2 stands below the other samples. Our tests show that it is not from the same population as 1 and 3. From our data we conclude that the samples are not all from a common population (based on the significant *F* value), and that samples 1 and 3 come from a common population but sample 2 comes from a different population. The last half of this conclusion could not have been drawn with confidence unless the Tukey test had been applied.

Exercises

13-1 The following data based on three groups of six high school students each are points made on a dartboard when throwing from different angles. Test H_0: $\mu_1 = \mu_2 = \mu_3$. Then test H_0: $\mu_i = \mu_j$ for all pairs of means and draw a conclusion about the samples based on these tests. Use the 95 percent confidence level.

90°	45° left	45° right
10	6	4
14	4	3
12	2	5
8	3	2
16	5	4
13	3	6

13-2 Three age groups of children tested for errors in |r| allophones had 10 children per group. The number of errors was recorded for each child in each age group. The following analysis of variance was done on the errors:

Source	SS	df	MS	F
Between groups	159.29	2	79.64	8.85
Within groups	243.00	27	9.00	
Total	402.29	29		
	$\bar{X}_1 = 14.6$ $\bar{X}_2 = 10.1$ $\bar{X}_3 = 9.4$			

Test the difference between the means in each pair, $\bar{X}_1 - \bar{X}_2$, $\bar{X}_1 - \bar{X}_3$, $\bar{X}_2 - \bar{X}_3$, at the 99 percent level. What conclusion do you draw from these tests?

The Scheffé Test

Because the Tukey test requires that sample sizes be equal, its application is somewhat limited. Many situations arise in research where samples of unequal size must be used. If samples are very nearly equal in size, cases from the larger samples can be randomly deleted before the F test is done, but care must be taken to ensure that the deletions are strictly random.

An alternative to random deletions is the Scheffé test. After noting that our general F test is significant, we proceed by calculating a test between each pair of means by

$$S_{i-j} = \frac{\bar{X}_i - \bar{X}_j}{\sqrt{MS_{wg}(1^2/n_i + 1^2/n_j)}} \qquad (13\text{-}4)$$

where $\bar{X}_i$, $\bar{X}_j$ = means of samples being compared

MS_{wg} = mean square within groups from overall analysis of variance

n_i, n_j = number of cases in sample i and j

Suppose we have the analysis of variance in Table 13-2. Since the overall F is significant, we begin our data snooping. To test the difference between the pairs of means we apply formula (13-4)

$$S_{1-2} = \frac{15.6 - 10.1}{\sqrt{9.93(1^2/12 + 1^2/10)}} = 4.08$$

Table 13-2 Analysis-of-Variance Data

Source	SS	df	MS	F
Between groups	371.40	2	185.70	18.70
Within groups	268.12	27	9.93	

	Sample 1	Sample 2	Sample 3
$\bar{X}_i$	15.6	10.1	9.4
n_i	12	10	8

$$S_{1-3} = \frac{15.6 - 9.4}{\sqrt{9.93(1^2/12 + 1^2/8)}} = 4.31$$

$$S_{2-3} = \frac{10.1 - 9.4}{\sqrt{9.93(1^2/10 + 1^2/8)}} = .47$$

Now we need a criterion against which to evaluate the S figures. We calculate this criterion as

$$C = \sqrt{(k - 1)F_{\alpha:(k-1)(N-k)}} \qquad (13\text{-}5)$$

where $F_{\alpha:(k-1)(N-k)}$ is the F-table value (from Tables C-4 to C-7) at the specified .05 or .01 criterion for the degrees of freedom $k - 1$ and $N - k$.

In our problem we have $3 - 1$ degrees of freedom for the between-group mean square and $30 - 3$ for the within-group mean square. Taking these to Table C-4, we find at the .05 level the value 3.3541. Substituting these data into formula (13-5) gives

$$C = \sqrt{(3 - 1)(3.3541)} = 2.59$$

This value is next compared with each of our computed S values. If S is greater than C, the difference between the means is significant at the .05 level. Two of our S values exceed 2.59, namely, S_{1-2} and S_{1-3}. The remaining one, S_{2-3}, does not reach our criterion of 2.59. The conclusion is that samples 2 and 3 do not appear to be from different populations, but they are not from the same population as sample 1. We have rejected H_0: $\mu_1 = \mu_2$, $\mu_1 = \mu_3$ and have accepted H_0: $\mu_2 = \mu_3$.

The Scheffé test can also be used when we combine samples and test the combination against other samples or other combinations. For example, in the above problem suppose we wished to combine samples

2 and 3 and test the combination against sample 1. The general procedure is

$$S_{i-(j+k)} = \frac{\bar{X}_i - (\bar{X}_j + \bar{X}_k)/2}{\sqrt{MS_{wg}(1^2/n_i + .5^2/n_j + .5^2/n_k)}}$$

Inserting the data from the above problem, we get

$$S_{1-(2+3)} = \frac{15.6 - (10.1 + 9.4)/2}{\sqrt{9.93(1^2/12 + .5^2/10 + .5^2/8)}} = 4.97$$

Once again we compare this computed S value with the C criterion and reject the null hypothesis. The combined samples 2 and 3 are probably not from the same population as sample 1.

Occasionally an S value will be negative. This is of no consequence since all evaluations are based on the absolute value of S and a value of 4.97 means the same as if it were -4.97, etc.

The Scheffé test can also be used with studies where sample sizes are equal. Since it is a more conservative test than Tukey's, i.e., less likely to find a difference significant, many investigators have chosen to use Tukey's test when sample sizes are equal, or choose a larger rejection area (e.g., .10) when using Scheffé.

Exercises

13-3 A questionnaire assessing attitudes toward a common salary schedule for teachers was circulated among teachers, administrators, and parents. An analysis of variance was done on the results. Test H_0: $\mu_1 = \mu_2$, $\mu_1 = \mu_3$, $\mu_2 = \mu_3$, with the .05 criterion. Then combine samples 2 and 3 and test them against sample 1.

Source	SS	df	MS	F
Between groups	481.32	2	240.66	9.57
Within groups	1182.36	47	25.16	
Total	1663.68	69		

	Administrators	Teachers	Parents
$\bar{X}_i$	21.2	14.3	12.8
n_i	10	22	18

13-4 An investigator is assessing speed of perception in four groups of college freshmen. The groups come from humanities, engineering, social science, and music majors. The analysis of variance on the data is below. Using the .01 criterion, assess the significance of the difference between all pairs of means. Then combine humanities and social science and test them against a combination of engineering and music.

Source	SS	df	MS	F
Between groups	2261.50	3	753.83	29.65
Within groups	1271.22	50	25.42	

	Humanities	Engineering	Social science	Music
$\bar{X}_i$	14.4	25.1	12.7	27.3
n_i	10	14	18	12

Summary

The analysis of variance tells us whether or are not our samples are from a common population. If we reject H_0: $\mu_1 = \mu_2 = \cdots = \mu_k$, we do not know which samples cluster into which population. The next job is to apply a post hoc test to sort out samples to see which ones come from different populations. We apply the Tukey or the Scheffé test to do this sorting.

The Tukey test requires that samples be equal in size. It calculates a standard error of the difference from the mean square within groups, the best estimate of the population variance. It then assesses differences between pairs of means, using this standard error as its yardstick. The difference between two means divided by the standard error is taken to the Studentized range table to identify the .05 or .01 criterion.

The Scheffé test does not require that groups be of equal size. It can also be applied to combine groups and compare this combination with another sample or combination of samples. It has the problem, however, of being a conservative test, not rejecting the null hypothesis as often as would Tukey's on the same data (assuming equal n's).

Scheffé has devised a procedure for calculating the criterion value rather than taking it directly from a table of critical values. This procedure is tied to the more rigorous rejection rate. Because of this rigor some investigators select a larger rejection area with Scheffé's test, e.g., the 10 percent level rather than the 5 percent level, but tables of the 10 percent criteria are not readily available.

The general process of applying post hoc tests is shown in the flowchart in Fig. 13-3.

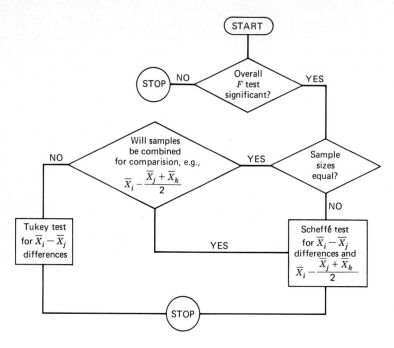

Figure 13-3

Key Terms

post hoc	Q_{i-j}
a posteriori	S_{i-j}
Tukey HSD	C
Scheffé test	

Problems

13-1 The following data collected on four age groups of women represent a 1-to-10 rating of a popular women's magazine. Do an analysis of variance and Tukey tests between all pairs of means. Use the .01 criterion for all tests.

20–29	30–39	40–49	50–59
10	9	4	3
8	7	5	4
9	8	3	3
7	7	2	5
9	8	3	4

13-2 The following analysis of variance was based on strength (in ounces) of right-hand grip of children in three groups of readers: A is five-year-olds who do not have a 20-word sight vocabulary, B is five-year-olds who do have a 20-word sight vocabulary, and C is six-year-olds who do not have a 20-word sight vocabulary. Do a Tukey test for all pairs of means. Use the .05 critical point.

Source	SS	df	MS	F
Between groups	141.87	2	70.93	5.91
Within groups	324.11	27	12.00	
Total	465.98	29		

$\bar{X}_A = 11.2 \quad \bar{X}_B = 15.6 \quad \bar{X}_C = 16.0$

Note: Sample sizes can be figured from the degrees of freedom.

13-3 In a study of opinions about labor unions, a researcher collected data from four groups of people: laboring men, laboring women, management men, and management women. The analysis of variance produced a significant *F* value on the attitude inventory, with $MS_{wg} = 10.43$. There were 16 cases in three groups, 18 in the fourth. Two cases were randomly deleted from the fourth group and Tukey tests between pairs of means are to be calculated. The means are laboring men 25.6, laboring women 20.3, management men 14.2, management women 18.7. Use the .05 criterion.

13-4 Three groups of college freshmen were timed while running the 50-yard dash. Their times are below. Calculate the overall *F* and do a Scheffé test for all possible pairs of means. Then combine business majors with music majors and test them against chemistry majors. Use .01 for the overall criterion; .05 for the Scheffé tests.

Business	Music	Chemistry
8	11	7
7	13	5
12	10	6
14	15	8
10	12	
10		

13-5 A biologist was testing the strength of various antitoxins against a disease in mice. He treated three groups of mice, all of whom had the disease, and then measured the time in minutes it took before the animal's temperature was back to normal. The MS_{wg} was 10.64; the overall F was beyond the .05 criterion. Test the differences between all pairs of means at the .05 critical value and then combine group A with C and compare this combination with B.

	A	B	C
$\bar{X}_i$	30.2	48.1	26.7
n_i	46	40	37

13-6 In a study of different types of learning techniques a psychologist had four groups of college sophomores. Three groups got special training in memorizing; one was the control. The overall F was significant at the .01 level. The MS_{wg} was 14.24. Test the differences between all pairs of means; then combine the control with group 2 and test this combination against the combined groups 3 and 4. Use the .05 criterion.

	Control	*2*	*3*	*4*
$\bar{X}_i$	17.3	19.1	23.6	21.9
n_i	14	10	11	12

13-7 List each null hypothesis for all pairs of means in a three-sample problem. Now list each null hypothesis for all pairs of means for a five-sample problem. As k increases, what happens to the number of comparisons that must be made?

14

Chi Square and Other Nonparametric Procedures

In testing hypotheses in previous chapters, we computed characteristics of a sample or samples in order to draw some conclusions about characteristics of the population; i.e., we dealt with statistics as estimates of parameters. These procedures are characteristically called *parametric tests*. In many situations our data clearly do not fit the assumptions necessary for parametric tests. For these situations methods have been developed which are free of the assumptions characteristic of operations such as t and F. Such procedures, called *nonparametric tests*, are carried out with their own assumptions. This chapter deals with three of these methods, chi square, the sign test, and the median test. These three, along with rank-order correlation, can be applied to most situations requiring nonparametric analysis.

Chi Square

A number of studies in the social sciences deal with counting of individuals who appear in various categories. We often wish to compare this count with the number of individuals that an a priori hypothesis says should appear in these categories. For example, suppose we are studying the political affiliations of women who belong to the Read Along Book Club. We find that there are 23 Republicans, 18 Democrats, and 5 women who belong to political parties other than the two major ones. In Table 14-1, these data are our observed frequencies, i.e., our actual count within the categories. We have noted that in the last election in this book club's county, 45 percent of the registered voters were Democrats, 40 percent were Republicans, and

Table 14-1   Fictitious Political Affiliations of Women in a Club

Parties	Observed frequency	Expected frequency
Republican	23	18.4
Democrat	18	20.7
Other	5	6.9
Total	46	46.0

15 percent were of other parties. If the ladies in the book club were distributed among the political parties in the same way as the total registered electorate, we would have 40 percent of the 46 members (18.4) affiliated with the Republicans, 45 percent (20.7) with the Democrats, and 15 perent (6.9) with other parties.

Now we ask: Are the political affiliations of these ladies typical of any sample from the county population, except for sampling error, or do their affiliations deviate so widely from the expected affiliations that we must conclude that it is not a chance deviation?

Fitting a Theoretical Distribution

To find out how much the observed frequencies f_o deviate from the expected f_e, we could subtract the expected frequency of each party from the actual frequency $f_o - f_e$ and add these differences together for all parties. However, if deviations from the expected value in one party is a large positive and in another a large negative (just as many differences can be positive as negative) we may well end up with a sum of $f_o - f_e$ equal to zero, even though a considerable difference existed between expected and observed frequencies in our data. In fact, the sum of random fluctuations from the expected values should be zero. Therefore, our problem is that in adding deviations, a deviation of actual frequencies from the expected value in one cell could be obscured by a similar deviation in the opposite direction in another cell. We get around this problem in chi square by squaring for each cell in the table the difference between the observed and expected frequencies, $(f_o - f_e)^2$. Thus, all values are positive.

A second problem is tied to the magnitude of an observed minus expected difference relative to the number of frequencies involved. For example, the 10-point difference in A of Table 14-2 is small in proportion to the number of individuals involved, but the same difference in B is proportionately great. In other words, the importance of a given $f_o - f_e$ is relative to the size of the expected group.

Therefore, we divide the squared difference between the actual and expected frequencies by the expected frequency, $(f_o - f_e)^2/f_e$. This puts the squared difference in proportion to the number of cases expected in the cell. We then add up for all cells these values of the squared differences divided by the expected frequency and this is our chi square. These steps are combined in the formula

$$\chi^2 = \sum \frac{(f_o - f_e)^2}{f_e}$$

(14-1)

and an example is worked in Table 14-3.

Significance of Chi Square

Chi square is based on the idea that if the hypothesis upon which the expected frequencies are computed is correct, deviations of actual frequencies from the expected ones will be random fluctuations only. From our work in previous chapters we know that when there are random fluctuations around any point, it is mathematically possible to compute the proportions of cases that deviate various amounts from that point. With these proportions we could build frequency curves which, although they may be unlike the binomial distribution in shape, can be used, in relation to degrees of freedom, to determine the likelihood of getting scores which deviate various amounts by chance alone. Then if our obtained deviations could be achieved only rarely by chance, we may wish to conclude that something other than chance was operating. The chi-square test, being tied to deviations, operates in a similar manner.

How do we determine the degrees of freedom in a problem such as the example in Table 14-3? Until now degrees of freedom have been associated with the number of individuals in a sample and the number

---

Table 14-2   Illustration of the Importance of Differences Relative to the Size of the Expected Frequencies

Category	A	B
Observed frequency	1510	12
Expected frequency	1500	22
Difference	10	10

Table 14-3 Procedure for Computing Chi Square

Formula (14-1)	What it says to do
$$\chi^2 = \sum \frac{(f_o - f_e)^2}{f_e}$$ where f_o = actual (observed) frequency for given cell f_e = expected frequency for that cell	1 Determine the number of individuals f_o that fall in each category being observed 2 With an a priori hypothesis determine the number of individuals f_e expected to fall in each category 3 For each cell subtract f_e from f_o, square the difference $(f_o - f_e)^2$, and divide the result by f_e 4 Sum the results of step 3 for all cells $$\chi^2 = \frac{(f_{o_1} - f_{e_1})^2}{f_{e_1}} + \frac{(f_{o_2} - f_{e_2})^2}{f_{e_2}} + \cdots + \frac{(f_{o_k} - f_{e_k})^2}{f_{e_k}}$$ where subscripts $1, 2, \ldots, k$ refer to various classified groups or cells in a table such as Table 14-1

Example We are studying color preferences of women college freshmen. We have begun with the basic colors of red, blue, yellow and have asked 60 women to select the color they like best from a card containing a 1-inch square of each of the three colors. We get the following distribution of choices:

Red	Blue	Yellow
13	27	20

We test the hypothesis that the observed choices do not differ from a random selection and make the test as follows. The above data are the observed frequencies, and our hypothesis is a random distribution of selections, or a third of the total responses in each category.

f_o	f_e	$\dfrac{(f_o - f_e)^2}{f_e}$
13	20	2.45
27	20	2.45
20	20	.00
		$\chi^2 = 4.90$

of restrictions imposed by such a condition as the population mean. In chi square, degrees of freedom are tied to the number of classifications into which we have sorted our individuals, and the number of restrictions imposed is determined by the total observations in the categories. In Table 14-1, we have three categories which must add up to 46 cases. We are free to vary the frequencies in two of the categories; but when the cases in two categories are established, the frequency in the third category is also determined, since it is the total number of cases less the frequencies of the "free" cells. We, therefore, have $k - 1$ degrees of freedom, where k is the number of categories into which we have segmented the classification variable. In the example in Table 14-3 we had three color categories for sorting, and so degrees of freedom are $3 - 1 = 2$.

The procedure shown in Table 14-3 for computing degrees of freedom pertains to problems where the data have been sorted on a single variable only, e.g., political party. Later we shall classify observations under two variables, such as income and political party, simultaneously. We shall take a second look at degrees of freedom at that point.

The significance of the chi-square value can be determined by consulting Table C-8 which lists for various degrees of freedom the chi-square values beyond which various percentages of chi squares will fall by chance alone. As in previous tests, we typically set the 5 or 1 percent critical values as our criterion of significance and accept or reject the null hypothesis in relation to these values of chi square.

For the problem in Table 14-3, we have 2 degrees of freedom. In Table C-8 we see that a chi-square value of 5.99 is necessary for significance at the 95 percent level of confidence, i.e., at the 5 percent critical point. Since our value of 4.90 is less than the table value, we accept the hypothesis that freshman women have a preference for color which does not differ from a random arrangement.

In the example of Table 14-3 we found our expected frequencies by hypothesizing a random distribution of the variable being observed, but any hypothesis that has a foundation can be used to establish expected frequencies. For example, we may have reason to believe that one out of three men prefers convertible cars to nonconvertibles, whereas only one out of five women prefers convertibles. This a priori proposal then could be the basis for determining how many men and how many women from a given group should be expected to choose a convertible.

A second example of this approach is found in analyzing items on certain tests. Suppose we have four alternatives on a multiple-choice test. Did our class respond to a given item on a random basis? If so, the responses of the group should be equally distributed among the four choices. Our expected frequency would be $.25N$ for each of the four alternatives to the test item. We could then compare these expected

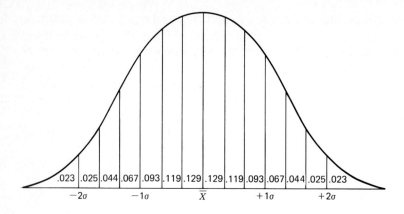

Figure 14-1

frequencies with actual student responses to see if the class did indeed respond to the item essentially on a random basis.

Other hypotheses may also be appropriate. For example, we may have reason to believe that the second of four alternatives to a test item should get as many choices as all other alternatives combined and that alternatives 1, 3, and 4 are equally attractive; i.e., the remaining half of the responses are equally spread over these three choices. In this case, the expected frequency for the second alternative would be $\frac{1}{2}N$, and the expected frequency for all other alternatives would be $\frac{1}{6}N$ each.

Fitting a Curve

In statistics we use many kinds of distributions of data, probably the most common of which is the normal curve. Suppose I have a set of data and I wish to know whether my frequency distribution fits a given theoretical distribution. Since the characteristics of these distributions are known, we can divide these distributions into segments along the base line and identify the portion of the curve in each segment. Then we can divide our observed frequency distribution at the same points and relate the portion of our data in the various segments to the portions in the theoretical distribution.

Let us demonstrate that last paragraph with an example using the normal curve as our theoretical distribution. Suppose I have a set of IQ data taken on 300 children. I wonder whether my distribution of IQ fits the normal curve. I can lay out a normal curve, beginning at the mean, and moving above and below it in standard-deviation metrics. The theoretical curve, divided up in units of .333 standard deviation, would look like Fig. 14-1. Here we have divided up the normal curve into segments and listed the portion of the curve in each segment.*

---

* These segments are based on z and areas of the normal curve in Table C-2.

Now we must compute the mean and standard deviation for our 300 student IQs and lay them out into segments of $.333\sigma$. My data look like Table 14-4 with a mean of 102.1 and a standard deviation of 12.0. The f_o column represents the actual number of cases that fall within given segments $.333\sigma$ wide. The f_e column is the theoretical number of cases out of 300 that would fall within those segments if my data fit the normal curve exactly. These come from multiplying the portion of the curve in a given segment times 300 cases. The portions of the normal curve can be calculated from Table C-2, as we did when we worked with z scores.

The remainder of the problem proceeds just like earlier χ^2 problems relating actual data to a theoretical distribution. For each segment of the curve we get the $f_o - f_e$ difference, square it, and divide by f_e. Summing across all segments produces the χ^2 value. This value is referred to Table C-8, with $k - 1 = 13$ degrees of freedom. The .05 criterion for 13 degrees of freedom is 22.362. Since this is much larger than our computed value, we conclude that our data do not depart from a normal distribution.

In testing our data against a theoretical distribution, we usually must have large samples, i.e., 200 or more cases. In fitting our data to a curve, unless we divide the curve into more than a few segments, considerable deviation of the data from the theoretical curve within a segment can go undetected. The extreme of this is shown in Fig. 14-2, where we have divided the curve into two segments, with exactly the

Table 14-4

$.333\sigma$ segments	f_o	f_e	$f_o - f_e$	$\dfrac{(f_o - f_e)^2}{f_e}$
126.1–130.0	9	6.9	2.1	.639
122.1–126.0	8	7.5	.5	.033
118.1–122.0	11	13.2	−2.2	.367
114.1–118.0	18	20.1	−2.1	.219
110.1–114.0	22	27.9	−5.9	1.248
106.1–109.0	38	35.7	2.3	.148
102.1–106.0	43	38.7	4.3	.478
98.1–102.0	35	38.7	−3.7	.353
94.1–98.0	34	35.7	−1.7	.080
90.1–94.0	32	27.9	4.1	.603
86.1–90.0	23	20.1	2.9	.418
82.1–86.0	12	13.2	−1.2	.109
78.1–82.0	9	7.5	1.5	.300
74.1–78.0	6	6.9	− .9	.117
	300	300		$\chi^2 = 5.114$

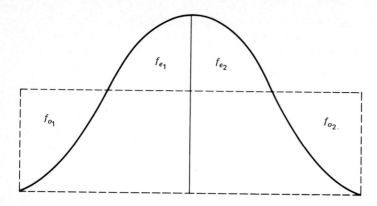

Figure 14-2

same number of cases in f_o as in f_e. This will produce a nonsignificant chi square, but the gross segments clearly do not fit the normal curve. When we divide the curve into many segments, it takes a fairly large N for there to be a sufficient number of cases in each segment.

A Test of Independence with Two Variables of Classification

Sometimes we wish to classify our observations under each of two different conditions simultaneously. For example, we may wish to consider urban-rural residence and preference for the three basic colors. Here we would be asking whether one variable of classification is independent of the other. (By independence we mean that knowledge of one characteristic for an individual tells us nothing about the other characteristic for that person.) Double-classification problems such as this require a second procedure for determining the expected frequencies. Let us look at the following table to see whether we can come up with a method for establishing these expected values. In the data below, we are comparing regular college students with part-time students on their feelings about the adequacy of social events at college Eks. The row and column totals are called *marginal totals*. A table laid out like the one below is called a *contingency table* because it reflects the extent to which frequencies in one variable (student status) are contingent upon, i.e., dependent upon, frequencies in the other variable (attitude).

	Satisfied	Dissatisfied	Total
Regular	40	30	70
Part-time	40	10	50
Total	80	40	120

In the data above, we look first at the marginal totals. Out of the 120 students in all, 70 were regular students and 50 were part-time. Also, 80 were satisfied, and 40 were dissatisfied with social events. Now if membership in a group is independent of attitudes, the proportion of regular students who are satisfied with social events will be the same as the proportion of part-time students who are satisfied, and similarly, the proportion of regular dissatisfied students will be the same as the proportion of part-time dissatisfied students. Therefore, of the total of 70 who are regularly enrolled, 80/120 will be expected to appear in the satisfied category and 40/120 in the dissatisfied category, or

$$f_e \text{ for regular satisfied students} = 70\left(\frac{80}{120}\right) = 46.67$$

$$f_e \text{ for regular dissatisfied students} = 70\left(\frac{40}{120}\right) = 23.33$$

In a like manner, we can also determine the expected frequencies for part-time students. There were 50 in all, and if the categories are independent, these 50 people will be divided between the satisfied and dissatisfied categories in proportion to the marginal totals for those two categories:

$$f_e \text{ for part-time satisfied students} = 50\left(\frac{80}{120}\right) = 33.33$$

$$f_e \text{ for part-time dissatisfied students} = 50\left(\frac{40}{120}\right) = 16.67$$

We now generalize this procedure for computing expected frequencies when we are testing the hypothesis that the categories are independent. First we get row and column totals and the grand total. The expected frequency for any given cell in our contingency table is then found by multiplying that cell's row and column totals and dividing this product by the grand total. In the table below the expected frequencies are found as listed.

A	B	C	T_1
D	E	F	T_2

T_3 T_4 T_5 T_g

$$f_{e_A} = \frac{T_1 T_3}{T_g} \qquad f_{e_B} = \frac{T_1 T_4}{T_g} \qquad f_{e_C} = \frac{T_1 T_5}{T_g}$$

$$f_{e_D} = \frac{T_2 T_3}{T_g} \qquad f_{e_E} = \frac{T_2 T_4}{T_g} \qquad f_{e_F} = \frac{T_2 T_5}{T_g}$$

where A, ... , F = classification categories
T_1, T_2 = total frequencies in rows 1 and 2
T_3, T_4, T_5 = column totals
T_g = grand total of all frequencies in table

Let us return to the problem above to see whether student status at Eks university is independent of attitude about social events. We can now check the calculations of expected frequencies which we made above. In chi square the sum of the expected frequencies must equal the sum of the actual frequencies. Our actual frequencies totaled 120, and our expected frequencies, rounded to two decimal places, also total to 120. With this knowledge we can proceed with increased confidence in our computations. The chi square becomes

f_o	f_e	$\dfrac{(f_o - f_e)^2}{f_e}$
40	46.67	.95
30	23.33	1.91
40	33.33	1.33
10	16.67	2.67
		$\chi^2 = 6.86$

Finding the degrees of freedom for a two-variable classification problem follows essentially the same procedure as a one-variable classification situation, but we have added restrictions because both row and column totals impose limitations. We noted earlier that the degrees of freedom in a single row (or column) arrangement are 1 less than the number of classification cells. However, in a two-way classification, each cell in a row (restricted by the row total) is also a cell in a column (restricted by a column total). The operation of such restrictions can be illustrated best by an example (Table 14-5).

Once the frequencies for cells A and B have been determined, cell C is not free to vary if the row is to add up to 20. Likewise, once cell A is established, cell D is not free to vary if the column is to add up to 10. Following this procedure, we find that once any two of the cells are set, no other cell frequency in the table is free to vary. In other words, we have only 2 degrees of freedom in the above situation. The results of this logical procedure for arriving at degrees of freedom are duplicated in the formula

$$df = (r - 1)(c - 1)$$

Table 14-5 Data for Illustrating Degrees of Freedom in Chi Square

Group	Class 1	Class 2	Class 3	Total
I	A	B	C	20
II	D	E	F	30
Total	10	26	14	50

where r is the number of rows and c the number of columns in the table.

For our problem relating enrollment status to satisfaction to social events, we have $(r - 1)(c - 1) = (2 - 1)(2 - 1) = 1$ degree of freedom. We consult Table C-9 and find that a chi square of 6.86 with 1 degree of freedom is significant at the 99 percent level of confidence. This tells us that the way frequencies spread themselves for one variable (student status) depends upon how the frequencies are dispersed for the other variable (attitude). In other words, if a chi-square test rejects the hypothesis of independence, we accept the hypothesis of dependence. Our knowledge of X does indeed tell us something about Y.

Correction for Small Frequencies in a 2 × 2 Table

When we have a small sample of cases, the usual computation of chi square gives us an overestimate of the true value. As a result, we reject some hypotheses which in fact should be accepted. We can avoid this problem in 2 × 2 tables, however, by applying *Yates' correction* for continuity. This correction is needed because chi square is based on frequencies. Frequencies move from unit to unit in discrete steps, whereas the chi-square table is based on a continuous distribution. When frequencies are large, the correction has very little effect on the chi-square value, but when frequencies are small, the correction makes a real difference. Therefore, we apply it only with small frequencies.

The procedure is a simple matter of subtracting .5 from the absolute value of the $f_o - f_e$ values in each of the four cells of the table. For example, if our table had frequencies like

3	9	12
6	6	12
9	15	24

we could calculate the expected frequencies by our usual method and find them to be

$$f_{e_1} = \frac{9(12)}{24} = 4.50 \qquad f_{e_2} = \frac{15(12)}{24} = 7.50$$

$$f_{e_3} = \frac{9(12)}{24} = 4.50 \qquad f_{e_4} = \frac{15(12)}{24} = 7.50$$

Then we would compute our chi square (the vertical rules mean the absolute value of the difference, regardless of sign):

$$\frac{(|3 - 4.5| - .5)^2}{4.5} = .22$$

$$\frac{(|6 - 4.5| - .5)^2}{4.5} = .22$$

$$\frac{(|9 - 7.5| - .5)^2}{7.5} = .13$$

$$\frac{(|6 - 7.5| - .5)^2}{7.5} = .13$$

$$\chi^2 = \overline{.70}$$

If Yates' correction had *not* been applied in this calculation, the chi square would have been

$$\frac{(3 - 4.5)^2}{4.5} = .50$$

$$\frac{(6 - 4.5)^2}{4.5} = .50$$

$$\frac{(9 - 7.5)^2}{7.5} = .30$$

$$\frac{(6 - 7.5)^2}{7.5} = .30$$

$$\chi^2 = \overline{1.60}$$

Yates' correction is applied when the least expected frequency in any cell is less than 5. If expected values are 5 or greater, the uncorrected procedure gives a reasonably accurate estimate. Yates' correction is suitable for tables which have only 1 degree of freedom, but suppose we have more than a single degree of freedom. Then if all but one of the cell expectations are large, we can achieve a reasonably close approximation of chi square with calculations in the usual manner without Yates' procedure.

We can occasionally avoid the problem of small frequencies in cells by combining categories. For example, suppose we have taken an opinion survey to find out how people feel about a proposed tax bill; we compare Republicans with Democrats on five categories of an agreement scale and get the following frequencies:

	Strongly disagree	Disagree	Undecided	Agree	Strongly agree
Democrats	2	4	25	14	10
Republicans	8	7	12	9	3

By condensing our scale to include agreeing, undecided, and disagreeing categories, we would have six cells all of which would be large enough for computing chi square directly with reasonable assurance that our chi square was quite accurate. The table then would look like this:

	Disagree	Undecided	Agree
Democrats	6	25	24
Republicans	15	12	12

When computing chi square, one is often faced with the dilemma of proceeding with a small cell frequency and maintaining categories which have research significance, or combining cells and achieving a more accurate estimate of chi square but losing categories. The choice is one which must be made for each individual problem.

Assumptions and Limitations in Chi Square

Like all statistical analyses, chi square is based on certain assumptions which must be met if the analysis is to produce dependable results. The following essential limitations on the use of chi square arise out of the basic assumptions:

1   Individual observations must be independent of each other. The response that subject A gives to a questionnaire should have no influence on the

response of subject B. The fact that X chose red as her favorite color should not influence Y's choice, etc. Independence of individual observations also means that a given individual can be represented only once in an analysis. Two encounters of the same person in the data cannot, of course, be independent observations.

2   Chi square must be limited to frequency (or counting) kinds of data. Sometimes a category is defined by a measurement, e.g., all students between IQs 100 and 110, but these people within a category are counted for the data that are to be analyzed. Measurements themselves cannot be analyzed by chi square. For example, it is *not* legitimate to establish $f_o - f_e$ by comparing children's IQs with the average for a defined class.

3   The sum of the expected frequencies must equal the sum of the actual frequencies.

4   As noted earlier, with 1 degree of freedom no expected frequency should be less than 5 unless Yates' correction is applied. If we have more than 1 degree of freedom, one small cell may not distort the results markedly. If cell categories can be reasonably combined to eliminate small cell expectancies, this alternative may be considered. However, a posteriori manipulations tend to destroy experimental sophistication.

Exercises

14-1   A school social worker wishing to see whether absence from class is tied to socioeconomic standing went to the school attendance records for the semester and made up the following table of persons who had been absent:

	1 day	2–5 days	Over 5 days
Middle income	25	12	5
Lower income	20	10	8

At the 99 percent level of confidence are socioeconomic standing and attendance independent?

14-2   A very obscure item was placed in a history test. The correct answer to the question could be found only in a footnote of a reserve book. The instructor decided that if the class had not read the footnote, and marked the item randomly, their responses would be equally distributed across the four multiple-choice alternatives. If the class had read the footnote, the distribution of frequencies would not be random. The number of times each alternative was chosen is given below. Decide whether the class might have read the footnote (use 5 percent critical value, i.e., the 95 percent level of confidence).

Alternative	A	B	C	D
Number of times chosen	7	17	9	11

14-3 A student group wishes to poll the student body to see whether students generally prefer lecture or discussion classes. Since they believe that high achievers may look at the problem differently from low achievers, they divide the group on the basis of the previous semester's grade-point average. The obtained data are as follows:

Achiever status	Prefer lecture	Prefer discussion
Above 2.5	10	17
Below 2.5	6	3

Is preference for type of class independent of academic achievement at the 99 perent level of confidence?

14-4 A counselor wishing to know whether television viewing preferences are different for bright and dull children took a sample of 40 bright children (IQ 115 and up) and a second sample of 40 children who were less bright (IQ 85 or less) and asked them to identify their favorite TV show. Shows were then classified into categories. Test the hypothesis that preferences are independent of intelligence at the 95 percent level of confidence.

	Cartoons	Documentary	Drama	Misc.
Bright	5	21	10	4
Less bright	14	11	9	6

14-5 A student in city planning read that 33 percent of all cars in his state were compact and 28 percent subcompact size. He went to a popular street corner in his town during the morning rush hour and tabulated the types of cars that passed by in a 30-minute period. Are the cars in his sample distributed the way they are in the state? Use the .05 criterion.

Full-sized	Compact	Sub-compact
29	51	54

14-6 A student who has measured the height (in inches) of a group of children in a learning disabilities class wishes to know whether the heights of these students fit a normal distribution. Her data are below. Test them against a normal curve at the 99 percent confidence level. (For practice use categories 1 standard deviation wide; we would want more categories in an actual problem.) Here $\bar{X} = 58.9$, $s = 3.0$.

X	f
66	2
65	2
64	6
63	7
62	13
61	11
60	16
59	24
58	17
57	15
56	12
55	9
54	4
53	5
52	1

The Sign Test

An English teacher has divided his class into two groups, matching individuals in group A with those in group B on a pretest of knowledge of parts of speech. Group A then practices diagramming sentences while group B writes stories and evaluates each other's themes. At the end of a 6-week period a posttest on parts of speech is again given. Does one group now achieve better than the other?

This problem looks like a correlated t-test situation, but if we cannot meet the basic parametric assumptions of t, we can apply the sign test. If the groups are equal in achievement, for each member of A who surpasses his/her matched partner, there should also be a member of B who surpasses his/her partner. The sign test then merely counts the number of cases in one group who exceed their matched partners and compares this with the number of persons in the second group who exceed their matched partners. A's who surpass B's or B's who surpass

A's will be a randomly determined event, like heads or tails on a coin toss.

The basic procedure for the sign test is as follows, using the data in Table 14-6 for illustration. Here we see scores for groups A and B, arranged so that matched partners are side by side. Then for each pair we subtract B from A (we get the same results if we subtract A from B); if A is larger than B, we assign that pair of individuals a plus; if B is larger than A, the pair gets a minus. If our groups have changed about equally, the pluses and minuses will be randomly distributed around a median of zero. Our null hypothesis is therefore that the median difference is zero. If there are considerably more of one sign than the other, the distribution of differences is clearly not random and the hypothesis of equal change in the two groups must be rejected.

When there are 10 or fewer cases the null hypothesis is tested by use of the binomial expansion (Chap. 9) with $p = .5$ and N equal to the number of pairs observed. In Table 14-6 we had 10 matched pairs; 5 of these were assigned plus; 3, minus; and 2, zero. We drop the zero pairs and deal with an N of 8.

By chance alone we would expect 4 pluses and 4 minuses from the 8 pairs with differences beyond zero. We now ask: Do the 5 pluses differ significantly from our chance value of 4? We now find the probability of getting 5 pluses in a binomial expansion, where p is .5. We can do this by expanding the binomial $(p + q)^8$, or we can read the values from Pascal's triangle, Table 9-2, and compute the probability of getting 5 or more pluses. The line in Pascal's triangle where N is 8 reads 1, 8, 28, 56,

Table 14-6 Posttest Results of an English Test for 12 Pairs of Subjects in Grade 12 Who Were Matched on a Pretest

Group A	Group B	Sign (A − B)
21	16	+
10	14	−
14	8	+
21	13	+
28	10	+
19	19	0
14	17	−
12	11	+
11	13	−
18	18	0

70, 56, 28, 8, 1, which sums to 256. Since the median (70) is the point where p is .50, that is, 4 pluses out of 8, we move to the right and find that 56 would represent the number of times out of 256 we would expect to get 5 pluses out of 8; 28 times we should get 6 out of 8, 8 times we should get 7 pluses, and 1 time we should get 8 pluses out of 8 chances. Thus, to determine the probability of getting 5 or more pluses out of 8, we would add $56 + 28 + 8 + 1$ and divide this sum by the total of 256. The resulting probability would be .36. If we set our criterion points, as we have in previous tests, at the .05 or .01 point, we see that our value of .36 is far from significant. We thus accept the null hypothesis and in fact say that we very easily could have obtained 5 pluses out of 8 on a chance basis alone.* Therefore, we have no basis for concluding that one group progressed farther than the other during the training weeks.

We agreed that we would use the binomial expansion to determine significance if the number of pairs being observed is 10 or less. If there are more than 10 pairs, we can use the normal curve as an approximation of the probabilities. The necessary z values are found with a mean of $.5N$ and a standard deviation of $\sqrt{N(.25)}$. The z value for a given number of pluses X would then be

$$z = \frac{(X \mp .5) - .5N}{\sqrt{N(.25)}} \tag{14-2}$$

If the number of pluses is *more* than $.5N$, we use $X - .5$; if the pluses are *less* than $.5N$, we use $X + .5$ in computing z. This procedure corrects for the discontinuity of the data. For example, the lower limit of the interval that represents 5 pluses out of 8 is actually 4.5. Thus $5 - .5$ includes the entire interval represented by 5.

Having computed our z value, we can consult a table of areas under the normal curve to determine the proportion of samples that would have more pluses than our obtained number. Either a one- or two-tailed test may be applied, depending upon the nature of our hypothesis.

Exercises

14-7   Fourteen students were given a test and 30 minutes later were given the same test again to see whether practice actually influenced their scores. The results of the pretests and posttests are below. Using a two-tailed

---

* The value 5 is actually the midpoint of the interval 4.5–5.5. Precisely, our conclusion is that .36 of the group lies beyond 4.5 or more pluses.

test at the 95 percent level of confidence, apply the sign test of significance to these data.

Pretest	109	108	108	103	103	99	94	94	94	91	91	89	88	71
Posttest	149	107	108	138	101	98	129	130	91	139	91	87	134	119

14-8 Using the data given in Exercise 14-7, compute $\Sigma(X_{pre} - X_{post})/N_{pairs}$ for the plus differences and again for the minus differences. Does the relative size of these two means compare favorably with your test of significance? What does this say about the strength of the sign test in reflecting magnitude of differences?

The Median Test

The median test is a nonparametric procedure which is used to test the hypothesis that two or more groups come from populations with the same median. Suppose we have two reading groups in Mrs. Dexter's fourth-grade class. We want to compare reading group A's reading ability with reading group B's ability. We give reading tests to both groups to determine their status. If groups A and B are both from populations with a common median, we can combine them to get the best estimate of that median; we shall call it the *grand median*. We can now actually count the number of cases above and the number below the grand median in A and the number of cases above and below the grand median in B. If A and B are indeed from populations with common medians, 50 percent of each group should fall above the grand median and 50 percent below. These are our expected frequencies for each group above and below the grand median. Applying the chi-square method, we can compare the actual frequencies in these segments with their expected values.

For example, let us return to Mrs. Dexter's reading groups and look at their records. We find that for the 6-week period the class has made the scores in reading comprehension recorded in Table 14-7 (page 318). The grand median is found by use of the frequency distribution in 14-7, which combines both groups. The grand median, computed as described in Chap. 3, is 18.5.

We now count the scores that appear above and below the grand median in group A and the scores above and below the grand median in group B. The result of this classification, a typical contingency table as used for chi-square analyses, would look like this:

	Group **A**	*Group* **B**
Above	3	9
Below	9	3

Table 14-7 Gain Scores for Reading Speed Shown First by Separate Groups of Students and Again in a Frequency Distribution Based on Combined Groups

| Separate groups | | Combined data | |
A	B	X	f
15	22	23	1
21	21	22	2
18	19	21	3
14	23	20	2
16	19	19	4
21	22	18	3
15	17	17	3
18	19	16	3
20	17	15	2
17	16	14	1
18	20		
16	19		

Applying chi square to these data, with an expected frequency of half of each of groups A and B above and half below the grand median, we have

$$\frac{(3 - 6)^2}{6} = 1.50$$

$$\frac{(9 - 6)^2}{6} = 1.50$$

$$\frac{(9 - 6)^2}{6} = 1.50$$

$$\frac{(3 - 6)^2}{6} = 1.50$$

$$\chi^2 = 6.00$$

With 1 degree of freedom, $(r - 1)(c - 1)$, we look this value up in Table C-8 and find that it is significant at the 95 percent level of confidence.

The median test can also be applied to more than two groups. The same general procedure applies. First we combine all groups and compute a grand median. Then for each group we count the number above and below the grand median and compute a chi-square test with 50 percent of each group expected above, and 50 percent below, the grand median. Although we had equal-sized groups in our example, this is not a requirement for the median test. We can proceed as usual even though the observed groups differ in number.

Exercises

14-9   A curriculum supervisor wished to test the advantage of a new procedure of teaching mathematics in grade 6. There are four rooms of children in the sixth grade; half are randomly assigned to take method A, half method B. After 6 weeks their scores on an arithmetic test are as follows. Test the hypothesis that the two groups are from populations with a common median (.05 criterion point).

A		B	
28	24	29	20
27	23	26	19
27	20	24	19
25	20	22	19
25	19	21	18
25	18	20	18
	15		17

14-10   A psychologist is studying a new drug called *Larnin* for its effect on the ability to learn. Thirty-two subjects were assigned to three groups. Group A got 1 grain of Larnin; B got 2 grains; and C got an inert powder. Each group was then asked to study a list of nonsense words for 10 minutes, after which they were tested for the number of words learned. Their scores are given below. Test the hypothesis that all three groups are from populations with a common median (.01 criterion point).

A		B		C	
10	6	11	5	12	8
9	6	11	5	12	7
8	5	10	4	11	5
8	3	10	4	11	5
7		8	3	9	5
6		5		8	

Summary

Since we are often faced with the problem of dealing with data which do not fit the assumptions necessary for using parametric tests such as t and F, a variety of nonparametric tests have been devised for the analysis of such data.

Three nonparametric procedures are adequate for dealing with most problems. Chi square handles frequency data in which we have sorted cases into defined categories and wish to see whether the observed frequencies depart significantly from frequencies that would be expected from a given hypothesis about the way cases will be sorted. The sign test deals with paired sets of data collected under different circumstances. If the differences between the two sets of data are randomly distributed, the median difference between the pairs of scores will be zero. Thus the sign test tests the hypothesis that the obtained pair differences are zero. The third procedure, the median test, says that if several samples of individuals all come from populations with a common median, each sample should have 50 percent of its cases above and 50 percent below the grand median of the combined groups. We then can count for each group the number of cases that actually do fall above and below the grand median and test these observed frequencies against the expected 50-50 split of frequencies in each group by the chi-square procedure.

Generally speaking parametric tests are more powerful than nonparametric tests in terms of accepting true hypotheses and rejecting false ones; i.e., conclusions based on parametric tests are more reliable than those based on nonparametric tests. For this reason when the data meet the assumptions of parametrics, parametric devices should be applied. However, for the many situations which do not fit these assumptions, nonparametric devices are very useful.

Key Terms

chi square	Yates' correction
expected frequencies	sign test
observed frequencies	median test
curve fitting	

Problems

In the following situations state which procedure—chi square, the sign test, or the median test—is the appropriate technique for analysis, assuming in each case that the data do not meet the assumptions of a parametric test.

14-1 I have a group of kindergarten children who are to be used to test some reading-readiness materials. I give them a pretest on reading readiness and a posttest. I want to see whether they have changed significantly from pretest to posttest.

14-2 I have randomly assigned three groups of emotionally disturbed children to three treatments. I have a psychiatrist rank the members of the total group on adjustment after 6 weeks of treatment. I wish to see whether the effects of the treatments are different.

14-3 I wish to know whether women's opinions about the death-penalty laws are independent of men's opinions. I ask a group of women and a group of men whether they are in favor of, opposed to, or undecided about the death-penalty law.

14-4 I wish to see whether there is a difference between fathers and their sons in the amount of anxiety they show in regard to test taking. I attach a galvanometer to the arm of each subject, then announce that I am going to give him an intelligence test. The galvanometer reading will be my measure of anxiety, and I will show one set of readings for fathers, one set for their sons.

A

*Review of Arithmetic and Algebraic Processes**

Since a basic knowledge of algebraic processes is essential to successful progress in statistical methods, a few basic rules for computation, with illustrations and exercises (answered in Appendix E), are presented.

A-1 Order of Operations

Suppose we have the following situation to deal with:

$$6 + 9 \div 2 \times 4 = ?$$

Is the answer 30, 7.125, 42, or 6.9? These answers result from completing the required operations in different sequences of adding, multiplying, and dividing. This section explains established procedures for carrying out basic arithmetic with numbers and fractions so that everyone will arrive at the same answer.

Rule A1-1 When both multiplication and addition (or subtraction) are called for in a computational procedure, the multiplication should be done first unless parentheses indicate otherwise. Therefore, $10 + 5 \times 4 = 30$, but $(10 + 5) \times 4 = 60$; $2 \times 10 - 5 = 15$, but $2 \times (10 - 5) = 10$.

Rule A1-2 When both division and addition (or subtraction) are called for, the division is done first, unless parentheses indicate otherwise. Therefore, $10 + 4 \div 2 = 12$ but $(10 + 4) \div 2 = 7$; $15 \div 5 - 3 = 0$, but $15 \div (5 - 3) = 7.5$.

Rule A1-3 When both multiplication and division are called for in one

---

* For a more detailed presentation see Helen Walker, "Mathematics Essential for Elementary Statistics," rev. ed., 382 pp., Holt, New York, 1951.

computational process, parentheses must be used to illustrate the order. Therefore, $20 \times 6 \div 2$ is not acceptable, but $20 \times (6 \div 2)$ or $(20 \times 6) \div 2$ is.

Rule A1-4   When an expression in parentheses is either multiplied or divided by a term outside the parentheses, the expression outside is distributed over all terms inside the parentheses according to the process (multiplication or division) indicated. This is called the *distributive law.* (Also note rule A2-3.) Therefore

$$4(7 + 3) = (28 + 12) = 40$$
$$\tfrac{1}{2}(8 - 4) = (4 - 2) = 2$$
$$X(Y + Z) = XY + XZ$$

$$\frac{1}{X}(Y + Z) = \frac{Y + Z}{X} = \frac{Y}{X} + \frac{Z}{X}$$

$$\left(\frac{\Sigma X}{N}\right)^2 = \frac{(\Sigma X)^2}{N^2}$$

Rule A1-5   Multiplying or dividing both the numerator and the denominator of a fraction by the same number does not change the relationship represented by the fraction; adding a number to, or subtracting it from, both the numerator and denominator does alter the basic relationship represented by the fraction. Therefore

$$\frac{1}{2} = \frac{2}{4} = \frac{3}{6} \qquad \text{and} \qquad \frac{x}{y} = \frac{ax}{ay} = \frac{azx}{azy}$$

where a and z are constants, but

$$\frac{1}{2} \neq \frac{2 + 1}{2 + 2} \qquad \text{and} \qquad \frac{x}{y} \neq \frac{a + x}{a + y}$$

Rule A1-6   To add or subtract two fractions, both fractions must be in the same denomination; that is, they must have common denominators. Therefore

$$\frac{1}{3} + \frac{1}{2} = \frac{1(2)}{3(2)} + \frac{1(3)}{2(3)} = \frac{5}{6} \qquad \text{and} \qquad \frac{a}{x} + \frac{b}{y} = \frac{ay}{xy} + \frac{bx}{xy} = \frac{ay + bx}{xy}$$

Rule A1-7   To multiply two fractions, we simply multiply the two numerators and then the two denominators. Therefore

$$\frac{2}{3}\left(\frac{2}{5}\right) = \frac{4}{15} \qquad \frac{a}{x}\left(\frac{b}{y}\right) = \frac{ab}{xy}$$

Rule A1-8   To divide one fraction by another, invert the divisor and multiply it times the dividend. Therefore

$$\frac{2}{3} \div \frac{2}{5} = \frac{2}{3}\left(\frac{5}{2}\right) = \frac{10}{6} = 1\tfrac{2}{3} \qquad \frac{a}{x} \div \frac{b}{y} = \frac{a}{x}\left(\frac{y}{b}\right) = \frac{ay}{bx}$$

Exercises

Complete the designated operations:

A1-1 $12 \times 7 + 4 \times 6 - 1 =$

A1-2 $5 + 12 \div 2 - 3 =$

A1-3 $(5 \times 6) \div 3 + 7 =$

A1-4 $3(4 + 6) =$

A1-5 $a(x + 2y) =$

A1-6 $i^2 \left[\Sigma f d^2 - \dfrac{(\Sigma f d)^2}{N} \right] =$

Which of the following terms stated as equal are in fact not equal?

A1-7 $(8 \times 4) + 3 = 8 \times (4 + 3)$

A1-8 $(7 \div 2)(4) = 7 \div (2 \times 4)$

A1-9 $(ab) + x = a(b + x)$

A1-10 $\dfrac{a}{b} = \dfrac{na}{nb}$

A1-11 $\dfrac{\Sigma x}{N} = \dfrac{N\Sigma x}{N^2}$

Complete the designated operations:

A1-12 $\frac{2}{3} \left(\frac{7}{8} \right) =$

A1-13 $\dfrac{a}{b} \left(\dfrac{x}{y} \right) =$

A1-14 $\dfrac{1}{N} \left(\dfrac{\Sigma X^2}{N} \right) =$

A1-15 $\frac{2}{3} \div \frac{1}{4} =$

A1-16 $\dfrac{x}{y} \div \dfrac{a}{b} =$

A1-17 $\frac{2}{3} + \frac{3}{4} =$

A1-18 $\dfrac{zx}{y} - \dfrac{a}{b}$

A-2 Symbols

This section reviews some common symbols and the operations they indicate. The symbols used in algebra and statistics are merely shorthand ways of stating quantitative values, processes, or relationships.

Rule A2-1 Alphabetic symbols usually stand for variables. N usually means the total number of observations made; however, if the total number of observations was created by combining several samples of observations, the lowercase n represents the number in each of the various samples. A single observation or raw score is usually designated X; if there is more than one observation per case, a second observation may be designated Y.

Rule A2-2   Differently based statistics are often distinguished by using a letter from the Latin alphabet for the sample statistic, says s, and the corresponding letter from the Greek alphabet for the population statistic, say σ.

Rule A2-3   Since there are only 26 letters in the Latin alphabet (52 if we count capitals and small letters), we occasionally run out of letters to represent numbers, We save letters by *subscripting*. A subscript is a symbol's number which tells us "which one" of the quantities in a given class is being considered. Therefore X_1 would indicate the first number in the group of quantities called X; X_2 would be the second one, etc. For example, let us use IQ as the condition to be labeled X and suppose we have five children's IQs to deal with. Then X_1 would be the first child's IQ, X_2 would be the second child's IQ; and X_5 would be the fifth child's IQ. Subscripts are adjectives telling which one.

Rule A2-4   Signs tell us what is to be done with numbers or indicate a relationship between numbers. The signs for adding, subtracting, multiplying and dividing are familiar, but signs showing relationships other than the equals sign are not so common. Some are shown in Table A-1.

Rule A2-5   Parentheses ( ), brackets [ ], and braces { } are used to set off operations that can be thought of as a single value; as a result they indicate the order of many processes. Therefore

$$(5 + 4)(6) + 1 = 9(6) + 1 = 55$$
$$= 30 + 24 + 1 = 55$$

It is clear that if addition (or subtraction) is the process within the parentheses, it may be done first, or the multiplication (or division) may be done first as long as we follow the distributive law stated in Rule A1-4. Thus

$$2(4 + 1) = 2(5) = 10$$
$$= 8 + 2 = 10$$

and
$$\frac{X_1 + X_2 + X_3}{N} = \frac{X_1}{N} + \frac{X_2}{N} + \frac{X_3}{N}$$

Table A-1   Mathematical Signs and Their Meaning

Sign	Meaning	Example
$<$	Less than	$a < b$ means "a is less than b"
$>$	More than	$a > b$ means "a is more than b"
$\leq$	Equal to or less than	$a \leq b$ means "a is equal to or less than b"
$\geq$	Equal to or more than	$a \geq b$ means "a is equal to or more than b"
$\neq$	Not equal	$a \neq b$ means "a and b are not equal"
Σ	The sum of	ΣX means "the sum of all the X values"

and
$$\frac{a(X_1 + X_2)}{N} = \frac{aX_1 + aX_2}{N} = \frac{aX_1}{N} + \frac{aX_2}{N}$$

Summation signs are often used with parentheses. The expression
$$\Sigma(X^2)$$
means that each X value is squared and all the squared values are added up. The expression
$$\Sigma(X - Y)$$
says to subtract $X - Y$ for all pairs of X and Y and then sum all these differences. If

$$X_1 = 4 \quad X_2 = 3 \quad X_3 = 1$$
and
$$Y_1 = 2 \quad Y_2 = 6 \quad Y_3 = 7$$
then
$$\Sigma(X - Y) = (4 - 2) + (3 - 6) + (1 - 7) = -7$$

The expression $\Sigma(X - Y)$ can also be written $\Sigma X - \Sigma Y$. If we sum the X values and sum the Y values and then subtract the sums, the answer is the same as if we first subtracted $X - Y$ and then summed the differences, as above. Thus

$$\Sigma X = 4 + 3 + 1 = 8$$
$$\Sigma Y = 2 + 6 + 7 = 15$$
and, as above,
$$8 - 15 = -7$$

Exercises

A2-1 We have the following IQs for four boys: Joe 105, John 110, Dick 100, Tom 107.
  (a) If we let N stand for the number of observations, what does N equal?
  (b) What is the value of X_3 and X_1?
  (c) What is the value of $\Sigma X/N$?
  (d) What is the meaning of the following statements?
    (1) $X_2 > X_1$ (2) $X_3 < X_1$
    (3) $(X_1 + X_2) > (X_3 + X_4)$ (4) $X_1 \neq X_2$
    (5) $A = \Sigma X^2$

Which of the following statements are correct?

A2-2 $(a + b)(c) = a + bc$ **A2-3** $\dfrac{X - Y}{2} = \dfrac{X}{2} - \dfrac{Y}{2}$

A2-4 $(N - 1)(N) = N^2 - N$ **A2-5** $\Sigma[X^2 - 2XM + M^2]$ $= \Sigma X^2 - 2\Sigma XM + \Sigma M^2$

State in words what each of the following equations says in symbols. X stands for IQ scores; N is the number of observations; C is a constant equal to 50; $\bar{X}$ is the average.

A2-6 $\bar{X} = \dfrac{\Sigma X}{N}$ **A2-7** $\dfrac{\Sigma(X - C)}{N} = \bar{X} - C$

A2-8 $\dfrac{\Sigma(XC)}{N} = \bar{X}C$ **A2-9** $\dfrac{\Sigma(X - \bar{X})}{N} = 0$

Put the following statements into algebraic shorthand.

A2-10 The sum of the ages of 10 girls divided by the number of girls observed is the average age for these girls.

A2-11 If we divide each of four numbers by a fifth number then add these four results, we have a quantity which is equal to the sum of the four numbers divided by the fifth one.

A-3 Positive and Negative Numbers

This section reviews computations with positive and negative numbers.

Rule A3-1 If numbers are to be added and have the same sign, add the absolute value of the numbers and sign the answer with the common sign.

Rule A3-2 If two numbers to be added have unlike signs, subtract the smaller from the larger and sign the result with the sign of the larger number.

Rule A3-3 When we subtract one number from another number, we change the sign of the number to be subtracted and proceed as in Rule A3-1 or A3-2 accordingly. Therefore,

$$7 - (-4) = 7 + 4 = 11 - 10 - (-3) = -7$$

Rule A3-4 If we multiply a positive number by a negative number, the answer is a negative number. Therefore

$$+ 8\,(-4) = -32 \quad -7(6) = -42$$

Rule A3-5 When we multiply two numbers with like signs, whether positive or negative, the answer should be signed positively. Therefore,

$$-7(-3) = 21 \quad 7(3) = 21$$

Rule A3-6 The rules for multiplication apply also to division; unlike signs between divisor and dividend produce a negative quotient, whereas like signs between a divisor and dividend produce a positive quotient.

Exercises

A3-1 $(-8) + (+4) - (-1) =$		**A3-2** $(+7) - (+3) + (-6) =$
A3-3 $(-6) + (-2) - (+1) =$		**A3-4** $(-7) - (+4) + (-6) =$
A3-5 $-2(+6) =$		**A3-6** $-4(-7) =$
A3-7 $(-18) \div (-6) =$		**A3-8** $(+24) \div (-4) =$

A-4 Functions

This section illustrates methods of dealing with two conditions when one of them changes in a prescribed manner with every given change in the other. When this happens, we say one condition is a function of the other.

Rule A4-1 A variable is a quantity which can take any one of a set of values. This set is called the *range of the variable.*

Rule A4-2 One variable is called a *function* of the other if for every value of the one there are one or more corresponding values of the other. In the equation

$$Y = bX$$

X and Y are variables and Y is a function of X; that is, $Y = f(X)$. Y is called the *dependent variable*, X the *independent variable*, and b is a constant; i.e., it takes only one value.

Rule A4-3 Functions can be graphed by arbitrarily inserting values for the independent variable X and solving the equation of the dependent variable Y. Therefore, if $Y = bX$ and $b = 2$,

X is	0	1	2	3	4
Y is	0	2	4	6	8

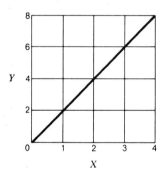

When the graph of a function is a straight line, we call it a *linear function*. When both variables appear to the first power, i.e., are neither squared, cubed, nor raised to a higher power, the function is linear. Therefore

$$Y = 3X + 2$$
$$Y = .7X - 3$$
$$Y = 3X$$

are all linear functions.

Rule A4-4 The horizontal axis on a bivariate (two-variable) graph is often called the X *axis* or the *abscissa*; the vertical axis is the Y *axis* or *ordinate*. Linear functions cross the X axis at the point where X is zero; this point is called the Y *intercept*. In the graph shown in Rule A4-3 the Y intercept is zero, but it need not always be.

Rule A4-5 When the independent variable X is multiplied by a constant, such as $Y = 2X$, where the constant is 2, that constant indicates the slope of the line in a bivariate graph and indicates how many units Y increases for each unit increase in X. If the slope is positive, the line will run from lower left to upper right; if the slope is negative, the line will run from upper left to lower right.

Exercises

A4-1   Given $Y = 3X + 2$.

(a) What is the independent variable?

(b) what is the dependent variable?

(c) What is the slope?

(d) What is the value of the Y intercept (when $X = 0$, $Y = ?$)?

Which of the following show a linear rather than curvilinear function?

A4-2 $Y = 6X$ **A4-3** $m = 2n^2 + 1$

A4-4 $Y^2 = 4X - 3$ **A4-5** $r = 2s + 5$

For each of the following pairs of equations choose the one for which the slope of the line in relation to the abscissa is the greater.

A4-6 $Y = 2X$ **A4-7** $Y = 6X - 1$

       $Y = 4X + 1$ $Y = 3X + 6$

A4-8 $Y = 4X - 5$

       $Y = 5X + 2$

A-5 Managing Linear Equations

This section describes the basic rules for handling linear equations.

Rule A5-1   The basic equality of any equation is not destroyed if the same number is added to each side of the equation or if the same number is subtracted from each side. Therefore if

$$4x = 7$$

then

$$4x + 2 = 7 + 2 \quad \text{and} \quad 4x - 2 = 7 - 2$$

A special case of this rule can be illustrated by the following problems. If

$$2x + 4 = 6$$

then (subtracting 4 from both sides of the equation)

$$2x + 4 - 4 = 6 - 4$$
$$2x = 6 - 4$$

In effect we have transferred the 4 in the original equation across the equals sign, changing its sign to minus. Again if

$$3y - 2 = 4$$

then (adding 2 to each side)

$$3y - 2 + 2 = 4 + 2$$
$$3y = 4 + 2$$

Once more we have in effect simply moved a quantity from one side of the equation to the other by changing its sign. It is generally true, then, that any term in an equation can be transferred from one side of the equals sign to the other if the sign of that term is changed.

Rule A5-2   The basic equality of any equation is not destroyed if each side of the equation is multiplied by the same number or if each side is divided by the same number. If

$$4x = 7$$

then

$$2(4x) = 2(7)$$

and

$$\frac{4x}{2} = \frac{7}{2}$$

Exercises

A5-1 $3x = 8$; $2(3x) =$ **A5-2** $4y - 5 = 12$; $(4y - 5)/2 =$

A5-3 $3a = 4b + 1$; $3a + 4 =$ **A5-4** $7m + 3 = 14$; $7m =$

A5-5 $a = y^2/x$; $ax =$ **A5-6** $NS_1S_2r = \Sigma xy$; $r =$

A-6   Multiplying Binomials

This section describes multiplication of the binomial by another binomial.

Rule A6-1   To square a given value we merely multiply it by itself. Therefore, to square a binomial, such as $x + y$, we multiply it by the same quantity

$$(x + y)(x + y)$$

To multiply one binomial by another, we first multiply each term in one by each term in the other and then collect terms:

$$
\begin{array}{r}
x + y \\
x + y \\
\hline
x(x) + xy \\
+ xy + y(y) \\
\hline
x^2 + 2xy + y^2
\end{array}
\qquad\qquad
\begin{array}{r}
x - 2 \\
x - 2 \\
\hline
x^2 - 2x \\
- 2x + 4 \\
\hline
x^2 - 4x + 4
\end{array}
$$

Rule A6-2   The quantities $x + y$ and $x + y$ are called the *factors* of $x^2 + 2xy + y^2$; that is they are the quantities which multiplied together yield the latter value. In this case $x + y$ is also the square root, but a factor is not always a square root. When we multiply $(x + 1)(x + 2)$ and get $x^2 + 3x + 2$, neither $x + 1$ nor $x + 2$ is a square root.

Exercises

A6-1 $(x + y)^2 =$

A6-3 $(2x - y)^2 =$

A6-5   What are the factors of $x^2 + 5x + 6$?

A6-2 $(x - 4)^2 =$

A6-4   What are the factors of $x^2 + 2xy + y^2$?

B

Mathematical Development of Statistical Procedures

B-1 The Raw-Score Formula for the Product-Moment Correlation r_{XY} Beginning with the Basic z-Score Formula

A group of scores X is to be correlated with a second group of scores Y. Since

$$r_{XY} = \frac{\Sigma z_X z_Y}{N} \qquad z_X = \frac{X - \bar{X}}{\sigma_X}; \qquad z_Y = \frac{Y - \bar{Y}}{\sigma_Y}$$

we have

$$r_{XY} = \frac{\Sigma(X - \bar{X})(Y - \bar{Y})}{N\sigma_X \sigma_Y}$$

and

$$\sigma_X = \sqrt{\frac{\Sigma X^2 - (\Sigma X)^2/N}{N}} \qquad \sigma_Y = \sqrt{\frac{\Sigma Y^2 - (\Sigma Y)^2/N}{N}}$$

Substituting gives

$$r_{XY} = \frac{\Sigma(X - \bar{X})(Y - \bar{Y})}{N\sqrt{\dfrac{[\Sigma X^2 - (\Sigma X)^2/N][\Sigma Y^2 - (\Sigma Y)^2/N]}{N^2}}} \tag{B-1}$$

Now let us deal only with the numerator of formula (B-1). For the development of the formula we look at the numerator

$$\Sigma(X - \bar{X})(Y - \bar{Y}) = \Sigma(XY - X\bar{Y} - \bar{X}Y + \bar{X}\bar{Y})$$
$$= \Sigma XY - \Sigma X\bar{Y} - \Sigma Y\bar{X} + \Sigma \bar{X}\bar{Y}$$

We already know that

$$\bar{X} = \frac{\Sigma X}{N} \qquad \bar{Y} = \frac{\Sigma Y}{N} \qquad \text{and} \qquad \Sigma \bar{X} \bar{Y} = N \frac{\Sigma X \Sigma Y}{N^2}$$

Substituting gives

$$\Sigma XY - \Sigma X\bar{Y} - \Sigma Y\bar{X} + \Sigma \bar{X}\bar{Y} = \Sigma XY - \frac{\Sigma X \Sigma Y}{N} - \frac{\Sigma Y \Sigma X}{N} + \frac{\Sigma X \Sigma Y}{N}$$

When we collect terms, the numerator of formula (B-1) becomes

$$\Sigma XY - \frac{\Sigma X \Sigma Y}{N}$$

Taking N^2 out from under the radical in the denominator of (B-1) makes it $1/N$, and we can cancel it with the N already outside the radical. With the new numerator and denominator formula (B-2) becomes

$$r_{XY} = \frac{\Sigma XY - \Sigma X \Sigma Y / N}{\sqrt{[\Sigma X^2 - (\Sigma X)^2/N][\Sigma Y^2 - (\Sigma Y)^2/N]}}$$

$$= \frac{\Sigma XY - \Sigma X \Sigma Y / N}{\sqrt{\Sigma X^2 - (\Sigma X)^2/N} \ \sqrt{\Sigma Y^2 - (\Sigma Y)^2/N}}$$

B-2 The Equation for Predicting a Y Score Once an X Score Is Known

The formula for the predicted score $\hat{Y}$ is

$$\hat{Y} = a + bX$$

where $a = Y$ when X is zero and b is the increase in Y with a unit increase in X. If raw scores are converted into deviation scores, that is, $y = Y - \bar{Y}$ and $x = X - \bar{X}$, the a value drops out since the mean of both x and y distributions becomes zero. Therefore, in deviation form the predicted score $\hat{y}$ becomes

$$\hat{y} = bx$$

We can now set up the function

$$f = \frac{\Sigma(y - \hat{y})^2}{N} = \frac{(y - bx)^2}{N}$$

where we have N deviations of the form

$$y - \hat{y} = y - bx$$

These values, when squared, added together, and divided by N, provide a variance of actual scores around the regression line. This variance can be minimized by a particular choice of b, the value of which can be found by the calculus:

$$f = \frac{\Sigma(y - bx)^2}{N}$$

differentiating with respect to b (x and y held constant). To illustrate with a general example, with $f = u^n$, the formulas for differentiation* show

$$df = nu^{n-1} \, du$$

Now if

$$f = (a - cx)^2$$
$$u = (a - cx)$$
$$du = 0 - c \, dx$$
$$n = 2$$
$$n - 1 = 1$$

substituting in $df = nu^{n-1} \, du$, we have

$$df = 2(a - cx)(0 - c \, dx)$$
$$= -2c(a - cx) \, dx$$

or

$$\frac{df}{dx} = -2c(a - cx)$$

With the above model the differentiation of the problem at hand is

$$f = \frac{1}{N} \Sigma(y - bx)^2$$

Referring to the above equation, $df = nu^{n-1} \, du$, placing the constant $1/N$ in evidence, since

$$d(cu) = c \, du$$
$$u = y - bx$$
$$du = 0 - x \, db$$
$$n = 2$$
$$n - 1 = 1$$

Substituting, we have

$$df = \frac{1}{N} \Sigma 2(y - bx)(0 - x \, db)$$

or

$$df = \frac{1}{N} \Sigma(-2x)(y - bx) \, db$$

---

* This procedure can be found in any elementary calculus book. The student who is not familiar with the calculus is asked to exercise faith at this point.

Rearranging gives

$$\frac{df}{db} = \frac{-2\Sigma x(y - bx)}{N}$$

When we set this equal to zero and divide by -2, we have

$$\frac{\Sigma x(y - bx)}{N} = 0$$

or

$$\frac{\Sigma xy - b\Sigma x^2}{N} = 0$$

Putting each term in the numerator over N gives

$$\frac{\Sigma xy}{N} - b\frac{\Sigma x^2}{N} = 0$$

Now if

$$\frac{\Sigma x^2}{N} = \sigma_X^2$$

and if

$$r_{XY} = \frac{\Sigma z_X z_Y}{N}$$

and

$$z_X = \frac{X - \bar{X}}{\sigma_X} = \frac{x}{\sigma_X} \qquad z_Y = \frac{Y - \bar{Y}}{\sigma_Y} = \frac{y}{\sigma_Y}$$

then

$$r_{XY} = \frac{\Sigma xy}{N\sigma_X \sigma_Y} \qquad \text{and} \qquad \frac{\Sigma xy}{N} = r_{XY}\sigma_X \sigma_Y$$

Thus

$$r_{XY}\sigma_X \sigma_Y - b\sigma_X^2 = 0$$

Factoring gives

$$\sigma_X(r_{XY}\sigma_y - b\sigma_X) = 0$$

and setting each factor equal to zero, we have

$$\sigma_X = 0 \qquad r_{XY}\sigma_Y - b\sigma_X = 0$$

Hence

$$b = r_{XY}\frac{\sigma_Y}{\sigma_X}$$

Substituting this value for b in the formula $\hat{y} = bx$, we have

$$\hat{y} = r_{XY}\frac{\sigma_Y}{\sigma_X}x$$

where y and x are deviation scores. Converting this formula to raw scores, $x = X - \bar{X}$ and $y = Y - \bar{Y}$, we have

$$\hat{Y} - \bar{Y} = r \frac{\sigma_Y}{\sigma_X} (X - \bar{X})$$

$$\hat{Y} = r \frac{\sigma_Y}{\sigma_X} (X - \bar{X}) + \bar{Y}$$

B-3   The Formula for the Standard Error of Estimate

The standard error of estimate $s_{Y.X}$ is the standard deviation of actual scores Y around their predicted scores $\hat{Y}$ in a regression situation. We begin with the variance (standard deviation squared), sometimes called the *residual variance*. It is found like other variances by squaring the deviations around a point, summing them, and dividing by the number N of deviations involved:

$$s_{Y.X}{}^2 = \frac{\Sigma(Y - \hat{Y})^2}{N}$$

If both actual and predicted scores are in deviation form, that is $Y - \bar{Y} = y$ and $\hat{Y} - \bar{Y} = \hat{y}$, then

$$s_{Y.X}{}^2 = \frac{\Sigma(y - \hat{y})^2}{N}$$

We just saw in the previous derivation that

$$\hat{y} = r \frac{\sigma_Y}{\sigma_X} x$$

We shall now apply this formula, but we shall use the sample estimates of σ, that is, s_X and s_Y. And so

$$s_{Y.X}{}^2 = \frac{\Sigma[y - r(s_Y/s_X)x]^2}{N - 1}$$

Squaring the numerator and putting each term over the denominator, we have

$$s_{Y.X}{}^2 = \frac{\Sigma y^2}{N - 1} - \frac{\Sigma xy}{N - 1} 2r \frac{s_Y}{s_X} + r^2 \frac{s_Y{}^2}{s_X{}^2} \frac{\Sigma x^2}{N - 1}$$

But since

$$\frac{\Sigma y^2}{N - 1} = s_Y{}^2 \qquad \frac{\Sigma x^2}{N - 1} = s_X{}^2 \qquad \frac{\Sigma xy}{N - 1} = r s_X s_Y$$

we can substitute these values into the previous formula to give

$$s_{Y.X}^2 = s_Y^2 - (rs_Xr_Y)(2r)\frac{s_Y}{s_X} + r^2\frac{s_Y^2}{s_X^2}s_X^2$$

$$= s_Y^2 - 2r^2s_Y^2 + r^2s_Y^2$$

$$= s_Y^2 - r^2s_Y^2$$

$$= s_Y^2(1 - r^2)$$

and the standard error is

$$s_{Y.X} = s_Y\sqrt{1 - r^2}$$

C

Tables

Suggestions for Expanding the Table of Squares and Square Roots

The student can increase the flexibility of Table C-1 by (1) using the square column n^2 as the number column n and the number n column as the square-root column $\sqrt{n}$ and (2) manipulating the decimal point in the terms in the n and the $\sqrt{n}$ columns.

If I multiply 4(4) and get 16, the number I began with, 4, is the square root of 16. Similarly, if I want the square root of a number beyond 10,000, I can find it, at least to the significant figures necessary for work in this book, by looking in the column labeled n^2. Then moving to the left I find the square root of this number in the column labeled n. For example, suppose I want the square root of 13,000. I move down the n^2 column until I find a number as near as possible to 13,000. It is 12,996. To the left of 12,996 is 114, which is almost exactly the square root of 13,000. The more precise student may wish to make interpolations to arrive at the exact values, but in the great majority of problems in this book interpolations will not alter the significant figures involved.

The square root of 13,000 can also be found by manipulating the decimal point in the number and in its square root. The value of 13,000 is equal to 130(100), and so its square root must be equal to $\sqrt{130}\ (\sqrt{100}) = \sqrt{130}\ (10)$.

Our table contains the square root of 130; it is 11.402. And if we multiply this value by 10, we have 114.02, which is the square root of 13,000.

Here is another example. What is the square root of 47,100? This number is equal to 471(100), and its square root is

$$\sqrt{471}\ (\sqrt{100}) = 21.703\ (10) = 217.03$$

What is the square root of 4.21? This number is equal to 421 $(\frac{1}{100})$ and its square root is

$$\sqrt{421} \; (1/\sqrt{100}) \text{ or } 20.518 \left(\tfrac{1}{10}\right) = 2.05$$

retaining only two decimal places.

Some practice with the above suggestions should make it unnecessary for students who do not have calculators to compute square roots by the long method. However, in all cases students should inspect their figures to see whether their square root when multiplied by itself would reasonably be equal to the number for which the root was taken.

Table C-1   Squares and Square Roots of Numbers from 1 to 9990

n	n^2	$\sqrt{n}$	$\sqrt{10n}$	$1000/n$	n	n^2	$\sqrt{n}$	$\sqrt{10n}$	$1000/n$
					50	2 500	7.0711	22.361	20.000
1	1	1.0000	3.1623	1000.0	51	2 601	7.1414	22.583	19.608
2	4	1.4142	4.4721	500.00	52	2 704	7.2111	22.804	19.231
3	9	1.7321	5.4772	333.33	53	2 809	7.2801	23.022	18.868
4	16	2.0000	6.3246	250.00	54	2 916	7.3485	23.238	18.519
5	25	2.2361	7.0711	200.00	55	3 025	7.4162	23.452	18.182
6	36	2.4495	7.7460	166.67	56	3 136	7.4833	23.664	17.857
7	49	2.6458	8.3666	142.86	57	3 249	7.5498	23.875	17.544
8	64	2.8284	8.9443	125.00	58	3 364	7.6158	24.083	17.241
9	81	3.0000	9.4868	111.11	59	3 481	7.6811	24.290	16.949
10	100	3.1623	10.000	100.00	60	3 600	7.7460	24.495	16.667
11	121	3.3166	10.488	90.909	61	3 721	7.8103	24.698	16.393
12	144	3.4641	10.954	83.333	62	3 844	7.8740	24.900	16.129
13	169	3.6056	11.402	76.923	63	3 969	7.9373	25.100	15.873
14	196	3.7417	11.832	71.429	64	4 096	8.0000	25.298	15.625
15	225	3.8730	12.247	66.667	65	4 225	8.0623	25.495	15.385
16	256	4.0000	12.649	62.500	66	4 356	8.1240	25.690	15.152
17	289	4.1231	13.038	58.824	67	4 489	8.1854	25.884	14.925
18	324	4.2426	13.416	55.556	68	4 624	8.2462	26.077	14.706
19	361	4.3589	13.784	52.632	69	4 761	8.3066	26.268	14.493
20	400	4.4721	14.142	50.000	70	4 900	8.3666	26.458	14.286
21	441	4.5826	14.491	47.619	71	5 041	8.4262	26.646	14.085
22	484	4.6904	14.832	45.455	72	5 184	8.4853	26.833	13.889
23	529	4.7958	15.166	43.478	73	5 329	8.5440	27.019	13.699
24	576	4.8990	15.492	41.667	74	5 476	8.6023	27.203	13.514
25	625	5.0000	15.811	40.000	75	5 625	8.6603	27.386	13.333
26	676	5.0990	16.125	38.462	76	5 776	8.7178	27.568	13.158
27	729	5.1962	16.432	37.037	77	5 929	8.7750	27.749	12.987
28	784	5.2915	16.733	35.714	78	6 084	8.8318	27.928	12.821
29	841	5.3852	17.029	34.483	79	6 241	8.8882	28.107	12.658
30	900	5.4772	17.321	33.333	80	6 400	8.9443	28.284	12.500
31	961	5.5678	17.607	32.258	81	6 561	9.0000	28.461	12.346
32	1 024	5.6569	17.889	31.250	82	6 724	9.0554	28.636	12.195
33	1 089	5.7446	18.166	30.303	83	6 889	9.1104	28.810	12.048
34	1 156	5.8310	18.439	29.412	84	7 056	9.1652	28.983	11.905
35	1 225	5.9161	18.708	28.571	85	7 225	9.2195	29.155	11.765
36	1 296	6.0000	18.974	27.778	86	7 396	9.2736	29.326	11.628
37	1 369	6.0828	19.235	27.027	87	7 569	9.3274	29.496	11.494
38	1 444	6.1644	19.494	26.316	88	7 744	9.3808	29.665	11.364
39	1 521	6.2450	19.748	25.641	89	7 921	9.4340	29.833	11.236
40	1 600	6.3246	20.000	25.000	90	8 100	9.4868	30.000	11.111
41	1 681	6.4031	20.248	24.390	91	8 281	9.5394	30.166	10.989
42	1 764	6.4807	20.494	23.810	92	8 464	9.5917	30.332	10.870
43	1 849	6.5574	20.736	23.256	93	8 649	9.6437	30.496	10.753
44	1 936	6.6333	20.976	22.727	94	8 836	9.6954	30.659	10.638
45	2 025	6.7082	21.213	22.222	95	9 025	9.7468	30.822	10.526
46	2 116	6.7823	21.448	21.739	96	9 216	9.7980	30.984	10.417
47	2 209	6.8557	21.679	21.277	97	9 409	9.8489	31.145	10.309
48	2 304	6.9282	21.909	20.833	98	9 604	9.8995	31.305	10.204
49	2 401	7.0000	22.136	20.408	99	9 801	9.9499	31.464	10.101

Table C-1   Squares and Square Roots of Numbers from 1 to 9990 (Continued)

n	n^2	$\sqrt{n}$	$\sqrt{10n}$	$1000/n$	n	n^2	$\sqrt{n}$	$\sqrt{10n}$	$1000/n$
100	10 000	10.000	31.623	10.000	150	22 500	12.247	38.730	6.6667
101	10 201	10.050	31.781	9.9010	151	22 801	12.288	38.859	6.6225
102	10 404	10.100	31.937	9.8039	152	23 104	12.329	38.987	6.5789
103	10 609	10.149	32.094	9.7087	153	23 409	12.369	39.115	6.5359
104	10 816	10.198	32.249	9.6154	154	23 716	12.410	39.243	6.4935
105	11 025	10.247	32.404	9.5238	155	24 025	12.450	39.370	6.4516
106	11 236	10.296	32.558	9.4340	156	24 336	12.490	39.497	6.4103
107	11 449	10.344	32.711	9.3458	157	24 649	12.530	39.623	6.3694
108	11 664	10.392	32.863	9.2593	158	24 964	12.570	39.749	6.3291
109	11 881	10.440	33.015	9.1743	159	25 281	12.610	39.875	6.2893
110	12 100	10.488	33.166	9.0909	160	25 600	12.649	40.000	6.2500
111	12 321	10.536	33.317	9.0090	161	25 921	12.689	40.125	6.2112
112	12 544	10.583	33.466	8.9286	162	26 244	12.728	40.249	6.1728
113	12 769	10.630	33.615	8.8496	163	26 569	12.767	40.373	6.1350
114	12 996	10.677	33.764	8.7719	164	26 896	12.806	40.497	6.0976
115	13 225	10.724	33.912	8.6957	165	27 225	12.845	40.620	6.0606
116	13 456	10.770	34.059	8.6207	166	27 556	12.884	40.743	6.0241
117	13 689	10.817	34.205	8.5470	167	27 889	12.923	40.866	5.9880
118	13 924	10.863	34.351	8.4746	168	28 224	12.961	40.988	5.9524
119	14 161	10.909	34.496	8.4034	169	28 561	13.000	41.110	5.9172
120	14 400	10.954	34.641	8.3333	170	28 900	13.038	41.231	5.8824
121	14 641	11.000	34.785	8.2645	171	29 241	13.077	41.352	5.8480
122	14 884	11.045	34.929	8.1967	172	29 584	13.115	41.473	5.8140
123	15 129	11.091	35.071	8.1301	173	29 929	13.153	41.593	5.7803
124	15 376	11.136	35.214	8.0645	174	30 276	13.191	41.713	5.7471
125	15 625	11.180	35.355	8.0000	175	30 625	13.229	41.833	5.7143
126	15 876	11.225	35.496	7.9365	176	30 976	13.267	41.952	5.6818
127	16 129	11.269	35.637	7.8740	177	31 329	13.304	42.071	5.6497
128	16 384	11.314	35.777	7.8125	178	31 684	13.342	42.190	5.6180
129	16 641	11.358	35.917	7.7519	179	32 041	13.379	42.308	5.5866
130	16 900	11.402	36.056	7.6923	180	32 400	13.416	42.426	5.5556
131	17 161	11.446	36.194	7.6336	181	32 761	13.454	42.544	5.5249
132	17 424	11.489	36.332	7.5758	182	33 124	13.491	42.661	5.4945
133	17 689	11.533	36.469	7.5188	183	33 489	13.528	42.779	5.4645
134	17 956	11.576	36.606	7.4627	184	33 856	13.565	42.895	5.4348
135	18 225	11.619	36.742	7.4074	185	34 225	13.601	43.012	5.4054
136	18 496	11.662	36.878	7.3529	186	34 596	13.638	43.128	5.3763
137	18 769	11.705	37.014	7.2993	187	34 969	13.675	43.244	5.3476
138	19 044	11.747	37.148	7.2464	188	35 344	13.711	43.359	5.3191
139	19 321	11.790	37.283	7.1942	189	35 721	13.748	43.474	5.2910
140	19 600	11.832	37.417	7.1429	190	36 100	13.784	43.589	5.2632
141	19 881	11.874	37.550	7.0922	191	36 481	13.820	43.704	5.2356
142	20 164	11.916	37.683	7.0423	192	36 864	13.856	43.818	5.2083
143	20 449	11.958	37.815	6.9930	193	37 249	13.892	43.932	5.1813
144	20 736	12.000	37.947	6.9444	194	37 636	13.928	44.045	5.1546
145	21 025	12.042	38.079	6.8966	195	38 025	13.964	44.159	5.1282
146	21 316	12.083	38.210	6.8493	196	38 416	14.000	44.272	5.1020
147	21 609	12.124	38.341	6.8027	197	38 809	14.036	44.385	5.0761
148	21 904	12.166	38.471	6.7568	198	39 204	14.071	44.497	5.0505
149	22 201	12.207	38.601	6.7114	199	39 601	14.107	44.609	5.0251

Table C-1   Squares and Square Roots of Numbers from 1 to 9990 (Continued)

n	n^2	$\sqrt{n}$	$\sqrt{10n}$	$1000/n$	n	n^2	$\sqrt{n}$	$\sqrt{10n}$	$1000/n$
200	40 000	14.142	44.721	5.0000	250	62 500	15.811	50.000	4.0000
201	40 401	14.177	44.833	4.9751	251	63 001	15.843	50.100	3.9841
202	40 804	14.213	44.944	4.9505	252	63 504	15.875	50.200	3.9683
203	41 209	14.248	45.056	4.9261	253	64 009	15.906	50.299	3.9526
204	41 616	14.283	45.166	4.9020	254	64 516	15.937	50.398	3.9370
205	42 025	14.318	45.277	4.8780	255	65 025	15.969	50.498	3.9216
206	42 436	14.353	45.387	4.8544	256	65 536	16.000	50.596	3.9063
207	42 849	14.387	45.497	4.8309	257	66 049	16.031	50.695	3.8911
208	43 264	14.422	45.607	4.8077	258	66 564	16.062	50.794	3.8760
209	43 681	14.457	45.717	4.7847	259	67 081	16.093	50.892	3.8610
210	44 100	14.491	45.826	4.7619	260	67 600	16.125	50.990	3.8462
211	44 521	14.526	45.935	4.7393	261	68 121	16.155	51.088	3.8314
212	44 944	14.560	46.043	4.7170	262	68 644	16.186	51.186	3.8168
213	45 369	14.595	46.152	4.6948	263	69 169	16.217	51.284	3.8023
214	45 796	14.629	46.260	4.6729	264	69 696	16.248	51.381	3.7879
215	46 225	14.663	46.368	4.6512	265	70 225	16.279	51.478	3.7736
216	46 656	14.697	46.476	4.6296	266	70 756	16.310	51.575	3.7594
217	47 089	14.731	46.583	4.6083	267	71 289	16.340	51.672	3.7453
218	47 524	14.765	46.690	4.5872	268	71 824	16.371	51.769	3.7313
219	47 961	14.799	46.797	4.5662	269	72 361	16.401	51.865	3.7175
220	48 400	14.832	46.904	4.5455	270	72 900	16.432	51.962	3.7037
221	48 841	14.866	47.011	4.5249	271	73 441	16.462	52.058	3.6900
222	49 284	14.900	47.117	4.5045	272	73 984	16.492	52.154	3.6765
223	49 729	14.933	47.223	4.4843	273	74 529	16.523	52.249	3.6630
224	50 176	14.967	47.329	4.4643	274	75 076	16.553	52.345	3.6496
225	50 625	15.000	47.434	4.4444	275	75 625	16.583	52.440	3.6364
226	51 076	15.033	47.539	4.4248	276	76 176	16.613	52.536	3.6232
227	51 529	15.067	47.645	4.4053	277	76 729	16.643	52.631	3.6101
228	51 984	15.100	47.749	4.3860	278	77 284	16.673	52.726	3.5971
229	52 441	15.133	47.854	4.3668	279	77 841	16.703	52.820	3.5842
230	52 900	15.166	47.958	4.3478	280	78 400	16.733	52.915	3.5714
231	53 361	15.199	48.062	4.3290	281	78 961	16.763	53.009	3.5587
232	53 824	15.232	48.166	4.3103	282	79 524	16.793	53.104	3.5461
233	54 289	15.264	48.270	4.2918	283	80 089	16.823	53.198	3.5336
234	54 756	15.297	48.374	4.2735	284	80 656	16.852	53.292	3.5211
235	55 225	15.330	48.477	4.2553	285	81 225	16.882	53.385	3.5088
236	55 696	15.362	48.580	4.2373	286	81 796	16.912	53.479	3.4965
237	56 169	15.395	48.683	4.2194	287	82 369	16.941	53.572	3.4843
238	56 644	15.427	48.785	4.2017	288	82 944	16.971	53.666	3.4722
239	57 121	15.460	48.888	4.1841	289	83 521	17.000	53.759	3.4602
240	57 600	15.492	48.990	4.1667	290	84 100	17.029	53.852	3.4483
241	58 081	15.524	49.092	4.1494	291	84 681	17.059	53.944	3.4364
242	58 564	15.556	49.194	4.1322	292	85 264	17.088	54.037	3.4247
243	59 049	15.588	49.295	4.1152	293	85 849	17.117	54.129	3.4130
244	59 536	15.621	49.396	4.0984	294	86 436	17.146	54.222	3.4014
245	60 025	15.652	49.497	4.0816	295	87 025	17.176	54.314	3.3898
246	60 516	15.684	49.598	4.0650	296	87 616	17.205	54.406	3.3784
247	61 009	15.716	49.699	4.0486	297	88 209	17.234	54.498	3.3670
248	61 504	15.748	49.800	4.0323	298	88 804	17.263	54.589	3.3557
249	62 001	15.780	49.900	4.0161	299	89 401	17.292	54.681	3.3445

Table C-1   Squares and Square Roots of Numbers from 1 to 9900 (Continued)

n	n²	√n	√10n	1000/n	n	n²	√n	√10n	1000/n
300	90 000	17.321	54.772	3.3333	350	122 500	18.708	59.161	2.8571
301	90 601	17.349	54.863	3.3223	351	123 201	18.735	59.245	2.8490
302	91 204	17.378	54.955	3.3113	352	123 904	18.762	59.330	2.8409
303	91 809	17.407	55.045	3.3003	353	124 609	18.788	59.414	2.8329
304	92 416	17.436	55.136	3.2895	354	125 316	18.815	59.498	2.8249
305	93 025	17.464	55.227	3.2787	355	126 025	18.841	59.582	2.8169
306	93 636	17.493	55.317	3.2680	356	126 736	18.868	59.666	2.8090
307	94 249	17.521	55.408	3.2573	357	127 449	18.894	59.749	2.8011
308	94 864	17.550	55.498	3.2468	358	128 164	18.921	59.833	2.7933
309	95 481	17.578	55.588	3.2362	359	128 881	18.947	59.917	2.7855
310	96 100	17.607	55.678	3.2258	360	129 600	18.974	60.000	2.7778
311	96 721	17.635	55.767	3.2154	361	130 321	19.000	60.083	2.7701
312	97 344	17.664	55.857	3.2051	362	131 044	19.026	60.166	2.7624
313	97 969	17.692	55.946	3.1949	363	131 769	19.053	60.249	2.7548
314	98 596	17.720	56.036	3.1847	364	132 496	19.079	60.332	2.7473
315	99 225	17.748	56.125	3.1746	365	133 225	19.105	60.415	2.7397
316	99 856	17.776	56.214	3.1646	366	133 956	19.131	60.498	2.7322
317	100 489	17.804	56.303	3.1546	367	134 689	19.157	60.581	2.7248
318	101 124	17.833	56.391	3.1447	368	135 424	19.183	60.663	2.7174
319	101 761	17.861	56.480	3.1348	369	136 161	19.209	60.745	2.7100
320	102 400	17.889	56.569	3.1250	370	136 900	19.235	60.828	2.7027
321	103 041	17.916	56.657	3.1153	371	137 641	19.261	60.910	2.6954
322	103 684	17.944	56.745	3.1056	372	138 384	19.287	60.992	2.6882
323	104 329	17.972	56.833	3.0960	373	139 129	19.313	61.074	2.6810
324	104 976	18.000	56.921	3.0864	374	139 876	19.339	61.156	2.6738
325	105 625	18.028	57.009	3.0769	375	140 625	19.365	61.237	2.6667
326	106 276	18.055	57.096	3.0675	376	141 376	19.391	61.319	2.6596
327	106 929	18.083	57.184	3.0581	377	142 129	19.416	61.400	2.6525
328	107 584	18.111	57.271	3.0488	378	142 884	19.442	61.482	2.6455
329	108 241	18.138	57.359	3.0395	379	143 641	19.468	61.563	2.6385
330	108 900	18.166	57.446	3.0303	380	144 400	19.494	61.644	2.6316
331	109 561	18.193	57.533	3.0211	381	145 161	19.519	61.725	2.6247
332	110 224	18.221	57.619	3.0120	382	145 924	19.545	61.806	2.6178
333	110 889	18.248	57.706	3.0030	383	146 689	19.570	61.887	2.6110
334	111 556	18.276	57.793	2.9940	384	147 456	19.596	61.968	2.6042
335	112 225	18.303	57.879	2.9851	385	148 225	19.621	62.048	2.5974
336	112 896	18.330	57.966	2.9762	386	148 996	19.647	62.129	2.5907
337	113 569	18.358	58.052	2.9674	387	149 769	19.672	62.209	2.5840
338	114 244	18.385	58.138	2.9586	388	150 544	19.698	62.290	2.5773
339	114 921	18.412	58.224	2.9499	389	151 321	19.723	62.370	2.5707
340	115 600	18.439	58.310	2.9412	390	152 100	19.748	62.450	2.5641
341	116 281	18.466	58.395	2.9326	391	152 881	19.774	62.530	2.5575
342	116 964	18.493	58.481	2.9240	392	153 664	19.799	62.610	2.5510
343	117 649	18.520	58.566	2.9155	393	154 449	19.824	62.690	2.5445
344	118 336	18.547	58.652	2.9070	394	155 236	19.849	62.769	2.5381
345	119 025	18.574	58.737	2.8986	395	156 025	19.875	62.849	2.5316
346	119 716	18.601	58.822	2.8902	396	156 816	19.900	62.929	2.5253
347	120 409	18.628	58.907	2.8818	397	157 609	19.925	63.008	2.5189
348	121 104	18.655	58.992	2.8736	398	158 404	19.950	63.087	2.5126
349	121 801	18.682	59.076	2.8653	399	159 201	19.975	63.166	2.5063

Table C-1 Squares and Square Roots of Numbers from 1 to 9990 (Continued)

n	n²	√n	√10n	1000/n	n	n²	√n	√10n	1000/n
400	160 000	20.000	63.246	2.5000	450	202 500	21.213	67.082	2.2222
401	160 801	20.025	63.325	2.4938	451	203 401	21.237	67.157	2.2173
402	161 604	20.050	63.403	2.4876	452	204 304	21.260	67.231	2.2124
403	162 409	20.075	63.482	2.4814	453	205 209	21.284	67.305	2.2075
404	163 216	20.100	63.561	2.4752	454	206 116	21.307	67.380	2.2026
405	164 025	20.125	63.640	2.4691	455	207 025	21.331	67.454	2.1978
406	164.836	20.149	63.718	2.4631	456	207 936	21.354	67.528	2.1930
407	165 649	20.174	63.797	2.4570	457	208 849	21.378	67.602	2.1882
408	166 464	20.199	63.875	2.4510	458	209 764	21.401	67.676	2.1834
409	167 281	20.224	63.953	2.4450	459	210 681	21.424	67.750	2.1786
410	168 100	20.248	64.031	2.4390	460	211 600	21.448	67.823	2.1739
411	168 921	20.273	64.109	2.4331	461	212 521	21.471	67.897	2.1692
412	169 744	20.298	64.187	2.4272	462	213 444	21.494	67.971	2.1645
413	170 569	20.322	64.265	2.4213	463	214 369	21.517	68.044	2.1598
414	171 396	20.347	64.343	2.4155	464	215 296	21.541	68.118	2.1552
415	172 225	20.372	64.420	2.4096	465	216 225	21.564	68.191	2.1505
416	173 056	20.396	64.498	2.4038	466	217 156	21.587	68.264	2.1459
417	173 889	20.421	64.576	2.3981	467	218 089	21.610	68.337	2.1413
418	174 724	20.445	64.653	2.3923	468	219 024	21.633	68.411	2.1368
419	175 561	20.469	64.730	2.3866	469	219 961	21.656	68.484	2.1322
420	176 400	20.494	64.807	2.3810	470	220 900	21.679	68.557	2.1277
421	177 241	20.518	64.885	2.3753	471	221 841	21.703	68.629	2.1231
422	178 084	20.543	64.962	2.3697	472	222 784	21.726	68.702	2.1186
423	178 929	20.567	65.038	2.3641	473	223 729	21.749	68.775	2.1142
424	179 776	20.591	65.115	2.3585	474	224 676	21.772	68.848	2.1097
425	180 625	20.616	65.192	2.3529	475	225 625	21.794	68.920	2.1053
426	181 476	20.640	65.269	2.3474	476	226 576	21.817	68.993	2.1008
427	182 329	20.664	65.345	2.3419	477	227 529	21.840	69.065	2.0964
428	183 184	20.688	65.422	2.3364	478	228 484	21.863	69.138	2.0921
429	184 041	20.712	65.498	2.3310	479	229 441	21.886	69.210	2.0877
430	184 900	20.736	65.574	2.3256	480	230 400	21.909	69.282	2.0833
431	185 761	20.761	65.651	2.3202	481	231 361	21.932	69.354	2.0790
432	186 624	20.785	65.727	2.3148	482	232 324	21.955	69.426	2.0747
433	187 489	20.809	65.803	2.3095	483	233 289	21.977	69.498	2.0704
434	188 356	20.833	65.879	2.3041	484	234 256	22.000	69.570	2.0661
435	189 225	20.857	65.955	2.2989	485	235 225	22.023	69.642	2.0619
436	190 096	20.881	66.030	2.2936	486	236 196	22.045	69.714	2.0576
437	190 969	20.905	66.106	2.2883	487	237 169	22.068	69.785	2.0534
438	191 844	20.928	66.182	2.2831	488	238 144	22.091	69.857	2.0492
439	192 721	20.952	66.257	2.2779	489	239 121	22.113	69.929	2.0450
440	193 600	20.976	66.333	2.2727	490	240 100	22.136	70.000	2.0408
441	194 481	21.000	66.408	2.2676	491	241 081	22.159	70.071	2.0367
442	195 364	21.024	66.483	2.2624	492	242 064	22.181	70.143	2.0325
443	196 249	21.048	66.558	2.2573	493	243 049	22.204	70.214	2.0284
444	197 136	21.071	66.633	2.2523	494	244 036	22.226	70.285	2.0243
445	198 025	21.095	66.708	2.2472	495	245 025	22.249	70.356	2.0202
446	198 916	21.119	66.783	2.2422	496	246 016	22.271	70.427	2.0161
447	199 809	21.142	66.858	2.2371	497	247 009	22.294	70.498	2.0121
448	200 704	21.166	66.933	2.2321	498	248 004	22.316	70.569	2.0080
449	201 601	21.190	67.007	2.2272	499	249 001	22.338	70.640	2.0040

Table C-1   Squares and Square Roots of Numbers from 1 to 9990 (Continued)

n	n^2	$\sqrt{n}$	$\sqrt{10n}$	$1000/n$	n	n^2	$\sqrt{n}$	$\sqrt{10n}$	$1000/n$
500	250 000	22.361	70.711	2.0000	550	302 500	23.452	74.162	1.8182
501	251 001	22.383	70.781	1.9960	551	303 601	23.473	74.229	1.8149
502	252 004	22.405	70.852	1.9920	552	304 704	23.495	74.297	1.8116
503	253 009	22.428	70.922	1.9881	553	305 809	23.516	74.364	1.8083
504	254 016	22.450	70.993	1.9841	554	306 916	23.537	74.431	1.8051
505	255 025	22.472	71.063	1.9802	555	308 025	23.558	74.498	1.8018
506	256 036	22.494	71.134	1.9763	556	309 136	23.580	74.565	1.7986
507	257 049	22.517	71.204	1.9724	557	310 249	23.601	74.632	1.7953
508	258 064	22.539	71.274	1.9685	558	311 364	23.622	74.699	1.7921
509	259 081	22.561	71.344	1.9646	559	312 481	23.643	74.766	1.7889
510	260 100	22.583	71.414	1.9608	560	313 600	23.664	74.833	1.7857
511	261 121	22.605	71.484	1.9569	561	314 721	23.685	74.900	1.7825
512	262 144	22.627	71.554	1.9531	562	315 844	23.707	74.967	1.7794
513	263 169	22.650	71.624	1.9493	563	316 969	23.728	75.033	1.7762
514	264 196	22.672	71.694	1.9455	564	318 096	23.749	75.100	1.7731
515	265 225	22.694	71.764	1.9417	565	319 225	23.770	75.166	1.7699
516	266 256	22.716	71.833	1.9380	566	320 356	23.791	75.233	1.7668
517	267 289	22.738	71.903	1.9342	567	321 489	23.812	75.299	1.7637
518	268 324	22.760	71.972	1.9305	568	322 624	23.833	75.366	1.7606
519	269 361	22.782	72.042	1.9268	569	323 761	23.854	75.432	1.7575
520	270 400	22.804	72.111	1.9231	570	324 900	23.875	75.498	1.7544
521	271 441	22.825	72.180	1.9194	571	326 041	23.896	75.565	1.7513
522	272 484	22.847	72.250	1.9157	572	327 184	23.917	75.631	1.7483
523	273 529	22.869	72.319	1.9120	573	328 329	23.937	75.697	1.7452
524	274 576	22.891	72.388	1.9084	574	329 476	23.958	75.763	1.7422
525	275 625	22.913	72.457	1.9048	575	330 625	23.979	75.829	1.7391
526	276 676	22.935	72.526	1.9011	576	331 776	24.000	75.895	1.7361
527	277 729	22.956	72.595	1.8975	577	332 929	24.021	75.961	1.7331
528	278 784	22.978	72.664	1.8939	578	334 084	24.042	76.026	1.7301
529	279 841	23.000	72.732	1.8904	579	335 241	24.062	76.092	1.7271
530	280 900	23.022	72.801	1.8868	580	336 400	24.083	76.158	1.7241
531	281 961	23.043	72.870	1.8832	581	337 561	24.104	76.223	1.7212
532	283 024	23.065	72.938	1.8797	582	338 724	24.125	76.289	1.7182
533	284 089	23.087	73.007	1.8762	583	339 889	24.145	76.354	1.7153
534	285 156	23.108	73.075	1.8727	584	341 056	24.166	76.420	1.7123
535	286 225	23.130	73.144	1.8692	585	342 225	24.187	76.485	1.7094
536	287 296	23.152	73.212	1.8657	586	343 396	24.207	76.551	1.7065
537	288 369	23.173	73.280	1.8622	587	344 569	24.228	76.616	1.7036
538	289 444	23.195	73.348	1.8587	588	345 744	24.249	76.681	1.7007
539	290 521	23.216	73.417	1.8553	589	346 921	24.269	76.746	1.6978
540	291 600	23.238	73.485	1.8519	590	348 100	24.290	76.811	1.6949
541	292 681	23.259	73.553	1.8484	591	349 281	24.310	76.877	1.6920
542	293 764	23.281	73.621	1.8450	592	350 464	24.331	76.942	1.6892
543	294 849	23.302	73.689	1.8416	593	351 649	24.352	77.006	1.6863
544	295 936	23.324	73.756	1.8382	594	352 836	24.372	77.071	1.6835
545	297 025	23.345	73.824	1.8349	595	354 025	24.393	77.136	1.6807
546	298 116	23.367	73.892	1.8315	596	355 216	24.413	77.201	1.6779
547	299 209	23.388	73.959	1.8282	597	356 409	24.434	77.266	1.6750
548	300 304	23.409	74.027	1.8248	598	357 604	24.454	77.330	1.6722
549	301 401	23.431	74.095	1.8215	599	358 801	24.474	77.395	1.6694

Table C-1 Squares and Square Roots of Numbers from 1 to 9990 (Continued)

n	n²	√n	√10n	1000/n	n	n²	√n	√10n	1000/n
600	360 000	24.495	77.460	1.6667	650	422 500	25.495	80.623	1.5385
601	361 201	24.515	77.524	1.6639	651	423 801	25.515	80.685	1.5361
602	362 404	24.536	77.589	1.6611	652	425 104	25.534	80.747	1.5337
603	363 609	24.556	77.653	1.6584	653	426 409	25.554	80.808	1.5314
604	364 816	24.576	77.717	1.6556	654	427 716	25.573	80.870	1.5291
605	366 025	24.597	77.782	1.6529	655	429 025	25.593	80.932	1.5267
606	367 236	24.617	77.846	1.6502	656	430 336	25.613	80.994	1.5244
607	368 449	24.637	77.910	1.6474	657	431 649	25.632	81.056	1.5221
608	369 664	24.658	77.974	1.6447	658	432 964	25.652	81.117	1.5198
609	370 881	24.678	78.038	1.6420	659	434 281	25.671	81.179	1.5175
610	372 100	24.698	78.103	1.6393	660	435 600	25.690	81.240	1.5152
611	373 321	24.718	78.166	1.6367	661	436 921	25.710	81.302	1.5129
612	374 544	24.739	78.230	1.6340	662	438 244	25.729	81.363	1.5106
613	375 769	24.759	78.294	1.6313	663	439 569	25.749	81.425	1.5083
614	376 996	24.779	78.358	1.6287	664	440 896	25.768	81.486	1.5060
615	378 225	24.799	78.422	1.6260	665	442 225	25.788	81.548	1.5038
616	379 456	24.819	78.486	1.6234	666	443 556	25.807	81.609	1.5015
617	380 689	24.839	78.549	1.6207	667	444 889	25.826	81.670	1.4993
618	381 924	24.860	78.613	1.6181	668	446 224	25.846	81.731	1.4970
619	383 161	24.880	78.677	1.6155	669	447 561	25.865	81.792	1.4948
620	384 400	24.900	78.740	1.6129	670	448 900	25.884	81.854	1.4925
621	385 641	24.920	78.804	1.6103	671	450 241	25.904	81.915	1.4903
622	386 884	24.940	78.867	1.6077	672	451 584	25.923	81.976	1.4881
623	388 129	24.960	78.930	1.6051	673	452 929	25.942	82.037	1.4859
624	389 376	24.980	78.994	1.6026	674	454 276	25.962	82.098	1.4837
625	390 625	25.000	79.057	1.6000	675	455 625	25.981	82.158	1.4815
626	391 876	25.020	79.120	1.5974	676	456 976	26.000	82.219	1.4793
627	393 129	25.040	79.183	1.5949	677	458 329	26.019	82.280	1.4771
628	394 384	25.060	79.246	1.5924	678	459 684	26.038	82.341	1.4749
629	395 641	25.080	79.310	1.5898	679	461 041	26.058	82.401	1.4728
630	396 900	25.100	79.373	1.5873	680	462 400	26.077	82.462	1.4706
631	398 161	25.120	79.436	1.5848	681	463 761	26.096	82.523	1.4684
632	399 424	25.140	79.498	1.5823	682	465 124	26.115	82.583	1.4663
633	400 689	25.159	79.561	1.5798	683	466 489	26.134	82.644	1.4641
634	401 956	25.179	79.624	1.5773	684	467 856	26.153	82.704	1.4620
635	403 225	25.199	79.687	1.5748	685	469 225	26.173	82.765	1.4599
636	404 496	25.219	79.750	1.5723	686	470 596	26.192	82.825	1.4577
637	405 769	25.239	79.812	1.5699	687	471 969	26.211	82.885	1.4556
638	407 044	25.259	79.875	1.5674	688	473 344	26.230	82.946	1.4535
639	408 321	25.278	79.937	1.5649	689	474 721	26.249	83.006	1.4514
640	409 600	25.298	80.000	1.5625	690	476 100	26.268	83.066	1.4493
641	410 881	25.318	80.062	1.5601	691	477 481	26.287	83.126	1.4472
642	412 164	25.338	80.125	1.5576	692	478 864	26.306	83.187	1.4451
643	413 449	25.357	80.187	1.5552	693	480 249	26.325	83.247	1.4430
644	414 736	25.377	80.250	1.5528	694	481 636	26.344	83.307	1.4409
645	416 025	25.397	80.312	1.5504	695	483 025	26.363	83.367	1.4388
646	417 316	25.417	80.374	1.5480	696	484 416	26.382	83.427	1.4368
647	418 609	25.436	80.436	1.5456	697	485 809	26.401	83.487	1.4347
648	419 904	25.456	80.498	1.5432	698	487 204	26.420	83.546	1.4327
649	421 201	25.475	80.561	1.5408	699	488 601	26.439	83.606	1.4306

Table C-1   Squares and Square Roots of Numbers from 1 to 9990 (Continued)

n	n^2	$\sqrt{n}$	$\sqrt{10n}$	$1000/n$	n	n^2	$\sqrt{n}$	$\sqrt{10n}$	$1000/n$
700	490 000	26.458	83.666	1.4286	750	562 500	27.386	86.603	1.3333
701	491 401	26.476	83.726	1.4265	751	564 001	27.404	86.660	1.3316
702	492 804	26.495	83.785	1.4245	752	565 504	27.423	86.718	1.3298
703	494 209	26.514	83.845	1.4225	753	567 009	27.441	86.776	1.3280
704	495 616	26.533	83.905	1.4205	754	568 516	27.459	86.833	1.3263
705	497 025	26.552	83.964	1.4184	755	570 025	27.477	86.891	1.3245
706	498 436	26.571	84.024	1.4164	756	571 536	27.495	86.948	1.3228
707	499 849	26.589	84.083	1.4144	757	573 049	27.514	87.006	1.3210
708	501 264	26.608	84.143	1.4124	758	574 564	27.532	87.063	1.3193
709	502 681	26.627	84.202	1.4104	759	576 081	27.550	87.121	1.3175
710	504 100	26.646	84.262	1.4085	760	577 600	27.568	87.178	1.3158
711	505 521	26.665	84.321	1.4065	761	579 121	27.586	87.235	1.3141
712	506 944	26.683	84.380	1.4045	762	580 644	27.604	87.293	1.3123
713	508 369	26.702	84.439	1.4025	763	582 169	27.622	87.350	1.3106
714	509 796	26.721	84.499	1.4006	764	583 696	27.641	87.407	1.3089
715	511 225	26.739	84.558	1.3986	765	585 225	27.659	87.464	1.3072
716	512 656	26.758	84.617	1.3966	766	586 756	27.677	87.521	1.3055
717	514 089	26.777	84.676	1.3947	767	588 289	27.695	87.579	1.3038
718	515 524	26.796	84.735	1.3928	768	589 824	27.713	87.636	1.3021
719	516 961	26.814	84.794	1.3908	769	591 361	27.731	87.693	1.3004
720	518 400	26.833	84.853	1.3889	770	592 900	27.749	87.750	1.2987
721	519 841	26.851	84.912	1.3870	771	594 441	27.767	87.807	1.2970
722	521 284	26.870	84.971	1.3850	772	595 984	27.785	87.864	1.2953
723	522 729	26.889	85.029	1.3831	773	597 529	27.803	87.920	1.2937
724	524 176	26.907	85.088	1.3812	774	599 076	27.821	87.977	1.2920
725	525 625	26.926	85.147	1.3793	775	600 625	27.839	88.034	1.2903
726	527 076	26.944	85.206	1.3774	776	602 176	27.857	88.091	1.2887
727	528 529	26.963	85.264	1.3755	777	603 729	27.875	88.148	1.2870
728	529 984	26.981	85.323	1.3736	778	605 284	27.893	88.204	1.2853
729	531 441	27.000	85.382	1.3717	779	606 841	27.911	88.261	1.2837
730	532 900	27.019	85.440	1.3699	780	608 400	27.928	88.318	1.2821
731	534 361	27.037	85.499	1.3680	781	609 961	27.946	88.374	1.2804
732	535 824	27.056	85.557	1.3661	782	611 524	27.964	88.431	1.2788
733	537 289	27.074	85.615	1.3643	783	613 089	27.982	88.487	1.2771
734	538 756	27.092	85.674	1.3624	784	614 656	28.000	88.544	1.2755
735	540 225	27.111	85.732	1.3605	785	616 225	28.018	88.600	1.2739
736	541 696	27.129	85.790	1.3587	786	617 796	28.036	88.657	1.2723
737	543 169	27.148	85.849	1.3569	787	619 369	28.054	88.713	1.2706
738	544 644	27.166	85.907	1.3550	788	620 944	28.071	88.769	1.2690
739	546 121	27.185	85.965	1.3532	789	622 521	28.089	88.826	1.2674
740	547 600	27.203	86.023	1.3514	790	624 100	28.107	88.882	1.2658
741	549 081	27.221	86.081	1.3495	791	625 681	28.125	88.938	1.2642
742	550 564	27.240	86.139	1.3477	792	627 264	28.142	88.994	1.2626
743	552 049	27.258	86.197	1.3459	793	628 849	28.160	89.051	1.2610
744	553 536	27.276	86.255	1.3441	794	630 436	28.178	89.107	1.2594
745	555 025	27.295	86.313	1.3423	795	632 025	28.196	89.163	1.2579
746	556 516	27.313	86.371	1.3405	796	633 616	28.213	89.219	1.2563
747	558 009	27.331	86.429	1.3387	797	635 209	28.231	89.275	1.2547
748	559 504	27.350	86.487	1.3369	798	636 804	28.249	89.331	1.2531
749	561 001	27.368	86.545	1.3351	799	638 401	28.267	89.387	1.2516

Table C-1   Squares and Square Roots of Numbers from 1 to 9990 (Continued)

n	n²	√n	√10n	1000/n	n	n²	√n	√10n	1000/n
800	640 000	28.284	89.443	1.2500	850	722 500	29.155	92.195	1.1765
801	641 601	28.302	89.499	1.2484	851	724 201	29.172	92.250	1.1751
802	643 204	28.320	89.554	1.2469	852	725 904	29.189	92.304	1.1737
803	644 809	28.337	89.610	1.2453	853	727 609	29.206	92.358	1.1723
804	646 416	28.355	89.666	1.2438	854	729 316	29.223	92.412	1.1710
805	648 025	28.373	89.722	1.2422	855	731 025	29.240	92.466	1.1696
806	649 636	28.390	89.778	1.2407	856	732 736	29.257	92.520	1.1682
807	651 249	28.408	89.833	1.2392	857	734 449	29.275	92.574	1.1669
808	652 864	28.425	89.889	1.2376	858	736 164	29.292	92.628	1.1655
809	654 481	28.443	89.944	1.2361	859	737 881	29.309	92.682	1.1641
810	656 100	28.461	90.000	1.2346	860	739 600	29.326	92.736	1.1628
811	657 721	28.478	90.056	1.2330	861	741 321	29.343	92.790	1.1614
812	659 344	28.496	90.111	1.2315	862	743 044	29.360	92.844	1.1601
813	660 969	28.513	90.167	1.2300	863	744 769	29.377	92.898	1.1587
814	662 596	28.531	90.222	1.2285	864	746 496	29.394	92.952	1.1574
815	664 225	28.548	90.277	1.2270	865	748 225	29.411	93.005	1.1561
816	665 856	28.566	90.333	1.2255	866	749 956	29.428	93.059	1.1547
817	667 489	28.583	90.388	1.2240	867	751 689	29.445	93.113	1.1534
818	669 124	28.601	90.443	1.2225	868	753 424	29.462	93.167	1.1521
819	670 761	28.618	90.499	1.2210	869	755 161	29.479	93.220	1.1507
820	672 400	28.636	90.554	1.2195	870	756 900	29.496	93.274	1.1494
821	674 041	28.653	90.609	1.2180	871	758 641	29.513	93.327	1.1481
822	675 684	28.671	90.664	1.2165	872	760 384	29.530	93.381	1.1468
823	677 329	28.688	90.719	1.2151	873	762 129	29.547	93.434	1.1455
824	678 976	28.705	90.774	1.2136	874	763 876	29.563	93.488	1.1442
825	680 625	28.723	90.830	1.2121	875	765 625	29.580	93.541	1.1429
826	682 276	28.740	90.885	1.2107	876	767 376	29.597	93.595	1.1416
827	683 929	28.758	90.940	1.2092	877	769 129	29.614	93.648	1.1403
828	685 584	28.775	90.995	1.2077	878	770 884	29.631	93.702	1.1390
829	687 241	28.792	91.049	1.2063	879	772 641	29.648	93.755	1.1377
830	688 900	28.810	91.104	1.2048	880	774 400	29.665	93.808	1.1364
831	690 561	28.827	91.159	1.2034	881	776 161	29.682	93.862	1.1351
832	692 224	28.844	91.214	1.2019	882	777 924	29.698	93.915	1.1338
833	693 889	28.862	91.269	1.2005	883	779 689	29.715	93.968	1.1325
834	695 556	28.879	91.324	1.1990	884	781 456	29.732	94.021	1.1312
835	697 225	28.896	91.378	1.1976	885	783 225	29.749	94.074	1.1299
836	698 896	28.914	91.433	1.1962	886	784 996	29.766	94.128	1.1287
837	700 569	28.931	91.488	1.1947	887	786 769	29.783	94.181	1.1274
838	702 244	28.948	91.542	1.1933	888	788 544	29.799	94.234	1.1261
839	703 921	28.966	91.597	1.1919	889	790 321	29.816	94.287	1.1249
840	705 600	28.983	91.652	1.1905	890	792 100	29.833	94.340	1.1236
841	707 281	29.000	91.706	1.1891	891	793 881	29.850	94.393	1.1223
842	708 964	29.017	91.761	1.1876	892	795 664	29.866	94.446	1.1211
843	710 649	29.034	91.815	1.1862	893	797 449	29.883	94.499	1.1198
844	712 336	29.052	91.869	1.1848	894	799 236	29.900	94.552	1.1186
845	714 025	29.069	91.924	1.1834	895	801 025	29.917	94.604	1.1173
846	715 716	29.086	91.978	1.1820	896	802 816	29.933	94.657	1.1161
847	717 409	29.103	92.033	1.1806	897	804 609	29.950	94.710	1.1148
848	719 104	29.120	92.087	1.1792	898	806 404	29.967	94.763	1.1136
849	720 801	29.138	92.141	1.1779	899	808 201	29.983	94.816	1.1123

Table C-1 **Squares and Square Roots of Numbers from 1 to 9990 (Continued)**

n	n^2	$\sqrt{n}$	$\sqrt{10n}$	$1000/n$	n	n^2	$\sqrt{n}$	$\sqrt{10n}$	$1000/n$
900	810 000	30.000	94.868	1.1111	950	902 500	30.822	97.468	1.0526
901	811 801	30.017	94.921	1.1099	951	904 401	30.838	97.519	1.0515
902	813 604	30.033	94.974	1.1086	952	906 304	30.855	97.570	1.0504
903	815 409	30.050	95.026	1.1074	953	908 209	30.871	97.622	1.0493
904	817 216	30.067	95.079	1.1062	954	910 116	30.887	97.673	1.0482
905	819 025	30.083	95.131	1.1050	955	912 025	30.903	97.724	1.0471
906	820 836	30.100	95.184	1.1038	956	913 936	30.919	97.775	1.0460
907	822 649	30.116	95.237	1.1025	957	915 849	30.935	97.826	1.0449
908	824 464	30.133	95.289	1.1013	958	917 764	30.952	97.877	1.0438
909	826 281	30.150	95.341	1.1001	959	919 681	30.968	97.929	1.0428
910	828 100	30.166	95.394	1.0989	960	921 600	30.984	97.980	1.0417
911	829 921	30.183	95.446	1.0977	961	923 521	31.000	98.031	1.0406
912	831 744	30.199	95.499	1.0965	962	925 444	31.016	98.082	1.0395
913	833 569	30.216	95.551	1.0953	963	927 369	31.032	98.133	1.0384
914	835 396	30.232	95.603	1.0941	964	929 296	31.048	98.184	1.0373
915	837 225	30.249	95.656	1.0929	965	931 225	31.064	98.234	1.0363
916	839 056	30.265	95.708	1.0917	966	933 156	31.081	98.285	1.0352
917	840 889	30.282	95.760	1.0905	967	935 089	31.097	98.336	1.0341
918	842 724	30.299	95.812	1.0893	968	937 024	31.113	98.387	1.0331
919	844 561	30.315	95.864	1.0881	969	938 961	31.129	98.438	1.0320
920	846 400	30.332	95.917	1.0870	970	940 900	31.145	98.489	1.0309
921	848 241	30.348	95.969	1.0858	971	942 841	31.161	98.539	1.0299
922	850 084	30.364	96.021	1.0846	972	944 784	31.177	98.590	1.0288
923	851 929	30.381	96.073	1.0834	973	946 729	31.193	98.641	1.0277
924	853 776	30.397	96.125	1.0823	974	948 676	31.209	98.691	1.0267
925	855 625	30.414	96.177	1.0811	975	950 625	31.225	98.742	1.0256
926	857 476	30.430	96.229	1.0799	976	952 576	31.241	98.793	1.0246
927	859 329	30.447	96.281	1.0787	977	954 529	31.257	98.843	1.0235
928	861 184	30.463	96.333	1.0776	978	956 484	31.273	98.894	1.0225
929	863 041	30.480	96.385	1.0764	979	958 441	31.289	98.944	1.0215
930	864 900	30.496	96.437	1.0753	980	960 400	31.305	98.995	1.0204
931	866 761	30.512	96.488	1.0741	981	962 361	31.321	99.045	1.0194
932	868 624	30.529	96.540	1.0730	982	964 324	31.337	99.096	1.0183
933	870 489	30.545	96.592	1.0718	983	966 289	31.353	99.146	1.0173
934	872 356	30.561	96.644	1.0707	984	968 256	31.369	99.197	1.0163
935	874 225	30.578	96.695	1.0695	985	970 225	31.385	99.247	1.0152
936	876 096	30.594	96.747	1.0684	986	972 196	31.401	99.298	1.0142
937	877 969	30.610	96.799	1.0672	987	974 169	31.417	99.348	1.0132
938	879 844	30.627	96.850	1.0661	988	976 144	31.432	99.398	1.0121
939	881 721	30.643	96.902	1.0650	989	978 121	31.448	99.448	1.0111
940	883 600	30.659	96.954	1.0638	990	980 100	31.464	99.499	1.0101
941	885 481	30.676	97.005	1.0627	991	982 081	31.480	99.549	1.0091
942	887 364	30.692	97.057	1.0616	992	984 064	31.496	99.599	1.0081
943	889 249	30.708	97.108	1.0604	993	986 049	31.512	99.649	1.0070
944	891 136	30.725	97.160	1.0593	994	988 036	31.528	99.700	1.0060
945	893 025	30.741	97.211	1.0582	995	990 025	31.544	99.750	1.0050
946	894 916	30.757	97.263	1.0571	996	992 016	31.559	99.800	1.0040
947	896 809	30.773	97.314	1.0560	997	994 009	31.575	99.850	1.0030
948	898 704	30.790	97.365	1.0549	998	996 004	31.591	99.900	1.0020
949	900 601	30.806	97.417	1.0537	999	998 001	31.607	99.950	1.0010

Table C-2 Areas under the Normal Curve

⭣ z or $\dfrac{x}{\sigma}$ 1	Area Between Mean and z 2	Area Beyond z 3	z or $\dfrac{x}{\sigma}$ 1	Area Between Mean and z 2	Area Beyond z 3
0.00	.0000	.5000	0.35	.1368	.3632
0.01	.0040	.4960	0.36	.1406	.3594
0.02	.0080	.4920	0.37	.1443	.3557
0.03	.0120	.4880	0.38	.1480	.3520
0.04	.0160	.4840	0.39	.1517	.3483
0.05	.0199	.4801	0.40	.1554	.3446
0.06	.0239	.4761	0.41	.1591	.3409
0.07	.0279	.4721	0.42	.1628	.3372
0.08	.0319	.4681	0.43	.1664	.3336
0.09	.0359	.4641	0.44	.1700	.3300
0.10	.0398	.4602	0.45	.1736	.3264
0.11	.0438	.4562	0.46	.1772	.3228
0.12	.0478	.4522	0.47	.1808	.3192
0.13	.0517	.4483	0.48	.1844	.3156
0.14	.0557	.4443	0.49	.1879	.3121
0.15	.0596	.4404	0.50	.1915	.3085
0.16	.0636	.4364	0.51	.1950	.3050
0.17	.0675	.4325	0.52	.1985	.3015
0.18	.0714	.4286	0.53	.2019	.2981
0.19	.0753	.4247	0.54	.2054	.2946
0.20	.0793	.4207	0.55	.2088	.2912
0.21	.0832	.4168	0.56	.2123	.2877
0.22	.0871	.4129	0.57	.2157	.2843
0.23	.0910	.4090	0.58	.2190	.2810
0.24	.0948	.4052	0.59	.2224	.2776
0.25	.0987	.4013	0.60	.2257	.2743
0.26	.1026	.3974	0.61	.2291	.2709
0.27	.1064	.3936	0.62	.2324	.2676
0.28	.1103	.3897	0.63	.2357	.2643
0.29	.1141	.3859	0.64	.2389	.2611
0.30	.1179	.3821	0.65	.2422	.2578
0.31	.1217	.3783	0.66	.2454	.2546
0.32	.1255	.3745	0.67	.2486	.2514
0.33	.1293	.3707	0.68	.2517	.2483
0.34	.1331	.3669	0.69	.2549	.2451

Table C-2   *Areas under the Normal Curve (Continued)*

z or $\frac{x}{\sigma}$ 1	Area Between Mean and z 2	Area Beyond z 3	z or $\frac{x}{\sigma}$ 1	Area Between Mean and z 2	Area Beyond z 3
0.70	.2580	.2420	1.05	.3531	.1469
0.71	.2611	.2389	1.06	.3554	.1446
0.72	.2642	.2358	1.07	.3577	.1423
0.73	.2673	.2327	1.08	.3599	.1401
0.74	.2704	.2296	1.09	.3621	.1379
0.75	.2734	.2266	1.10	.3643	.1357
0.76	.2764	.2236	1.11	.3665	.1335
0.77	.2794	.2206	1.12	.3686	.1314
0.78	.2823	.2177	1.13	.3708	.1292
0.79	.2852	.2148	1.14	.3729	.1271
0.80	.2881	.2119	1.15	.3749	.1251
0.81	.2910	.2090	1.16	.3770	.1230
0.82	.2939	.2061	1.17	.3790	.1210
0.83	.2967	.2033	1.18	.3810	.1190
0.84	.2995	.2005	1.19	.3830	.1170
0.85	.3023	.1977	1.20	.3849	.1151
0.86	.3051	.1949	1.21	.3869	.1131
0.87	.3078	.1922	1.22	.3888	.1112
0.88	.3106	.1894	1.23	.3907	.1093
0.89	.3133	.1867	1.24	.3925	.1075
0.90	.3159	.1841	1.25	.3944	.1056
0.91	.3186	.1814	1.26	.3962	.1038
0.92	.3212	.1788	1.27	.3980	.1020
0.93	.3238	.1762	1.28	.3997	.1003
0.94	.3264	.1736	1.29	.4015	.0985
0.95	.3289	.1711	1.30	.4032	.0968
0.96	.3315	.1685	1.31	.4049	.0951
0.97	.3340	.1660	1.32	.4066	.0934
0.98	.3365	.1635	1.33	.4082	.0918
0.99	.3389	.1611	1.34	.4099	.0901
1.00	.3413	.1587	1.35	.4115	.0885
1.01	.3438	.1562	1.36	.4131	.0869
1.02	.3461	.1539	1.37	.4147	.0853
1.03	.3485	.1515	1.38	.4162	.0838
1.04	.3508	.1492	1.39	.4177	.0823

Table C-2   Areas under the Normal Curve (Continued)

z or $\frac{x}{\sigma}$ 1	Area Between Mean and z 2	Area Beyond z 3	z or $\frac{x}{\sigma}$ 1	Area Between Mean and z 2	Area Beyond z 3
1.40	.4192	.0808	1.75	.4599	.0401
1.41	.4207	.0793	1.76	.4608	.0392
1.42	.4222	.0778	1.77	.4616	.0384
1.43	.4236	.0764	1.78	.4625	.0375
1.44	.4251	.0749	1.79	.4633	.0367
1.45	.4265	.0735	1.80	.4641	.0359
1.46	.4279	.0721	1.81	.4649	.0351
1.47	.4292	.0708	1.82	.4656	.0344
1.48	.4306	.0694	1.83	.4664	.0336
1.49	.4319	.0681	1.84	.4671	.0329
1.50	.4332	.0668	1.85	.4678	.0322
1.51	.4345	.0655	1.86	.4686	.0314
1.52	.4357	.0643	1.87	.4693	.0307
1.53	.4370	.0630	1.88	.4699	.0301
1.54	.4382	.0618	1.89	.4706	.0294
1.55	.4394	.0606	1.90	.4713	.0287
1.56	.4406	.0594	1.91	.4719	.0281
1.57	.4418	.0582	1.92	.4726	.0274
1.58	.4429	.0571	1.93	.4732	.0268
1.59	.4441	.0559	1.94	.4738	.0262
1.60	.4452	.0548	1.95	.4744	.0256
1.61	.4463	.0537	1.96	.4750	.0250
1.62	.4474	.0526	1.97	.4756	.0244
1.63	.4484	.0516	1.98	.4761	.0239
1.64	.4495	.0505	1.99	.4767	.0233
1.65	.4505	.0495	2.00	.4772	.0228
1.66	.4515	.0485	2.01	.4778	.0222
1.67	.4525	.0475	2.02	.4783	.0217
1.68	.4535	.0465	2.03	.4788	.0212
1.69	.4545	.0455	2.04	.4793	.0207
1.70	.4554	.0446	2.05	.4798	.0202
1.71	.4564	.0436	2.06	.4803	.0197
1.72	.4573	.0427	2.07	.4808	.0192
1.73	.4582	.0418	2.08	.4812	.0188
1.74	.4591	.0409	2.09	.4817	.0183

Table C-2 Areas under the Normal Curve (Continued)

z or $\dfrac{x}{\sigma}$ 1	Area Between Mean and z 2	Area Beyond z 3	z or $\dfrac{x}{\sigma}$ 1	Area Between Mean and z 2	Area Beyond z 3
2.10	.4821	.0179	2.45	.4929	.0071
2.11	.4826	.0174	2.46	.4931	.0069
2.12	.4830	.0170	2.47	.4932	.0068
2.13	.4834	.0166	2.48	.4934	.0066
2.14	.4838	.0162	2.49	.4936	.0064
2.15	.4842	.0158	2.50	.4938	.0062
2.16	.4846	.0154	2.51	.4940	.0060
2.17	.4850	.0150	2.52	.4941	.0059
2.18	.4854	.0146	2.53	.4943	.0057
2.19	.4857	.0143	2.54	.4945	.0055
2.20	.4861	.0139	2.55	.4946	.0054
2.21	.4864	.0136	2.56	.4948	.0052
2.22	.4868	.0132	2.57	.4949	.0051
2.23	.4871	.0129	2.58	.4951	.0049
2.24	.4875	.0125	2.59	.4952	.0048
2.25	.4878	.0122	2.60	.4953	.0047
2.26	.4881	.0119	2.61	.4955	.0045
2.27	.4884	.0116	2.62	.4956	.0044
2.28	.4887	.0113	2.63	.4957	.0043
2.29	.4890	.0110	2.64	.4959	.0041
2.30	.4893	.0107	2.65	.4960	.0040
2.31	.4896	.0104	2.66	.4961	.0039
2.32	.4898	.0102	2.67	.4962	.0038
2.33	.4901	.0099	2.68	.4963	.0037
2.34	.4904	.0096	2.69	.4964	.0036
2.35	.4906	.0094	2.70	.4965	.0035
2.36	.4909	.0091	2.71	.4966	.0034
2.37	.4911	.0089	2.72	.4967	.0033
2.38	.4913	.0087	2.73	.4968	.0032
2.39	.4916	.0084	2.74	.4969	.0031
2.40	.4918	.0082	2.75	.4970	.0030
2.41	.4920	.0080	2.76	.4971	.0029
2.42	.4922	.0078	2.77	.4972	.0028
2.43	.4925	.0075	2.78	.4973	.0027
2.44	.4927	.0073	2.79	.4974	.0026

Table C-2 Areas under the Normal Curve (Continued)

z or $\frac{x}{\sigma}$ 1	Area Between Mean and z 2	Area Beyond z 3	z or $\frac{x}{\sigma}$ 1	Area Between Mean and z 2	Area Beyond z 3
2.80	.4974	.0026	3.15	.4992	.0008
2.81	.4975	.0025	3.16	.4992	.0008
2.82	.4976	.0024	3.17	.4992	.0008
2.83	.4977	.0023	3.18	.4993	.0007
2.84	.4977	.0023	3.19	.4993	.0007
2.85	.4978	.0022	3.20	.4993	.0007
2.86	.4979	.0021	3.21	.4993	.0007
2.87	.4979	.0021	3.22	.4994	.0006
2.88	.4980	.0020	3.23	.4994	.0006
2.89	.4981	.0019	3.24	.4994	.0006
2.90	.4981	.0019	3.30	.4995	.0005
2.91	.4982	.0018	3.40	.4997	.0003
2.92	.4982	.0018	3.50	.4998	.0002
2.93	.4983	.0017	3.60	.4998	.0002
2.94	.4984	.0016	3.70	.4999	.0001
2.95	.4984	.0016			
2.96	.4985	.0015			
2.97	.4985	.0015			
2.98	.4986	.0014			
2.99	.4986	.0014			
3.00	.4987	.0013			
3.01	.4987	.0013			
3.02	.4987	.0013			
3.03	.4988	.0012			
3.04	.4988	.0012			
3.05	.4989	.0011			
3.06	.4989	.0011			
3.07	.4989	.0011			
3.08	.4990	.0010			
3.09	.4990	.0010			
3.10	.4990	.0010			
3.11	.4991	.0009			
3.12	.4991	.0009			
3.13	.4991	.0009			
3.14	.4992	.0008			

Table C-3   Probability Distribution of t*

n	.1	.05	.02	.01	.001
1	6.314	12.706	31.821	63.657	636.619
2	2.920	4.303	6.965	9.925	31.598
3	2.353	3.182	4.541	5.841	12.924
4	2.132	2.776	3.747	4.604	8.610
5	2.015	2.571	3.365	4.032	6.869
6	1.943	2.447	3.143	3.707	5.959
7	1.895	2.365	2.998	3.499	5.408
8	1.860	2.306	2.896	3.355	5.041
9	1.833	2.262	2.821	3.250	4.781
10	1.812	2.228	2.764	3.169	4.587
11	1.796	2.201	2.718	3.106	4.437
12	1.782	2.179	2.681	3.055	4.318
13	1.771	2.160	2.650	3.012	4.221
14	1.761	2.145	2.624	2.977	4.140
15	1.753	2.131	2.602	2.947	4.073
16	1.746	2.120	2.583	2.921	4.015
17	1.740	2.110	2.567	2.898	3.965
18	1.734	2.101	2.552	2.878	3.922
19	1.729	2.093	2.539	2.861	3.883
20	1.725	2.086	2.528	2.845	3.850
21	1.721	2.080	2.518	2.831	3.819
22	1.717	2.074	2.508	2.819	3.792
23	1.714	2.069	2.500	2.807	3.767
24	1.711	2.064	2.492	2.797	3.745
25	1.708	2.060	2.485	2.787	3.725
26	1.706	2.056	2.479	2.779	3.707
27	1.703	2.052	2.473	2.771	3.690
28	1.701	2.048	2.467	2.763	3.674
29	1.699	2.045	2.462	2.756	3.659
30	1.697	2.042	2.457	2.750	3.646

* From Table III of Fisher and Yates, "Statistical Tables for Biological, Agricultural and Medical Research," published by Oliver and Boyd Ltd., Edinburgh and London, 1953; by permission of the authors and publishers.

Table C-4 Percentage Points of the F Distribution: 5 Percent Points*

df_2 \ df_1	1	2	3	4	5	6	7	8	9
1	161.45	199.50	215.71	224.58	230.16	233.99	236.77	238.88	240.54
2	18.513	19.000	19.164	19.247	19.296	19.330	19.353	19.371	19.385
3	10.128	9.5521	9.2766	9.1172	9.0135	8.9406	8.8868	8.8452	8.8123
4	7.7086	6.9443	6.5914	6.3883	6.2560	6.1631	6.0942	6.0410	5.9988
5	6.6079	5.7861	5.4095	5.1922	5.0503	4.9503	4.8759	4.8183	4.7725
6	5.9874	5.1433	4.7571	4.5337	4.3874	4.2839	4.2066	4.1468	4.0990
7	5.5914	4.7374	4.3468	4.1203	3.9715	3.8660	3.7870	3.7257	3.6767
8	5.3177	4.4590	4.0662	3.8378	3.6875	3.5806	3.5005	3.4381	3.3881
9	5.1174	4.2565	3.8626	3.6331	3.4817	3.3738	3.2927	3.2296	3.1789
10	4.9646	4.1028	3.7083	3.4780	3.3258	3.2172	3.1355	3.0717	3.0204
11	4.8443	3.9823	3.5874	3.3567	3.2039	3.0946	3.0123	2.9480	2.8962
12	4.7472	3.8853	3.4903	3.2592	3.1059	2.9961	2.9134	2.8486	2.7964
13	4.6672	3.8056	3.4105	3.1791	3.0254	2.9153	2.8321	2.7669	2.7144
14	4.6001	3.7389	3.3439	3.1122	2.9582	2.8477	2.7642	2.6987	2.6458
15	4.5431	3.6823	3.2874	3.0556	2.9013	2.7905	2.7066	2.6408	2.5876
16	4.4940	3.6337	3.2389	3.0069	2.8524	2.7413	2.6572	2.5911	2.5377
17	4.4513	3.5915	3.1968	2.9647	2.8100	2.6987	2.6143	2.5480	2.4943
18	4.4139	3.5546	3.1599	2.9277	2.7729	2.6613	2.5767	2.5102	2.4563
19	4.3808	3.5219	3.1274	2.8951	2.7401	2.6283	2.5435	2.4768	2.4227
20	4.3513	3.4928	3.0984	2.8661	2.7109	2.5990	2.5140	2.4471	2.3928
21	4.3248	3.4668	3.0725	2.8401	2.6848	2.5757	2.4876	2.4205	2.3661
22	4.3009	3.4434	3.0491	2.8167	2.6613	2.5491	2.4638	2.3965	2.3419
23	4.2793	3.4221	3.0280	2.7955	2.6400	2.5277	2.4422	2.3748	2.3201
24	4.2597	3.4028	3.0088	2.7763	2.6207	2.5082	2.4226	2.3551	2.3002
25	4.2417	3.3852	2.9912	2.7587	2.6030	2.4904	2.4047	2.3371	2.2821
26	4.2252	3.3690	2.9751	2.7426	2.5868	2.4741	2.3883	2.3205	2.2655
27	4.2100	3.3541	2.9604	2.7278	2.5719	2.4591	2.3732	2.3053	2.2501
28	4.1960	3.3404	2.9467	2.7141	2.5581	2.4453	2.3593	2.2913	2.2360
29	4.1830	3.3277	2.9340	2.7014	2.5454	2.4324	2.3463	2.2782	2.2229
30	4.1709	3.3158	2.9223	2.6896	2.5336	2.4205	2.3343	2.2662	2.2107
40	4.0848	3.2317	2.8387	2.6060	2.4495	2.3359	2.2490	2.1802	2.1240
60	4.0012	3.1504	2.7581	2.5252	2.3683	2.2540	2.1665	2.0970	2.0401
120	3.9201	3.0718	2.6802	2.4472	2.2900	2.1750	2.0867	2.0164	1.9588
∞	3.8415	2.9957	2.6049	2.3719	2.2141	2.0986	2.0096	1.9384	1.8799

* From Table V. of Fisher and Yates, "Statistical Tables for Biological, Agricultural and Medical Research," published by Oliver and Boyd Ltd., Edinburgh and London, 1953; by permission of the publishers.

Table C-4   Percentage Points of the F Distribution: 5 Percent Points (Continued)

d_{f_2} \\ d_{f_1}	10	12	15	20	24	30	40	60	120	∞
1	241.88	243.91	245.95	248.01	249.05	250.09	251.14	252.20	253.25	254.32
2	19.396	19.413	19.429	19.446	19.454	19.462	19.471	19.479	19.487	19.196
3	8.7855	8.7446	8.7029	8.6602	8.6385	8.6166	8.5944	8.5720	8.5491	8.5265
4	5.9644	5.9117	5.8578	5.8025	5.7744	5.7459	5 7170	5.6878	5.6581	5.6281
5	4.7351	4.6777	4.6188	4.5581	4.5272	4.4957	4.4638	4.4314	4.3984	4.3650
6	4.0600	3.9999	3.9381	3.8742	3.8415	3.8082	3.7743	3.7398	3.7047	3.6688
7	3.6365	3.5747	3.5108	3.4445	3.4105	3.3758	3.3404	3.3043	3.2674	3.2298
8	3.3472	3.2840	3.2184	3.1503	3.1152	3.0794	3.0428	3.0053	2.9669	2.9276
9	3.1373	3.0729	3.0061	2.9365	2.9005	2.8637	2.8259	2.7872	2.7475	2.7067
10	2.9782	2.9130	2.8450	2.7740	2.7372	2.6996	2.6609	2.6211	2.5801	2.5379
11	2.8536	2.7876	2.7186	2.6464	2.6090	2.5705	2.5309	2.4901	2.4480	2.4015
12	2.7534	2.6866	2.6169	2.5436	2.5055	2.4663	2.4259	2.3842	2.3410	2.2962
13	2.6710	2.6037	2.5331	2.4589	2.4202	2.3803	2.3392	2.2966	2.2524	2.2064
14	2.6021	2.5342	2.4630	2.3879	2.3487	2.3082	2.2664	2.2230	2.1778	2.1307
15	2.5437	2.4753	2.4035	2.3275	2.2878	2.2468	2.2043	2.1601	2.1141	2.0658
16	2.4935	2.4247	2.3522	2.2756	2.2354	2.1938	2.1507	2.1058	2.0589	2.0096
17	2.4499	2.3807	2.3077	2.2304	2.1898	2.1477	2.1040	2.0584	2.0107	1.9601
18	2.4117	2.3421	2.2686	2.1906	2.1497	2.1071	2.0629	2.0166	1.9681	1.9168
19	2.3779	2.3080	2.2341	2.1555	2.1141	2.0712	2.0264	1.9796	1.9302	1.8780
20	2.3479	2.2776	2.2033	2.1242	2.0825	2.0391	1.9938	1.9464	1.8963	1.8432
21	2.3210	2.2504	2.1757	2.0960	2.0540	2.0102	1.9645	1.9165	1.8657	1.8117
22	2.2967	2.2258	2.1508	2.0707	2.0283	1.9842	1.9380	1.8895	1.8380	1.7831
23	2.2747	2.2036	2.1282	2.0476	2.0050	1.9605	1.9139	1.8649	1.8128	1.7570
24	2.2547	2.1834	2.1077	2.0267	1.9838	1.9390	1.8920	1.8424	1.7897	1.7331
25	2.2365	2.1649	2.0889	2.0075	1.9643	1.9192	1.8718	1.8217	1.7684	1.7110
26	2.2197	2.1479	2.0716	1.9898	1.9464	1.9010	1.8533	1.8027	1.7488	1.6906
27	2.2043	2.1323	2.0558	1.9736	1.9299	1.8842	1.8361	1.7851	1.7307	1.6717
28	2.1900	2.1179	2.0411	1.9586	1.9147	1.8687	1.8203	1.7689	1.7138	1.6541
29	2.1768	2.1045	2.0275	1.9446	1.9005	1.8543	1.8055	1.7537	1.6981	1.6377
30	2.1646	2.0921	2.0148	1.9317	1.8874	1.8409	1.7918	1.7396	1.6835	1.6223
40	2.0772	2.0035	1.9245	1.8389	1.7929	1.7444	1.6928	1.6373	1.5766	1.5089
60	1.9926	1.9174	1.8364	1.7480	1.7001	1.6491	1.5943	1.5343	1.4673	1.3893
120	1.9105	1.8337	1.7505	1.6587	1.6084	1.5543	1.4952	1.4290	1.3519	1.2539
∞	1.8307	1.7522	1.6664	1.5705	1.5173	1.4591	1.3940	1.3180	1.2214	1.0000

Table C-5   Percentage Points of the F Distribution: 2.5 Percent Points

df_1 / df_2	1	2	3	4	5	6	7	8	9
1	647.79	799.50	864.16	899.58	921.85	937.11	948.22	956.66	963.28
2	38.506	39.000	39.165	39.248	29.298	39.331	39.355	39.373	39.387
3	17.443	16.044	15.439	15.101	14.885	14.735	14.624	14.540	14.473
4	12.218	10.649	9.9792	9.6045	9.3645	9.1973	9.0741	8.9796	8.9047
5	10.007	8.4336	7.7636	7.3879	7.1464	6.9777	6.8531	6.7572	6.6810
6	8.8131	7.2598	6.5988	6.2272	5.9876	5.8197	5.6955	5.5996	5.5234
7	8.0727	6.5415	5.8898	5.5226	5.2852	5.1186	4.9949	4.8994	4.8232
8	7.5709	6.0595	5.4160	5.0526	4.8173	4.6517	4.5286	4.4332	4.3572
9	7.2093	5.7147	5.0781	4.7181	4.4844	4.3197	4.1971	4.1020	4.0260
10	6.9367	5.4564	4.8256	4.4683	4.2361	4.0721	3.9498	3.8549	3.7790
11	6.7241	5.2559	4.6300	4.2751	4.0440	3.8807	3.7586	3.6638	3.5879
12	6.5538	5.0959	4.4742	4.1212	3.8911	3.7283	3.6065	3.5118	3.4358
13	6.4143	4.9653	4.3472	3.9959	3.7667	3.6043	3.4827	3.3880	3.3120
14	6.2979	4.8567	4.2417	3.8919	3.6634	3.5014	3.3799	3.2853	3.2093
15	6.1995	4.7650	4.1528	3.8043	3.5764	3.4147	3.2934	3.1987	3.1227
16	6.1151	4.6867	4.0768	3.7294	3.5021	3.3406	3.2194	3.1248	3.0488
17	6.0420	4.6189	4.0112	3.6648	3.4379	3.2767	3.1556	3.0610	2.9849
18	5.9781	4.5597	3.9539	3.6083	3.3820	3.2209	3.0999	3.0053	2.9291
19	5.9216	4.5075	3.9034	3.5587	3.3327	3.1718	3.0509	2.9563	2.8800
20	5.8715	4.4613	3.8587	3.5147	3.2891	3.1283	3.0074	2.9128	2.8365
21	5.8266	4.4199	3.8188	3.4754	3.2501	3.0895	2.9686	2.8740	2.7977
22	5.7863	4.3828	3.7829	3.4401	3.2151	3.0546	2.9338	2.8392	2.7628
23	5.7498	4.3492	3.7505	3.4083	3.1835	3.0232	2.9024	2.8077	2.7313
24	5.7167	4.3187	3.7211	3.3794	3.1548	2.9946	2.8738	2.7791	2.7027
25	5.6864	4.2909	3.6943	3.3530	3.1287	2.9685	2.8478	2.7531	2.6766
26	5.6586	4.2655	3.6697	3.3289	3.1048	2.9447	2.8240	2.7293	2.6528
27	5.6331	4.2421	3.6472	3.3067	3.0828	2.9228	2.8021	2.7074	2.6309
28	5.6096	4.2205	3.6264	3.2863	3.0625	2.9027	2.7820	2.6872	2.6106
29	5.5878	4.2006	3.6072	3.2674	3.0438	2.8840	2.7633	2.6686	2.5919
30	5.5675	4.1821	3.5894	3.2499	3.0265	2.8667	2.7460	2.6513	2.5746
40	5.4239	4.0510	3.4633	3.1261	2.9037	2.7444	2.6238	2.5289	2.4519
60	5.2857	3.9253	3.3425	3.0077	2.7863	2.6274	2.5068	2.4117	2.3344
120	5.1524	3.8046	3.2270	2.8943	2.6740	2.5154	2.3948	2.2994	2.2217
∞	5.0239	3.6889	3.1161	2.7858	2.5665	2.4082	2.2875	2.1918	2.1136

Table C-5   Percentage Points of the F Distribution: 2.5 Percent Points (Continued)

d_{f_2} \ d_{f_1}	10	12	15	20	24	30	40	60	120	∞
1	968.63	976.71	984.87	993.10	997.25	1001.4	1005.6	1009.8	1014.0	1018.3
2	39.398	39.415	39.431	39.448	39.456	39.465	39.473	39.481	39.490	39.498
3	14.419	14.337	14.253	14.167	14.124	14.081	14.037	13.992	13.947	13.902
4	8.8439	8.7512	8.6565	8.5599	8.5109	8.4613	8.4111	8.3604	8.3092	8.2573
5	6.6192	6.5246	6.4277	6.3285	6.2780	6.2269	6.1751	6.1225	6.0693	6.0153
6	5.4613	5.3662	5.2687	5.1684	5.1172	5.0652	5.0125	4.9589	4.9045	4.8491
7	4.7611	4.6658	4.5678	4.4667	4.4150	4.3624	4.3089	4.2544	4.1989	4.1423
8	4.2951	4.1997	4.1012	3.9995	3.9472	3.8940	3.8398	3.7844	3.7279	3.6702
9	3.9639	3.8682	3.7694	3.6669	3.6142	3.5604	3.5055	3.4493	3.3918	3.3329
10	3.7168	3.6209	3.5217	3.4186	3.3654	3.3110	3.2554	3.1984	3.1399	3.0798
11	3.5257	3.4296	3.3299	3.2261	3.1725	3.1176	3.0613	3.0035	2.9441	2.8828
12	3.3736	3.2773	3.1772	3.0728	3.0187	2.9633	2.9063	2.8478	2.7874	2.7249
13	3.2197	3.1532	3.0527	2.9477	2.8932	2.8373	2.7797	2.7204	2.6590	2.5955
14	3.1169	3.0501	2.9493	2.8437	2.7888	2.7324	2.6742	2.6142	2.5519	2.4872
15	3.0602	2.9633	2.8621	2.7559	2.7006	2.6437	2.5850	2.5242	2.4611	2.3953
16	2.9862	2.8890	2.7875	2.6808	2.6252	2.5678	2.5085	2.4471	2.3831	2.3163
17	2.9222	2.8249	2.7230	2.6158	2.5598	2.5021	2.4422	2.3801	2.3153	2.2474
18	2.8661	2.7689	2.6667	2.5590	2.5027	2.4445	2.3842	2.3214	2.2558	2.1869
19	2.8173	2.7196	2.6171	2.5089	2.4523	2.3937	2.3329	2.2695	2.2032	2.1333
20	2.7737	2.6758	2.5731	2.4645	2.4076	2.3486	2.2873	2.2234	2.1562	2.0853
21	2.7348	2.6368	2.5338	2.4247	2.3675	2.3082	2.2465	2.1819	2.1141	2.0422
22	2.6998	2.6017	2.4984	2.3890	2.3315	2.2718	2.2097	2.1446	2.0760	2.0032
23	2.6682	2.5699	2.4665	2.3567	2.2989	2.2389	2.1763	2.1107	2.0415	1.9677
24	2.6396	2.5412	2.4374	2.3273	2.2693	2.2090	2.1460	2.0799	2.0099	1.9353
25	2.6135	2.5149	2.4110	2.3005	2.2422	2.1816	2.1183	2.0517	1.9811	1.9055
26	2.5895	2.4909	2.3867	2.2759	2.2171	2.1565	2.0928	2.0257	1.9545	1.8781
27	2.5676	2.4688	2.3644	2.2533	2.1946	2.1334	2.0693	2.0018	1.9299	1.8527
28	2.5473	2.4484	2.3438	2.2324	2.1735	2.1121	2.0477	1.9796	1.9072	1.8291
29	2.5286	2.4295	2.3248	2.2131	2.1540	2.0923	2.0276	1.9591	1.8861	1.8072
30	2.5112	2.4120	2.3072	2.1952	2.1359	2.0739	2.0089	1.9400	1.8664	1.7867
40	2.3882	2.2882	2.1819	2.0677	2.0069	1.9429	1.8752	1.8028	1.7242	1.6371
60	2.2702	2.1692	2.0613	1.9445	1.8817	1.8152	1.7440	1.6668	1.5810	1.4822
120	2.1570	2.0548	1.9450	1.8249	1.7597	1.6899	1.6141	1.5299	1.4327	1.3104
∞	2.0483	1.9447	1.8326	1.7085	1.6402	1.5660	1.4835	1.3883	1.2684	1.0000

Table C-6 Percentage Points of the F Distribution: 1 Percent Points

d_{f_2} \ d_{f_1}	1	2	3	4	5	6	7	8	9
1	4052.2	4999.5	5403.3	5624.6	5763.7	5859.0	5928.3	5981.6	6022.5
2	98.503	99.000	99.166	99.249	99.299	99.332	99.356	99.374	99.388
3	34.116	30.817	29.457	28.710	28.237	27.911	27.672	27.489	27.345
4	21.198	18.000	16.694	15.977	15.522	15.207	14.976	14.799	14.659
5	16.258	13.274	12.060	11.392	10.967	10.672	10.456	10.289	10.158
6	13.745	10.925	9.7795	9.1483	8.7459	8.4661	8.2600	8.1016	7.9761
7	12.246	9.5466	8.4513	7.8467	7.4604	7.1914	6.9928	6.8401	6.7188
8	11.259	8.6491	7.5910	7.0060	6.6318	6.3707	6.1776	6.0289	5.9106
9	10.561	8.0215	6.9919	6.4221	6.0569	5.8018	5.6129	5.4671	5.3511
10	10.044	7.5594	6.5523	5.9943	5.6363	5.3858	5.2001	5.0567	4.9424
11	9.6460	7.2057	6.2167	5.6683	5.3160	5.0692	4.8861	4.7445	4.6315
12	9.3302	6.9266	5.9526	5.4119	5.0643	4.8206	4.6395	4.4994	4.3875
13	9.0738	6.7010	5.7394	5.2053	4.8616	4.6204	4.4410	4.3021	4.1911
14	8.8616	6.5119	5.5639	5.0354	4.6950	4.4558	4.2779	4.1399	4.0297
15	8.6831	6.3589	5.4170	4.8932	4.5556	4.3183	4.1415	4.0045	3.8948
16	8.5310	6.2262	5.2922	4.7726	4.4374	4.2016	4.0259	3.8896	3.7804
17	8.3997	6.1121	5.1850	4.6690	4.3359	4.1015	3.9267	3.7910	3.6822
18	8.2854	6.0129	5.0919	4.5790	4.2479	4.0146	3.8406	3.7054	3.5971
19	8.1850	5.9259	5.0103	4.5003	4.1708	3.9386	3.7653	3.6305	3.5225
20	8.0960	5.8189	4.9382	4.4307	4.1027	3.8714	3.6987	3.5644	3.4567
21	8.0166	5.7804	4.8740	4.3688	4.0421	3.8117	3.6396	3.5056	3.3981
22	7.9454	5.7190	4.8166	4.3134	3.9880	3.7583	3.5867	3.4530	3.3458
23	7.8811	5.6637	4.7649	4.2635	3.9392	3.7102	3.5390	3.4057	3.2986
24	7.8229	5.6136	4.7181	4.2184	3.8951	3.6667	3.4959	3.3629	3.2560
25	7.7698	5.5680	4.6755	4.1774	3.8550	3.6272	3.4568	3.3239	3.2172
26	7.7213	5.5263	4.6366	4.1400	3.8183	3.5911	3.4210	3.2884	3.1818
27	7.6767	5.4881	4.6009	4.1056	3.7848	3.5580	3.3882	3.2558	3.1494
28	7.6356	5.4529	4.5681	4.0740	3.7539	3.5276	3.3581	3.2259	3.1195
29	7.5976	5.4205	4.5378	4.0449	3.7254	3.4995	3.3302	3.1982	3.0920
30	7.5625	5.3904	4.5097	4.0179	3.6990	3.4735	3.3045	3.1726	3.0665
40	7.3141	5.1785	4.3126	3.8283	3.5138	3.2910	3.1238	2.9930	2.8876
60	7.0771	4.9774	4.1259	3.6491	3.3389	3.1187	2.9530	2.8233	2.7185
120	6.8510	4.7865	3.9493	3.4796	3.1735	2.9559	2.7918	2.6629	2.5586
∞	6.6349	4.6052	3.7816	3.3192	3.0173	2.8020	2.6393	2.5113	2.4073

Table C-6   Percentage Points of the F Distribution: 1 Percent Points (Continued)

d_{f_2} \ d_{f_1}	10	12	15	20	24	30	40	60	120	∞
1	6055.8	6106.3	6157.3	6208.7	6234.6	6260.7	6286.8	6313.0	6339.4	6366.0
2	99.399	99.416	99.432	99.449	99.458	99.466	99.474	99.483	99.491	99.501
3	27.229	27.052	26.872	26.690	26.598	26.505	26.411	26.316	26.221	26.125
4	14.546	14.374	14.198	14.020	13.929	13.838	13.745	13.652	13.558	13.463
5	10.051	9.8883	9.7222	9.5527	9.4665	9.3793	9.2912	9.2020	9.1118	9.0204
6	7.8741	7.7183	7.5590	7.3958	7.3127	7.2285	7.1432	7.0568	6.9690	6.8801
7	6.6201	6.4691	6.3143	6.1554	6.0743	5.9921	5.9084	5.8236	5.7372	5.6495
8	5.8143	5.6668	5.5151	5.3591	5.2793	5.1981	5.1156	5.0316	4.9460	4.8588
9	5.2565	5.1114	4.9621	4.8080	4.7290	4.6486	4.5667	4.4831	4.3978	4.3105
10	4.8492	4.7059	4.5582	4.4054	4.3269	4.2469	4.1653	4.0819	3.9965	3.9090
11	4.5393	4.3974	4.2509	4.0990	4.0209	3.9411	3.8596	3.7761	3.6904	3.6025
12	4.2961	4.1553	4.0096	3.8584	3.7805	3.7008	3.6192	3.5355	3.4494	3.3608
13	4.1003	3.9603	3.8154	3.6646	3.5868	3.5070	3.4253	3.3413	3.2548	3.1654
14	3.9394	3.8001	3.6557	3.5052	3.4274	3.3476	3.2656	3.1813	3.0942	3.0040
15	3.8049	3.6662	3.5222	3.3719	3.2940	3.2141	3.1319	3.0471	2.9595	2.8684
16	3.6909	3.5527	3.4089	3.2588	3.1808	3.1007	3.0182	2.9330	2.8447	2.7528
17	3.5931	3.4552	3.3117	3.1615	3.0835	3.0032	2.9205	2.8348	2.7459	2.6530
18	3.5082	3.3706	3.2273	3.0771	2.9990	2.9185	2.8354	2.7493	2.6597	2.5660
19	3.4338	3.2965	3.1533	3.0031	2.9249	2.8422	2.7608	2.67	2.5839	2.4893
20	3.3682	3.2311	3.0880	2.9377	2.8594	2.7785	2.6947	2.6077	2.5168	2.4212
21	3.3098	3.1729	3.0299	2.8796	2.8011	2.7200	2.6359	2.5484	2.4568	2.3603
22	3.2576	3.1209	2.9780	2.8274	2.7488	2.6675	2.5831	2.4951	2.4029	2.3055
23	3.2106	3.0740	2.9311	2.7805	2.7017	2.6202	2.5355	2.4471	2.3542	2.2559
24	3.1681	3.0316	2.8887	2.7380	2.6591	2.5773	2.4923	2.4035	2.3099	2.2107
25	3.1294	2.9931	2.8502	2.6993	2.6203	2.5383	2.4530	2.3637	2.2695	2.1694
26	3.0941	2.9579	2.8150	2.6640	2.5848	2.5026	2.4170	2.3273	2.2325	2.1315
27	3.0618	2.9256	2.7827	2.6316	2.5522	2.4699	2.3840	2.2938	2.1984	2.0965
28	3.0320	2.8959	2.7530	2.6017	2.5223	2.4397	2.3535	2.2629	2.1670	2.0642
29	3.0045	2.8685	2.7256	2.5742	2.4946	2.4118	2.3253	2.2344	2.1378	2.0342
30	2.9791	2.8431	2.7002	2.5487	2.4689	2.3860	2.2992	2.2079	2.1107	2.0062
40	2.8005	2.6648	2.5216	2.3689	2.2880	2.2034	2.1142	2.0194	1.9172	1.8047
60	2.6318	2.4961	2.3523	2.1978	2.1154	2.0285	1.9360	1.8363	1.7263	1.6006
120	2.4721	2.3363	2.1915	2.0346	1.9500	1.8600	1.7628	1.6557	1.5330	1.3805
∞	2.3209	2.1848	2.0385	1.8783	1.7908	1.6964	1.5923	1.4730	1.3246	1.0000

Table C-7 Percentage Points of the F Distribution: 0.5 Percent Points

d_{f_2} \ d_{f_1}	1	2	3	4	5	6	7	8	9
1	16211	20000	21615	22500	23056	23437	23715	23925	24091
2	198.50	199.00	199.17	199.25	199.30	199.33	199.36	199.37	199.39
3	55.552	49.799	47.467	46.195	45.392	44.838	44.434	44.126	43.882
4	31.333	26.284	24.259	23.155	22.456	21.975	21.622	21.352	21.139
5	22.785	18.314	16.530	15.556	14.940	14.513	14.200	13.961	13.772
6	18.635	14.544	12.917	12.028	11.464	11.073	10.786	10.566	10.391
7	16.236	12.404	10.882	10.050	9.5221	9.1554	8.8854	8.6781	8.5138
8	14.688	11.042	9.5965	8.8051	8.3018	7.9520	7.6942	7.4960	7.3386
9	13.614	10.107	8.7171	7.9559	7.4711	7.1338	6.8819	6.6933	6.5411
10	12.826	9.4270	8.0807	7.3128	6.8723	6.5446	6.3025	6.1159	5.9676
11	12.226	8.9122	7.6004	6.8809	6.4217	6.1015	5.8648	5.6821	5.5368
12	11.754	8.5096	7.2258	6.5211	6.0711	5.7570	5.5245	5.3451	5.2021
13	11.374	8.1865	6.9257	6.2335	5.7910	5.1819	5.2529	5.0761	4.9351
14	11.060	7.9217	6.6803	5.9984	5.5623	5.2574	5.0313	4.8566	4.7173
15	10.798	7.7008	6.4760	5.8029	5.3721	5.0708	4.8473	4.6743	4.5364
16	10.575	7.5138	6.3034	5.6378	5.2117	4.9134	4.6920	4.5207	4.3838
17	10.384	7.3536	6.1556	5.4967	5.0746	4.7789	4.5594	4.3893	4.2535
18	10.218	7.2148	6.0277	5.3746	4.9560	4.6627	4.4448	4.2759	4.1410
19	10.073	7.0935	5.9161	5.2681	4.8526	4.5614	4.3448	4.1770	4.0428
20	9.9439	6.9865	5.8177	5.1743	4.7616	4.4721	4.2569	4.0900	3.9564
21	9.8295	6.8914	5.7304	5.0911	4.6808	4.3931	4.1789	4.0128	3.8799
22	9.7271	6.8064	5.6524	5.0168	4.6088	4.3225	4.1094	3.9440	3.8116
23	9.6348	6.7300	5.5823	4.9500	4.5441	4.2591	4.0469	3.8822	3.7502
24	9.5513	6.6610	5.5190	4.8898	4.4857	4.2019	3.9905	3.8264	3.6949
25	9.4753	6.5982	5.4615	4.8351	4.4327	4.1500	3.9394	3.7758	3.6447
26	9.4059	6.5409	5.4091	4.7852	4.3844	4.1027	3.8928	3.7297	3.5989
27	9.3423	6.4885	5.3611	4.7396	4.3402	4.0594	3.8501	3.6875	3.5571
28	9.2838	6.4403	5.3170	4.6977	4.2996	4.0197	3.8110	3.6487	3.5186
29	9.2297	6.3958	5.2764	4.6591	4.2622	3.9830	3.7749	3.6130	3.4832
30	9.1797	6.3547	5.2388	4.6233	4.2276	3.9492	3.7416	3.5801	3.4505
40	8.8278	6.0664	4.9759	4.3738	3.9860	3.7129	3.5088	3.3498	3.2220
60	8.4946	5.7950	4.7290	4.1399	3.7600	3.4918	3.2911	3.1344	3.0083
120	8.1790	5.5393	4.4973	3.9207	3.5482	3.2849	3.0874	2.9330	2.8083
∞	7.8794	5.2983	4.2794	3.7151	3.3499	3.0913	2.8968	2.7414	2.6210

Table C-7 Percentage Points of the F Distribution: 0.5 Percent Points (Continued)

df_1 / df_2	10	12	15	20	24	30	40	60	120	∞
1	24224	24426	24630	24836	24940	25044	25148	25253	25359	25465
2	199.40	199.42	199.43	199.45	199.46	199.47	199.47	199.48	199.49	199.51
3	43.686	43.387	43.085	42.778	42.622	42.466	42.308	42.149	41.989	41.829
4	20.967	20.705	20.438	20.167	20.030	19.892	19.752	19.611	19.468	19.325
5	13.618	13.384	13.146	12.903	12.780	12.656	12.530	12.402	12.274	12.144
6	10.250	10.034	9.8140	9.5888	9.4741	9.3583	9.2408	9.1219	9.0015	8.8793
7	8.3803	8.1764	7.9678	7.7540	7.6450	7.5345	7.4225	7.3088	7.1933	7.0760
8	7.2107	7.0149	6.8143	6.6082	6.5029	6.3961	6.2875	6.1772	6.0619	5.9505
9	6.4171	6.2274	6.0325	5.8318	5.7292	5.6248	5.5186	5.4104	5.3001	5.1875
10	5.8167	5.6613	5.4707	5.2740	5.1732	5.0705	4.9659	4.8592	4.7501	4.6385
11	5.4182	5.2363	5.0489	4.8552	4.7557	4.6543	4.5508	4.4450	4.3367	4.2256
12	5.0855	4.9063	4.7214	4.5299	4.4315	4.3309	4.2282	4.1229	4.0149	3.9039
13	4.8199	4.6429	4.4600	4.2703	4.1726	4.0727	3.9704	3.8655	3.7577	3.6465
14	4.6034	4.4281	4.2468	4.0585	3.9614	3.8619	3.7600	3.6553	3.5473	3.4359
15	4.4236	4.2498	4.0698	3.8826	3.7859	3.6867	3.5850	3.4803	3.3722	3.2602
16	4.2719	4.0994	3.9205	3.7342	3.6378	3.5388	3.4372	3.3324	3.2240	3.1115
17	4.1423	3.9709	3.7929	3.6073	3.5112	3.4124	3.3107	3.2058	3.0971	2.9839
18	4.0305	3.8599	3.6827	3.4977	3.4017	3.3030	3.2014	3.0962	2.9871	2.8732
19	3.9329	3.7631	3.5866	3.4020	3.3062	3.2075	3.1058	3.0004	2.8908	2.7762
20	3.8470	3.6779	3.5020	3.3178	3.2220	3.1234	3.0215	2.9159	2.8058	2.6904
21	3.7709	3.6024	3.4270	3.2431	3.1474	3.0488	2.9467	2.8408	2.7302	2.6140
22	3.7030	3.5350	3.3600	3.1764	3.0807	2.9821	2.8799	2.7736	2.6625	2.5455
23	3.6420	3.4745	3.2999	3.1165	3.0208	2.9221	2.8198	2.7132	2.6016	2.4837
24	3.5870	3.4199	3.2456	3.0624	2.9667	2.8679	2.7654	2.6585	2.5463	2.4276
25	3.5370	3.3704	3.1963	3.0133	2.9176	2.8187	2.7160	2.6088	2.4960	2.3765
26	3.4916	3.3252	3.1515	2.9685	2.8728	2.7738	2.6709	2.5633	2.4501	2.3297
27	3.4499	3.2839	3.1104	2.9275	2.8318	2.7327	2.6296	2.5217	2.4078	2.2867
28	3.4117	3.2460	3.0727	2.8899	2.7941	2.6949	2.5916	2.4834	2.3689	2.2469
29	3.3765	3.2111	3.0379	2.8551	2.7594	2.6601	2.5565	2.4479	2.3330	2.2102
30	3.3440	3.1787	3.0057	2.8230	2.7272	2.6278	2.5241	2.4151	2.2997	2.1760
40	3.1167	2.9531	2.7811	2.5984	2.5020	2.4015	2.2958	2.1838	2.0635	1.9318
60	2.9042	2.7419	2.5705	2.3872	2.2898	2.1874	2.0789	1.9622	1.8341	1.6885
120	2.7052	2.5439	2.3727	2.1881	2.0890	1.9839	1.8709	1.7469	1.6055	1.4311
∞	2.5188	2.3583	2.1868	1.9998	1.8983	1.7891	1.6691	1.5325	1.3637	1.0000

Table C-8   Probability Distribution of χ^2*

n	.50	.30	.20	.10	.05	.02	.01	.001
1	.455	1.074	1.642	2.706	3.841	5.412	6.635	10.827
2	1.386	2.408	3.219	4.605	5.991	7.824	9.210	13.815
3	2.366	3.665	4.642	6.251	7.815	9.837	11.345	16.266
4	3.357	4.878	5.989	7.779	9.488	11.668	13.277	18.467
5	4.351	6.064	7.289	9.236	11.070	13.388	15.086	20.515
6	5.348	7.231	8.558	10.645	12.592	15.033	16.812	22.457
7	6.346	8.383	9.803	12.017	14.067	16.622	18.475	24.322
8	7.344	9.524	11.030	13.362	15.507	18.168	20.090	26.125
9	8.343	10.656	12.242	14.684	16.919	19.679	21.666	27.877
10	9.342	11.781	13.442	15.987	18.307	21.161	23.209	29.588
11	10.341	12.899	14.631	17.275	19.675	22.618	24.725	31.264
12	11.340	14.011	15.812	18.549	21.026	24.054	26.217	32.909
13	12.340	15.119	16.985	19.812	22.362	25.472	27.688	34.528
14	13.339	16.222	18.151	21.064	23.685	26.873	29.141	36.123
15	14.339	17.322	19.311	22.307	24.996	28.259	30.578	37.697
16	15.338	18.418	20.465	23.542	26.296	29.633	32.000	39.252
17	16.338	19.511	21.615	24.769	27.587	30.995	33.409	40.790
18	17.338	20.601	22.760	25.989	28.869	32.346	34.805	42.312
19	18.338	21.689	23.900	27.204	30.144	33.687	36.191	43.820
20	19.337	22.775	25.038	28.412	31.410	35.020	37.566	45.315
21	20.337	23.858	26.171	29.615	32.671	36.343	38.932	46.797
22	21.337	24.939	27.301	30.813	33.924	37.659	40.289	48.268
23	22.337	26.018	28.429	32.007	35.172	38.968	41.638	49.728
24	23.337	27.096	29.553	33.196	36.415	40.270	42.980	51.179
25	24.337	28.172	30.675	34.382	37.652	41.566	44.314	52.620
26	25.336	29.246	31.795	35.563	38.885	42.856	45.642	54.052
27	26.336	30.319	32.912	36.741	40.113	44.140	46.963	55.476
28	27.336	31.391	34.027	37.916	41.337	45.419	48.278	56.893
29	28.336	32.461	35.139	39.087	42.557	46.693	49.588	58.302
30	29.336	33.530	36.250	40.256	43.773	47.962	50.892	59.703

* From Table IV of Fisher and Yates, "Statistical Tables for Biological, Agricultural and Medical Research," published by Oliver and Boyd Ltd., Edinburgh and London, 1953; by permission of the publishers.

Table C-9 Percentage Points of the Studentized Range*

df	α	r = number of means or number of steps between ranked means									
		2	3	4	5	6	7	8	9	10	11
5	.05	3.64	4.60	5.22	5.67	6.03	6.33	6.58	6.80	6.99	7.17
	.01	5.70	6.98	7.80	8.42	8.91	9.32	9.67	9.97	10.24	10.48
6	.05	3.46	4.34	4.90	5.30	5.63	5.90	6.12	6.32	6.49	6.65
	.01	5.24	6.33	7.03	7.56	7.97	8.32	8.61	8.87	9.10	9.30
7	.05	3.34	4.16	4.68	5.06	5.36	5.61	5.82	6.00	6.16	6.30
	.01	4.95	5.92	6.54	7.01	7.37	7.68	7.94	8.17	8.37	8.55
8	.05	3.26	4.04	4.53	4.89	5.17	5.40	5.60	5.77	5.92	6.05
	.01	4.75	5.64	6.20	6.62	6.96	7.24	7.47	7.68	7.86	8.03
9	.05	3.20	3.95	4.41	4.76	5.02	5.24	5.43	5.59	5.74	5.87
	.01	4.60	5.43	5.96	6.35	6.66	6.91	7.13	7.33	7.49	7.65
10	.05	3.15	3.88	4.33	4.65	4.91	5.12	5.30	5.46	5.60	5.72
	.01	4.48	5.27	5.77	6.14	6.43	6.67	6.87	7.05	7.21	7.36
11	.05	3.11	3.82	4.26	4.57	4.82	5.03	5.20	5.35	5.49	5.61
	.01	4.39	5.15	5.62	5.97	6.25	6.48	6.67	6.84	6.99	7.13
12	.05	3.08	3.77	4.20	4.51	4.75	4.95	5.12	5.27	5.39	5.51
	.01	4.32	5.05	5.50	5.84	6.10	6.32	6.51	6.67	6.81	6.94
13	.05	3.06	3.73	4.15	4.45	4.69	4.88	5.05	5.19	5.32	5.43
	.01	4.26	4.96	5.40	5.73	5.98	6.19	6.37	6.53	6.67	6.79
14	.05	3.03	3.70	4.11	4.41	4.64	4.83	4.99	5.13	5.25	5.36
	.01	4.21	4.89	5.32	5.63	5.88	6.08	6.26	6.41	6.54	6.66
15	.05	3.01	3.67	4.08	4.37	4.59	4.78	4.94	5.08	5.20	5.31
	.01	4.17	4.84	5.25	5.56	5.80	5.99	6.16	6.31	6.44	6.55
16	.05	3.00	3.65	4.05	4.33	4.56	4.74	4.90	5.03	5.15	5.26
	.01	4.13	4.79	5.19	5.49	5.72	5.92	6.08	6.22	6.35	6.46
17	.05	2.98	3.63	4.02	4.30	4.52	4.70	4.86	4.99	5.11	5.21
	.01	4.10	4.74	5.14	5.43	5.66	5.85	6.01	6.15	6.27	6.38
18	.05	2.97	3.61	4.00	4.28	4.49	4.67	4.82	4.96	5.07	5.17
	.01	4.07	4.70	5.09	5.38	5.60	5.79	5.94	6.08	6.20	6.31
19	.05	2.96	3.59	3.98	4.25	4.47	4.65	4.79	4.92	5.04	5.14
	.01	4.05	4.67	5.05	5.33	5.55	5.73	5.89	6.02	6.14	6.25
20	.05	2.95	3.58	3.96	4.23	4.45	4.62	4.77	4.90	5.01	5.11
	.01	4.02	4.64	5.02	5.29	5.51	5.69	5.84	5.97	6.09	6.19
24	.05	2.92	3.53	3.90	4.17	4.37	5.54	4.68	4.81	4.92	5.01
	.01	3.96	4.55	4.91	5.17	5.37	5.54	5.69	5.81	5.92	6.02
30	.05	2.89	3.49	3.85	4.10	4.30	4.46	4.60	4.72	4.82	4.92
	.01	3.89	4.45	4.80	5.05	5.24	5.40	5.54	5.65	5.76	5.85
40	.05	2.86	3.44	3.79	4.04	4.23	4.39	4.52	4.63	4.73	4.82
	.01	3.82	4.37	4.70	4.93	5.11	5.26	5.39	5.50	5.60	5.69
60	.05	2.83	3.40	3.74	3.98	4.16	4.31	4.44	4.55	4.65	4.73
	.01	3.76	4.28	4.59	4.82	4.99	5.13	5.25	5.36	5.45	5.53
120	.05	2.80	3.36	3.68	3.92	4.10	4.24	4.36	4.47	4.56	4.64
	.01	3.70	4.20	4.50	4.71	4.87	5.01	5.12	5.21	5.30	5.37
∞	.05	2.77	3.31	3.63	3.86	4.03	4.17	4.29	4.39	4.47	4.55
	.01	3.64	4.12	4.40	4.60	4.76	4.88	4.99	5.08	5.16	5.23

Table C-9   Percentage Points of the Studentized Range (Continued)

r = number of means or number of steps between ranked means										
12	13	14	15	16	17	18	19	20	α	df
7.32	7.47	7.60	7.72	7.83	7.93	8.03	8.12	8.21	.05	5
10.70	10.89	11.08	11.24	11.40	11.55	11.68	11.81	11.93	.01	
6.79	6.92	7.03	7.14	.724	7.34	7.43	7.51	7.59	.05	6
9.48	9.65	9.81	9.95	10.08	10.21	10.32	10.43	10.54	.01	
6.43	6.55	6.66	6.76	6.85	6.94	7.02	7.10	7.17	.05	7
8.71	8.86	9.00	9.12	9.24	9.35	9.46	9.55	9.65	.01	
6.18	6.29	6.39	6.48	6.57	6.65	6.73	6.80	6.87	.05	8
8.18	8.31	8.44	8.55	8.66	8.76	8.85	8.94	9.03	.01	
5.98	6.09	6.19	6.28	6.36	6.44	6.51	6.58	6.64	.05	9
7.78	7.91	8.03	8.13	8.23	8.33	8.41	8.49	8.57	.01	
5.83	5.93	6.03	6.11	6.19	6.27	6.34	6.40	6.47	.05	10
7.49	7.60	7.71	7.81	7.91	7.99	8.08	8.15	8.23	.01	
5.71	5.81	5.90	5.98	6.06	6.13	6.20	6.27	6.33	.05	11
7.25	7.36	7.46	7.56	7.65	7.73	7.81	7.88	7.95	.01	
5.61	5.71	5.80	5.88	5.95	6.02	6.09	6.15	6.21	.05	12
7.06	7.17	7.26	7.36	7.44	7.52	7.59	7.66	7.73	.01	
5.53	5.63	5.71	5.79	5.86	5.93	5.99	6.05	6.11	.05	13
6.90	7.01	7.10	7.19	7.27	7.35	7.42	7.48	7.55	.01	
5.46	5.55	5.64	5.71	5.79	5.85	5.91	5.97	6.03	.05	14
6.77	6.87	6.96	7.05	7.13	7.20	7.27	7.33	7.39	.01	
5.40	5.49	5.57	5.65	5.72	5.78	5.85	5.90	5.96	.05	15
6.66	6.76	6.84	6.93	7.00	7.07	7.14	7.20	7.26	.01	
5.35	5.44	5.52	5.59	5.66	5.73	5.79	5.84	5.90	.05	16
6.56	6.66	6.74	6.82	6.90	6.97	7.03	7.09	7.15	.01	
5.31	5.39	5.47	5.54	5.61	5.67	5.73	5.79	5.84	.05	17
6.48	6.57	6.66	6.73	6.81	6.87	6.94	7.00	7.05	.01	
5.27	5.35	5.43	5.50	5.57	5.63	5.69	5.74	5.79	.05	18
6.41	6.50	6.58	6.65	6.73	6.79	6.85	6.91	6.97	.01	
5.23	5.31	5.39	5.46	5.53	5.59	5.65	5.70	5.75	.05	19
6.34	6.43	6.51	6.58	6.65	6.72	6.78	6.84	6.89	.01	
5.20	5.28	5.36	5.43	5.49	5.55	5.61	5.66	5.71	.05	20
6.28	6.37	6.45	6.52	6.59	6.65	6.71	6.77	6.82	.01	
5.10	5.18	1.25	5.32	5.38	5.44	5.49	5.55	5.59	.05	24
6.11	6.19	6.26	6.33	6.39	6.45	6.5	6.56	6.61	.01	
5.00	5.08	5.15	5.21	5.27	5.33	5.38	5.43	5.47	.05	30
5.93	6.01	6.08	6.14	6.20	6.26	6.31	6.36	6.41	.01	
4.90	4.98	5.04	5.11	5.16	5.22	5.27	5.31	5.36	.05	40
5.76	5.83	5.90	5.96	6.02	6.07	6.12	6.16	6.21	.01	
4.81	4.88	4.94	5.00	5.06	5.11	5.15	5.20	5.24	.05	60
5.60	5.67	5.73	5.78	5.84	5.89	5.93	5.97	6.01	.01	
4.71	4.78	4.84	4.90	4.95	5.00	5.04	5.09	5.13	.05	120
5.44	5.50	5.56	5.61	5.66	5.71	5.75	5.79	5.83	.01	
4.62	4.68	4.74	4.80	4.85	4.89	4.93	4.97	5.01	.05	∞
5.29	5.35	5.40	5.45	5.49	5.54	5.57	5.61	5.65	.01	

Table C-10   Transformation of r to z

r	z	r	z	r	z
.01	.010	.34	.354	.67	.811
.02	.020	.35	.366	.68	.829
.03	.030	.36	.377	.69	.848
.04	.040	.37	.389	.70	.867
.05	.050	.38	.400	.71	.887
.06	.060	.39	.412	.72	.908
.07	.070	.40	.424	.73	.929
.08	.080	.41	.436	.74	.950
.09	.090	.42	.448	.75	.973
.10	.100	.43	.460	.76	.996
.11	.110	.44	.472	.77	1.020
.12	.121	.45	.485	.78	1.045
.13	.131	.46	.497	.79	1.071
.14	.141	.47	.510	.80	1.099
.15	.151	.48	.523	.81	1.127
.16	.161	.49	.536	.82	1.157
.17	.172	.50	.549	.83	1.188
.18	.181	.51	.563	.84	1.221
.19	.192	.52	.577	.85	1.256
.20	.203	.53	.590	.86	1.293
.21	.214	.54	.604	.87	1.333
.22	.224	.55	.618	.88	1.376
.23	.234	.56	.633	.89	1.422
.24	.245	.57	.648	.90	1.472
.25	.256	.58	.663	.91	1.528
.26	.266	.59	.678	.92	1.589
.27	.277	.60	.693	.93	1.658
.28	.288	.61	.709	.94	1.738
.29	.299	.62	.725	.95	1.832
.30	.309	.63	.741	.96	1.946
.31	.321	.64	.758	.97	2.092
.32	.332	.65	.775	.98	2.298
.33	.343	.66	.793	.99	2.647

D

Introduction to Computers

Anyone who deals with data must inevitably consider using a computer. Computers are often believed to be designed for very complex problems, but many situations described in this book can be managed with rather rudimentary knowledge of computers. We shall see how this works.

A computer has essentially four parts: input mechanism, a processor, a storage mechanism, and an output mechanism (Fig. D-1). The input device reads instructions and data from computer cards, a keyboard terminal (much like a typewriter) or magnetic tape and sends this to storage. The processor calls information from storage, manages the computations or logical systems prescribed by the user, and sends data back and forth to storage. The output device typically takes results of the processing out of storage and prints them on paper, punches a new deck of cards, writes them on a cathode ray tube (like a TV screen), or writes a new magnetic tape.

Let us look at each of these operations in more detail. How do we get data into the computer? The most common way is to punch our data into computer cards, which can be "read" by the input device (here a card reader). The punching is typically done by a machine called a *keypunch*, which has a keyboard similar to that of a typewriter. When we press a key on the keyboard, the machine punches a rectangular hole in a location on a computer card reserved for the symbol on the key we pressed. A computer card is shown in Fig. D-2, with holes punched in the card for all digits 1 through 0 and all alphabetic characters. There are 80 columns in a card, so 80 single-digit numbers can be recorded in a single card.

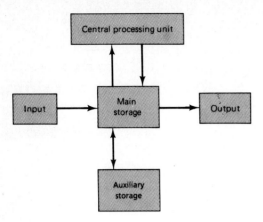

Figure D-1

Suppose we want to punch in two-digit numbers, like 56 or 21. Now we use two adjacent columns in the card. If we want to punch a 46 into the card we could punch the 4 into the first column and the 6 into the next column of the same card. This is shown in Fig. D-3.

Let us now suppose we have a study in which we have recorded children's weight in pounds and their height in inches. Jane weighed 75 pounds and was 56 inches tall. How do I put these figures into a computer card? I will put the 7 and 5 in adjacent columns and the 5 and 6 in adjacent columns, but just which of all the 80 columns I choose is not important as long as I use the same two pairs of columns for every child in my study. (Of course, I will need a separate card for each child.) Let us put the weight in columns 6 and 7, the height in columns 8 and 9. The data for Jane would look like Fig. D-4.

We would have a card for each child in our study: weight punched

Figure D-2

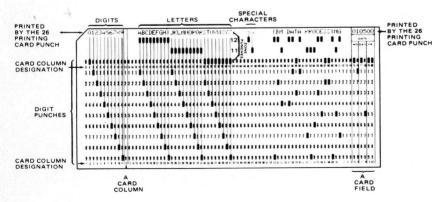

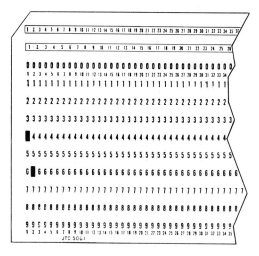

Figure D-3

into columns 6 and 7, height in columns 8 and 9. I left card columns 1, 2, 3, and 4 blank before I recorded the weight data, and I also left the rest of the card blank to the right of column 9. Now suppose I want to give each child an identification number. I could use the first three columns in the card for this number, taking me up to as many as 999 children. If I have fewer than 1000 children, the first three card columns will be sufficient for my identification (ID) number. I can start by calling the first child number 1, the second number 2, the third 3, etc. But since we have reserved three columns for identification, we can

Figure D-4

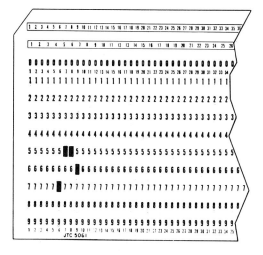

write these numbers as follows: 001, 002, 003, etc. Child number 12 would have an ID of 012, child number 43 would be 043, and child number 102 would be 102.

Here are the data on the first three children in our study, and Fig. D-5 shows how the data would appear in the data cards.

Name	ID	Height, in	Weight, lb
John Able	001	49	79
Bill Baker	002	53	84
Chuck Cart	003	43	76

Our next job is to tell the computer where to look on the cards to find the data on which we wish to do our calculations. We do this by writing a *format statement*. A format statement has three parts: (1) a number telling us how many *fields* the data are in, (2) a letter telling the computer about our numerical system, and (3) a second number telling the computer the maximum number of digits in the data on which we shall do calculations and where the decimal point is in these digits. Now let us see what this description means.

A *field* is simply a column or group of adjacent columns on a data

Figure D-5

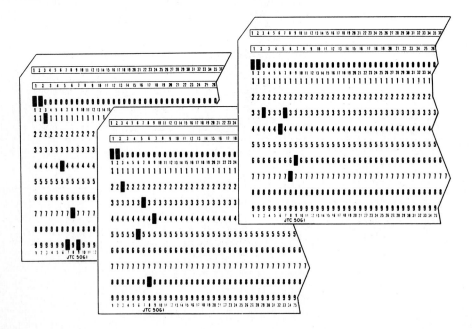

card reserved for recording the data on a given variable. For the above data we set aside columns 6 and 7 in which we then record the height data. This is a two-column field. Our ID data are in columns 1, 2, and 3. This is a three-column field.

Our format statement for the above data on height and weight can now be written to look like this:

(F3.0, 2X, 2F2.0)

The first term, F3.0 refers to the ID data. We usually have a number preceding the F to tell us how many successive, identical fields we have, but since we have only one three-column field here, we need not precede the F with a number. If we did have two or more successive, identical fields, we would precede the F with the number of fields.

The F tells the computer something about the nature of the numbers involved in our data, but that "something" is not sufficiently important to discuss at this point. (Other letters are used in programming, but the F formating serves our purpose here.) The 3.0 says that our first field on the card (ID) is a three-digit number. The zero to the right of the decimal says that in our three-digit number there are no digits to the right of the decimal. (If our data had represented money, with a three-column field we would write F 3.2, because we would have two digits—cents—to the right of the decimal in a three-digit number.)

Next on our cards we skipped two columns. We tell the computer this by writing 2X. The X tells the computer to skip columns; the 2 tells how many to skip.

Then we come to our variables, height and weight. These variables are identified by the term 2F2.0. The 2 to the left of the F in the term 2F2.0 tells the computer that there are two adjacent fields on the card that have the same number of card columns per field. The 2.0 to the right of the F tells the computer that each field is two digits wide, and that there are no digits to the right of the decimal.

Since we are not recording data in the remainder of the card to the right of column 9, we do not include the rest of the card in the format statement. The computer will disregard that portion of the card. However, if we leave the first few columns of the card blank, we must write this into the format. The computer looks at the first item in our format statement as beginning with column 1 on the card, so if we skip initial columns, we must tell the computer that we have done so. For example, if we had no ID on our children above, we would leave columns 1, 2, and 3 blank (plus columns 4 and 5 which we did not use above), and our format statement would read

(5X, 2F2.0)

instead of

(F3.0, 2X, 2F2.0)

Format statements are put into the computer by means of our input device, beginning in column 1 of the card. Since different computing systems require slight variations in how the formats are punched, students should inquire about use of parentheses, commas, etc., before employing a given system.

Now that the computer knows where to find our data on the card, it must next be told what to do with these figures. We tell the computer what operations to perform through use of a *program.*

A program is a set of mathematical and logical statements that direct the computer in completing the processing of our data. Some programs simply translate mathematical formulas into *machine language*, while others are complex logical systems which provide criteria for decisions and ask the computer not only to do mathematical computations but also to make specified decisions about what data are admissible and what is to be done with the results of computation. Fortunately for users of this book a number of *library* programs are already available for solving a wide variety of statistical problems. Library programs have already been composed by professional programmers and are available intact at almost every computing center. Manuals on the use of library packages are also available at computer centers.

Probably the most common packages are the "BMD Biomedical Computer Programs P Series" (BMDP), by W. J. Dixon and M. B. Brown, published by the University of California Press, Berkeley, 1981 (popularly called the Bimeds), and the "Statistical Package for the Social Sciences" (SPSS), combined edition, by Norman Nie, Dale Bent, and C. Hadlai Hull, published by McGraw-Hill Book Company, New York, 1981. For working with a computer both the Bimed manual and the SPSS manual are good places for the novice to begin. They include instruction on all phases of elementary use of the computer from format-statement writing to step-by-step procedures for running data on the computer via the library programs.

After the computer has called our data from storage and has done the computations the program has told it to do, it must then output the results. Typically we ask it to print the results on paper, but we may ask it to punch a new deck of cards, or write a new magnetic tape, or display it on the cathode ray tube (CRT) of a computer terminal. This operation completes our use of the computer for a given problem.

To get the computer to perform all these operations we must put the following items into the machine: system cards, format, program deck of cards, our data deck of cards, and a finish card. We have already talked about programs and data cards, but not systems cards.

Since systems cards differ with different computer systems, you must inquire at each computing center to see what is required. Systems

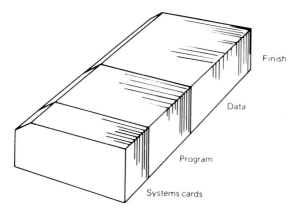

Figure D-6

cards contain signals to the computer that a new job is coming, and they lay out some basic parameters of the job.

A complete deck, ready for a computer run, might look like Fig. D-6. The person using the computer must get the deck in order. Here is what we must do:

1 We select the program that will give us the results we want. Assuming we shall use a library program, we consult a manual such as Bimed or SPSS, find the program, and note its title and number. The manual will tell us to punch several control cards that regulate how the program will be used.

2 Our data must be punched into cards. This means setting up a card format and having the data punched into cards according to that format. Data are punched on a keypunch machine, and anyone who has rudimentary typing skills can quickly learn to operate a keypunch at a minimum level of proficiency.

3 We set up our systems cards. Most computing centers have users' manuals, program consultants, or both, which tell us how to set up our systems cards and how to gain access to the computer. The student who is using the computing facility for the first time will want to exploit these sources fully.

4 Our deck of cards for the computer run is then arranged as shown in Fig. D-6, and we consult the users' manual or the program consultant to find the prescribed routine for getting our problem onto the computing system and submit the problem following the routine.

5 We pick up the results from the computer output.

6 If the problem has failed to run, we must go back and check each detail of the setup. Computers must receive instructions exactly as prescribed. Were our systems cards exact and in proper order? Was our format done exactly as prescribed for our computer system? Was our deck in correct order,

including a finish card? Did we include the control cards required by our program, and were they punched exactly the way the program stipulates? If we find no errors in our setup, we should take our problem to a program consultant who will help us find the "bug."

Sometimes, instead of cards, instructions and data are sent directly to the computer via a computer terminal. A terminal looks like a television screen with a typewriter keyboard attached. When the terminal is used to transmit instructions and data to the computer, the same general procedure is used as with cards. A set of systems instructions comes first, followed by a program, then the data, and a termination statement.

Initial work with the computer may seem time-consuming, but the savings in time and effort in future work will be a vast compensation for the initial effort.

E

Answers to Exercises and Problems

Chapter 1

Problems

1-2 (*a*) parameter   (*b*) statistic
       (*c*) statistic   (*d*) parameter

1-3 (*a*) discrete   (*b*) continuous
       (*c*) discrete   (*d*) continuous
       (*e*) discrete

1-4 (*a*) experimental group
       (*b*) control group
       (*c*) independent variable
       (*d*) conceptual dependent variable
       (*e*) anxiety-test scores

1-5 (*a*) small classes, large classes
       (*b*) class size
       (*c*) reading ability
       (*d*) scores on Metro Reading Readiness Test
       (*e*) inferential: data were generalized to the population of kindergarteners

Chapter 2

Exercises

2-1   (*a*) 1 IQ point   (*b*) 5 units per interval
          (*c*) 14 intervals   (*d*) 78–82
          (*e*) 77.5–82.5

2-2   (*a*) 7 units per interval   (*b*) 15 intervals

2-3 $\dfrac{144 - 80}{15} + 1 = 4.33$, interval $= 5$ units

X	f
140–144	1
135–139	1
130–134	1
125–129	3
120–124	3
115–119	3
110–114	5
105–109	3
100–104	3
95–99	2
90–94	2
84–89	1
80–84	1
	29

2-4

X	f
132–144	3
119–131	6
106–118	10
93–105	8
80–92	2
	29

2-5   Check work with instructor

2-6 (*c*) data not separated by discrete units

2-7 (*d*) smoothing reduces unique characteristics of samples

2-8

cf	cp
25	1.00
24	.96
23	.92
22	.88
20	.80
18	.72
15	.60
11	.44
8	.32
6	.24
5	.20
4	.16
2	.08
1	.04
1	.04

Problems

2-1 (*a*) .5–5.5   (*b*) 55.5–58.5
      (*c*) 27.5–32.5   (*d*) 100.5–103.5

2-2 (*a*) 3 points, 11   (*b*) 5 points, 14
      (*c*) 1 point, 20   (*d*) 7 points, 16 or 9 points, 12

2-3 (*c*) bimodal   (*e*) 52 percent, 48 percent

2-4 (*a*) positively skewed   (*b*) bimodal
      (*c*) reduce skewness, becomes unimodal

2-5 (*b*) between 2 and 3

2-6 (*a*) 20 percent   (*b*) about 4 percent

Chapter 3

Exercises

3-4	f_X	f_Y
21	1	
20	1	
19	1	
18	2	
17	1	
16	2	1
15	3	6
14	2	4
13	4	4
12	4	1
11	1	1
10	1	1
9	3	3
8	1	4
7	2	2
6	1	2
5		1

(a) 13.0, 12.5

(b) no, X has many more high ratings than Y

3-5 62.23, yes
3-6 106.38, yes
3-7 10.5
3-8 53.0 Mary; 53.0 George
3-9 211/30 = 7.03
3-10 545/50 = 10.9

Problems

3-1 (a) 23 (b) 22.0
    (c) 1054/50 = 21.08
3-2 (a) 690/30 = 23 (b) 21.38
    (c) 21.0
3-3 (a) 6.0

3-4 (a)

	Pine Crossing	Wolf Run	Culdesac
$\bar{X}$	44.27	46.88	42.50
Mdn	43.41	50.00	42.00
N	55	40	30

     (c) 44.68

3-6 c

3-7 a and d

3-8 (a) median
      (b) mean
      (c) median

3-9 (a) Y^2 (b) $(X - \bar{X})^2$
      (c) XY (d) $(X - \bar{X})(Y - \bar{Y})$

3-10 104, 1080

3-11 19.4

3-12 10.08

3-13 14.13

3-14 6.12

Chapter 4

Exercises

4-1 (a) percentile, percentile rank   (b) percentile, percentile rank

4-2 (a) 32.48   (b) 25.33

4-3 (a) 92d percentile rank   (b) 48th percentile rank
      (c) 27.05   (d) 82d percentile rank

Problems

4-1 (a) 12, 32
      (b) $X_{20} = 35.50$, $X_{67} = 53.62$, $X_{90} = 61.50$
      (c) $P_{36} = 21$, $P_{55} = 70.67$, $P_{91} = 30.33$
      (d) $D_3 = 40.93$
      (e) $Q_2 = 47.50 = $ median

4-2 Compared with norms group, she ranks above almost four-fifths in hand-eye coordination score; he ranks above 78 percent.

4-3 (a) 87th   (b) 52.36
      (c) 38.65   (d) 22.84
      (e) Verbal 49.72 Numerical 32.63   (f) 36

Chapter 5

Exercises

5-4 2.48
5-5 .63
5-6 Exercise 5-5 harder because of decimal points
5-7 (a) 3.46   (b) 3.65
5-8 3.30
5-9 4.38
5-10 4.16
5-11 (a) 11.34   (b) 11.44

Problems

5-1 (a) 10.88   (b) 10.96
          (c) 9.0
5-2 (a) 7.34   (b) 7.34
          (d) 7.53
5-3 between 1 standard deviation below the mean and 1 standard deviation above the mean
5-4 (a) class I, wider spread upward ($3\sigma + \bar{X} = 36 + 101 = 136$; 3σ for class II $= 18 + 105 = 123$)
          (b) class I, wider spread downward ($-3\sigma + \bar{X} = -36 + 101 = 65$; -3σ for class II $= 87$)
5-5 (a) $\sigma_A = 1.90$, $\sigma_B = 3.64$
          (b) $QD_A = 1.17$, $QD_B = 1.63$
5-6 6.81 and 6.99 pushup data; 1.49 and 1.53 time
5-7 pushups 14.94–28.56; runs 5.36–8.34
5-8 (a) 28   (b) 4.25
          (c) 6.40   (d) 6.53

Chapter 6

Exercises

6-1 (a) 84.13   (b) 97.72
6-2 68 $\pm$ 4 = 64 to 72 inches
6-3 (a) .75
          (b) ordinate to right of $\bar{X}$, 77.34 percent below, 22.66 percent above
6-4 (a) −.50
          (b) 30.85 percent below, 69.15 percent above
6-5 (a) −.33
          (b) spatial-relations
          (c) 37.07 percent below

6-6 (*a*) learning
(*b*) 37.79 percent
(*c*) 25.14 percent

6-7

Raw score	$X - \bar{X}$	z	T	% below
132	22	1.1	61	86
126	16	.8	58	79
121	11	.55	56	71
90	−20	−1.0	40	16
86	−24	−1.2	38	12
70	−40	−2.0	30	2

6-9 (*a*) $z = .88$, $c = 117.5$ (*b*) 588
(*c*) 294 (*d*) 58.75

6-10 80

Problems

6-1

X	z	T
91	−.5	45
119	1.83	68
110	1.08	61
86	−0.92	41
97	0	50

6-2 Joe A, Mary B, Bill B
6-3 Joe 483, Mary 608, Bill 425
6-4 Joe 9.2
Mary 13.4
Bill 9.0

6-5

z	$\bar{X}$ to z	Area in larger	Area in smaller
1.20	.38	.88	.12
−.65	.24	.74	.26
.75	.27	.77	.23
−.10	.04	.54	.46

6-6

Score	z	Area to z	Total area above z
32	.67	.2486	.2514
36	1.33	.4082	.0918
21	−1.17	.3790	.8790
16	−2.00	.4772	.9772
24	−.67	.2486	.7486

6-7 (a) $z = 1.2$, 11.51 percent

(b) $z = -.20$, $z = .67$, area between is 32.79 percent

(c) $z = -.67$, $z = -.20$, area between is 16.93 percent

6-8 (a) $z_A = 1.4 = (X_S - 18.4)/4$, $X_S = 24$ spelling words

(b) 61.06

(c) $.4452 - .20 = .2452 = z$ of $-.66$; $-.66 = (X - 18.4)/4 = 16$ spelling words

6-9 (a) 5.48 percent   (b) 70.29 percent

6-10 (a) arithmetic 1.38; reading .26; science $-.13$

(b) arithmetic 91.62 percent; reading 60.26 percent; science 44.83 percent

(c) science

(d) 25.34 percent

(e) 29.96 percent

(f) arithmetic 269; reading 213; science 193.5

Chapter 7

Exercises

7-1 (a) positive, moderate   (b) positive, moderate

(c) zero   (d) negative, low

(e) negative, low   (f) positive, high

7-2 (a) .92   (b) same, (7-3)

7-3 .79

7-4 .79

7-5 (a) .04, .16, .25, .36, .49, .64, .81, 1.00

7-6 (a)

r_{XY}	r_{XY}^2	$\sqrt{1 - r_{XY}^2}$
.20	.04	.98
.40	.16	.92
.60	.36	.80
.80	.64	.60
1.00	1.00	.00

(*b*) increases; no

(*c*) decreases; no

7-7 (*a*)

r_{XY}	z
.11	.110
.78	1.045
.53	.590
.24	.245
.98	2.298

(*b*)

z	r_{XY}
.060	.060
.245	.240
.604	.540
.282	.275
1.832	.950

7-8 $\bar{z} = .4351$, $r_{XY} = .41$

7-9 .485

7-10 (*a*) no, not homoscedastic and may not be rectilinear

(*b*) probably met

(*c*) no, not homoscedastic

(*d*) probably met

(*e*) no, not rectilinear

7-11 .83

7-12 −.74

7-13 .25

Problems

7-1 .69

7-2 −.86

7-3 .81

7-4 .25

7-5 .69

7-6 −.77

7-7 (*a*) .85   (*b*) .15   (*c*) .25

7-8 .43 ($\sigma = .51$)

7-9 −.77 ($\sigma = 79.93$)

7-10 .39

7-11 −.24

Chapter 8

Exercises

8-1 $a = 6.0 - (.56)\, 9.7 = .57$

$b = .56$

$\hat{Y} = .57 + .56X$

John: $\hat{Y} = .57 + .56\,(6) = 3.93$

Joe: $\hat{Y} = .57 + .56\,(14) = 8.41$

8-2 $b = .611$

$a = 5.75 - (.611)\ 10.58 = -.71$

$\hat{Y}_Q = -.71 + (.611)\ 12 = 6.62$

$\hat{Y}_R = -.71 + (.611)\ 8 = 4.18$

8-3 $\hat{Y} = .63\,\frac{3.0}{1.8}\,(X - 5.2) + 14.5$

Student	A	B	C	D	E
$\hat{Y}$	17.44	9.04	16.39	11.14	14.50

8-4 $\hat{Y} = .71\,\frac{12}{15.1}\,(X - 110) + 108.5$

Student	F	G	H	I	J
$\hat{Y}$	111.3	101.73	105.79	118.09	109.06

8-5 equal $\bar{Y}$

8-6 (a) 10.56%   (b) 30.85%; .62%

8-7 26.43

Problems

8-1 (a) $b_{YX} = .52$ (b) 5.14   (c) 13.98, 9.3

8-2 14.00, 9.30 (difference tied to rounding of $\bar{X}$'s, standard deviation)

8-3 (a) .22   (b) −15.47   (c) 5.21 (IQ 94), 10.93 (IQ 120)

8-4 (a) $9.84 (IQ) 115); $7.85 (IQ 106)

(b) .22 or 22 cents

(c) −15.47 (the intercept with these data)

(d) 22.36 percent ($z = .76$)

8-5 (a) 28.10% ($z = .58$)

(b) 19.22% ($z = -.87$)

(c) 18 ± 1.96 (3.44) = 11.26 to 24.74

8-6 (a) $s_{YX} = .58$, $z = .34$, 36.69 percent

(b) $z = -.86$, 19.49 percent

Chapter 9

Exercises

9-1 (a) .125   (b) .375   (c) .375   (d) .125

9-2 (a) .0039   (b) .0312   (c) .273

9-3 (a) .0547   (b) .2734   (c) .0078

9-4 (a) .0256   (b) .3456

9-5 (a) .76   (b) .64   (c) .60

9-6 (a) .875   (b) .375   (c) .750

9-7 (a) .6154   (b) .5385

9-8 (a) .6154   (b) .8846

9-9 (a) .24   (b) .1474

Problems

9-1 (*a*) .0313   (*b*) .1563   (*c*) .3125   (*d*) .3125
     (*e*) .1563   (*f*) .0313   (*g*) yes

9-2 .1115

9-3 .1641

9-4 (*a*) .0156, .0938, .2344, .3125, .2344, .0938, .0156
     (*b*) .0938, .0156, .1094

9-5 (*a*) .50   (*b*) .2308   (*c*) .0769   (*d*) .1538
     (*e*) .6154

9-6 (*a*) .0278   (*b*) .0278

9-7 (*a*) .50 (.50) = .25
     (*b*) .50 $(\frac{9}{19})$ = .2368

9-8 .725: $P(C) + P(F) - P(C \text{ and } F) = .50 + .475 - .25$

9-9 .75; $P(\text{Min}) + P(\text{Male}) - P(\text{Min and Male}) = .50 + .525 - .275$

Chapter 10

Exercises

10-1 -3.5, reject H_0 (-3.5 is beyond -1.96)

10-2 3.8, reject H_0 (3.8 is beyond 1.96)

10-3 4.78, reject H_0 (-4.78 is beyond -2.58)

10-4 Type II (accept H_0 when sample is from other population)

10-5 Type I (reject H_0 when sample is from own population)

10-6 Type II (accept H_0 when sample is from other population)

10-7 (*a*) 28.40–25.60   (*b*) 28.84–25.16

10-8 (*a*) 37.31–34.69   (*b*) 37.72–34.28

10-9 (*a*) 2.31   (*b*) 2.18
      (*c*) 2.09   (*d*) 2.05

10-10 gets smaller; can reject H_0 with smaller t

10-11 $t = -1.05$, accept H_0

10-12 6.63, reject H_0

10-13 63.11 to 58.89

10-14 19.35 to 15.85

10-15 2.62, reject H_0

10-16 5.49, reject H_0

10-17 1.66, accept H_0

10-18 1.66, accept H_0

10-19 (*a*) .890   (*b*) 1.334   (*c*) 1.664

Problems

10-1 2.1, reject H_0

10-2 12.14, reject H_0

10-3 .122 accept H_0

10-4 -1.74 accept H_0

10-5 -1.85, accept H_0; $112 \pm 3.179 = 115.18$ to 108.82

10-6 -3.16, reject H_0; $7 \pm 3.250(.632) = 9.06$ to 4.95

10-7 30.29–21.71

10-8 -2.91, reject H_0, probably not from population with a mean of 102 months MA.

10-9 2.65, reject H_0 (one-tailed test, use 1.65 standard errors; see Table C-2)

10-10 -1.78 (one-tailed test table value is 1.708, reject H_0)

Chapter 11

Exercises

11-1

$\bar{X}_i - \bar{X}_j$	f
5	1
4	1
3	2
2	2
1	2
0	4
−1	3
−2	1
−3	2
−4	1
−5	1

(a) Yes, 11 of 20 have differences of at least 2 points.

(b) No, in fact none of the pairs has an $\bar{X}_1 - \bar{X}_2$ differences as great as ± 6.

11-2 (a) 2.71

11-3 (a) experimental is programmed text group; control is regular text-lecture

(b) 2.00, reject H_0; $2.00 > 1.96$.

(c) no

11-4 (a) 3.13, reject H_0 ($3.13 > 2.58$)

11-5 (a) 2.65, reject H_0

11-6 (a) 1.00   (b) 1.00

11-7 (a) $F = 1.40$, accept H_0: $\sigma_1^2 = \sigma_2^2$

(b) $t = .54$, accept H_0: $\mu_1 = \mu_2$

11-8 $t = 2.97$, reject H_0: $\mu_1 = \mu_2$

11-9 $t = .20$, accept H_0

11-10 $t = .45$, accept H_0

11-11 7.02, reject H_0

11-12 2.50, reject H_0

11-13 2.19, reject H_0

11-14 4.26, reject H_0

Problems

11-1 2.10, accept H_0

11-2 1.24, accept H_0

11-3 1.62, accept H_0

11-4 .85, accept H_0

11-5 (a) 1.08, accept H_0: $\sigma_1^2 = \sigma_2^2$

(b) 2.40, accept H_0: $\mu_1 = \mu_2$

11-6 2.11, accept H_0: $\mu_1 = \mu_2$

11-7 1.59, accept H_0: $\mu_1 = \mu_2$

11-8 $F = 5.57$, reject H_0: $\sigma_1^2 = \sigma_2^2$;

$t = 1.97$, calculated critical value is 2.262 accept H_0: $\mu_1 = \mu_2$

11-9 2.53, reject H_0: $\mu_1 = \mu_2$

11-10 2.54, reject H_0: $\mu_1 = \mu_2$

Chapter 12

Exercises

12-1 3.45, accept H_0: $\mu_1 = \mu_2 = \mu_3$

12-2 should be equal

12-3 2.31, accept H_0: $\mu_1 = \mu_2 = \mu_3$

Problems

12-1 $F = 6.26$, reject H_0 at .05

12-2 $F = .03$, accept H_0 at .05

12-3 $F = 1.18$, accept H_0 at .01

12-4 (a) yes   (b) yes, $SS_{bg} = SS_t - SS_{wg}$

12-5 $F = 29.48$, reject H_0

12-6 $\bar{X}_1 = 26.63$, $\bar{X}_2 = 22.43$, $\bar{X}_3 = 10.0$; high males appear lower in job satisfaction than others

12-7 $F = 4.24$, reject H_0

12-8 $F = 5.96$, accept H_0

12-9 $F = 12.67$, reject H_0

12-10 $F = 4.58$, accept H_0

Chapter 13

Exercises

13-1 $F = 33.13$, reject H_0: $\mu_1 = \mu_2 = \mu_3$

$Q_{1-2} = 10.07$, reject H_0: $\mu_1 = \mu_2$

$Q_{1-3} = 9.87$, reject H_0: $\mu_1 = \mu_3$

$Q_{2-3} = -.21$, accept H_0: $\mu_2 = \mu_3$

sample 1 not from same population as samples 2 and 3

13-2 $Q_{1-2} = 4.74$, reject H_0: $\mu_1 = \mu_2$

$Q_{1-3} = 5.48$, reject H_0: $\mu_1 = \mu_3$

$Q_{2-3} = .74$, accept H_0: $\mu_2 = \mu_3$

sample 1 not from same population as samples 2 and 3

13-3* $S_{1-2} = 3.61$
$S_{1-3} = 4.24$
$S_{2-3} = .94$
$S_{1-(2+3)} = 4.31$
$C = \sqrt{(3-1)(3.23)} = 2.54$
reject H_0: $\mu_1 = \mu_2$
reject H_0: $\mu_1 = \mu_3$
accept H_0: $\mu_2 = \mu_3$

reject H_0: $\mu_1 = \dfrac{\mu_2 + \mu_3}{2}$

13-4* $S_{1-2} = -5.12$
$S_{1-3} = .85$
$S_{1-4} = -5.97$
$S_{2-3} = 6.89$
$S_{2-4} = -1.11$
$S_{3-4} = -7.77$
$S_{(1+3)-(2+4)} = 9.04$
$C = \sqrt{(4-1)(4.3126)} = 3.60$
Reject all null hypotheses except: $\mu_1 = \mu_3$, $\mu_2 = \mu_4$

Problems

13-1* $F = 35.93$, reject H_0
Tukey $s_q = .45$
$Q_{1-2} = 1.78$, accept H_0 $Q_{2-3} = 9.842$, reject H_0
$Q_{1-3} = 11.63$, reject H_0 $Q_{2-4} = 8.94$, reject H_0
$Q_{1-4} = 10.73$, reject H_0 $Q_{3-4} = -.89$, accept H_0
Samples 1 and 2 from common population; samples 3 and 4 from common population

13-2* $s_q = 1.10$
$Q_{A-B} = -4.00$, reject H_0
$Q_{A-C} = -4.36$, reject H_0
$Q_{B-C} = -.36$, accept H_0
Sample A not from same population as B and C, but B and C from common population

13-3* $s_q = .81$
$Q_{1-2} = 6.57$, reject H_0 $Q_{2-3} = 7.56$, reject H_0
$Q_{1-3} = 14.13$, reject H_0 $Q_{2-4} = 1.98$, accept H_0
$Q_{1-4} = 8.55$, reject H_0 $Q_{3-4} = 5.57$, reject H_0
Samples 2 and 4 from a common population, but all others from different populations

13-4* $F = 36.55/4.39 = 8.33$
$S_{1-2} = -1.60$
$S_{1-3} = 2.70$
$S_{2-3} = 4.04$
$S_{(1+2)-3} = 3.83$
$C = \sqrt{(3-1)(3.8853)} = 2.79$
Samples 2 and 3 not from common population; sample 1 could be in

*Answers may vary slightly depending on rounding of decimals.

same population as 2 or 3; samples 1 and 2 combined not from same
population as sample 3

13-5* $S_{A-B} = -25.38$

$S_{A-C} = 4.86$

$S_{B-C} = 28.76$

$S_{B-(A,C)} = 31.24$

$C = \sqrt{(3-1)(3.0718)} = 2.48$

No pairs of samples from common populations

13-6* $S_{1-2} = -1.15$ $\qquad\qquad$ $S_{2-3} = -2.73$

 $S_{1-3} = -4.14$ $\qquad\qquad$ $S_{2-4} = -1.73$

 $S_{1-4} = -3.12$ $\qquad\qquad$ $S_{3-4} = 1.08$

 $S_{(1,2)-(3,4)} = -4.10$

 $C = \sqrt{(4-1)(2.8387)} = 2.92$

All from same population, except 1 and 3, 1 and 4

13-7 | *Three samples* | | *Five samples* | | | |
|---|---|---|---|---|---|
| $\mu_1 - \mu_2$ | $\mu_1 - \mu_2$ | $\mu_2 - \mu_3$ | $\mu_3 - \mu_4$ | $\mu_4 - \mu_5$ |
| $\mu_1 - \mu_3$ | $\mu_1 - \mu_3$ | $\mu_2 - \mu_4$ | $\mu_3 - \mu_5$ | |
| $\mu_2 - \mu_3$ | $\mu_1 - \mu_4$ | $\mu_2 - \mu_5$ | | |
| | $\mu_1 - \mu_5$ | | | |

Chapter 14

Exercises

14-1 $\chi^2 = 1.24$, attendance is independent of socioeconomic standing

14-2 $\chi^2 = 5.09$, does not differ from random

14-3 $\chi^2 = 1.35$ (with Yates correction) does not differ from random

14-4 $\chi^2 = 7.84$, significant relationship between intelligence and preference
for programs

14-5 $\chi^2 = 18.63$, sample not distributed like population

14-6 $\chi^2 = 4.59$, does not depart from normality

14-7 .29, pretest not significantly different from posttest

14-8 not particularly powerful in detecting differences

14-9 1.38, A's not different in achievement from B's

14-10 1.31, A's, B's, C's all from same population

Problems

14-1 sign test

14-2 median test

14-3 chi square

14-4 sign test

*Answers may vary slightly depending on rounding of decimals.

Appendix A

Exercises

A1-1	107	**A1-2**	8
A1-3	17	**A1-4**	30
A1-5	$ax + 2ay$	**A1-6**	$i^2 \Sigma f d^2 - i^2 \dfrac{(\Sigma f d)^2}{N}$
A1-7	unequal	**A1-8**	unequal
A1-9	unequal	**A1-10**	equal
A1-11	equal	**A1-12**	.538
A1-13	ax/by	**A1-14**	$\Sigma X^2/N^2$
A1-15	2.67	**A1-16**	xb/ya
A1-17	1.42	**A1-18**	$(bzx - ay)/yb$

A2-1   (a) 4, (b) $X_3 = 100$, $X_1 - 105$, (c) 105.5, (d) (1) X_2 is more than X_1, (2) X_3 is less than X_1, (3) $X_1 + X_2$ is more than $X_3 + X_4$, (4) X_1 is not equal to X_2, (5) $A = X_1^2 + X_2^2 + \cdots + X_N^2 = $ the sum of the squared X scores

A2-2	incorrect	**A2-3**	correct
A2-4	correct	**A2-5**	correct

A2-6   The average is the sum of all scores divided by the number of scores.

A2-7   If we subtract a constant from each score, add up all the scores after this subtraction, and divide this sum by the number of scores we have, the result will be the average minus the constant.

A2-8   If we multiply each score by a constant, add up the resulting products, and divide by the number of scores we have, the result is the mean multiplied by the constant.

A2-9   If we subtract the average from each score, summate the resulting differences, and divide them by the number of scores we have, the result will be zero.

A2-10   The value of $X_1 + X_2 + X_3 + X_4$ is not equal to the value of $X_5 + X_6 + X_7 + X_8$.

A2-11 $\dfrac{X_1}{X_5} + \dfrac{X_2}{X_5} + \dfrac{X_3}{X_5} + \dfrac{X_4}{X_5} = \dfrac{(X_1 + X_2 + X_3 + X_4)}{X_5}$

A3-1	-3	**A3-2**	-2
A3-3	-9	**A3-4**	-17
A3-5	-12	**A3-6**	28
A3-7	3	**A3-8**	-6

A4-1 *(a) X (b) y,*
(c) 3 (d) 2

A4-2 linear

A4-3 curvilinear

A4-4 curvilinear

A4-5 linear

A4-6 $y = 4x + 1$

A4-7 $y = 6x - 1$

A4-8 $y = -5x + 2$

A5-1 16

A5-2 6

A5-3 $4b + 5$

A5-4 11

A5-5 y^2

A5-6 $\dfrac{\Sigma xy}{Ns_1 s_2}$

A6-1 $x^2 + 2xy + y^2$

A6-2 $x^2 - 8x + 16$

A6-3 $4x^2 - 4xy + y^2$

A6-4 $(x + y)(x - y)$

A6-5 $(x + 3)(x + 2)$

7

Index